Topics in Atmospheric and Oceanographic Sciences

Editors: Michael Ghil Jürgen Sündermann Robert Sadourny

Intense Atmospheric Vortices

Proceedings of the Joint Symposium (IUTAM/IUGG) held at Reading (United Kingdom) July 14–17, 1981

Edited by
L. Bengtsson and J. Lighthill

With 195 Figures

Springer-Verlag
Berlin Heidelberg New York 1982

Dr. LENNART BENGTSSON, European Centre for Medium Range Weather Forecasts, Shinfield Park, Reading, Berkshire RG2 9AX / England

Sir JAMES LIGHTHILL, University College London
Gower Street, London WC1E 6BT / England

Series Editors:

Professor Dr. MICHAEL GHIL
Courant Institute of Mathematical Sciences, New York University
251 Mercer Street, New York, NY 10012 / USA

Professor Dr. JÜRGEN SÜNDERMANN
Universität Hamburg, Institut für Meereskunde
Heimhuder Straße 71, 2000 Hamburg 13 / FRG

Professor Dr. ROBERT SADOURNY
Laboratoire de Météorologie Dynamique, Ecole Normale Supérieure
24 rue Lhomond, 75231 Paris Cedex 05 / France

ISBN-13: 978-3-540-11657-8 e-ISBN-13: 978-3-642-81866-0
DOI: 10.1007/ 978-3-642-81866-0

Library of Congress Cataloging in Publication Data. Main entry under title: Intense atmospheric vortices. (Topics in atmospheric and oceanographic sciences) Symposium sponsored by the International Union of Theoretical and Applied Mechanics and the International Union of Geodesy and Geophysics, and held at the European Centre for Medium Range Weather Forecasts. Bibliography: p. 1. Vortex-motion--Congresses. 2. Tropics--Cyclones--Congresses. 3. Tornadoes--Congresses. I. Bengtsson, Lennart. II. Lighthill, M. J., Sir. III. International Union of Theoretical and Applied Mechanics. IV. International Union of Geodesy and Geophysics. V. Series. QC880.4.V6I57 1982 551.5'53 82-10365

2132/3130-543210

Preface

The concept of vorticity is of central importance in fluid mechanics and the change and variability of atmospheric flow is dominated by transient vortices of different time- and space scales. Of particular importance are the most intense vortices such as hurricanes, typhoons and tornadoes which are associated with extreme and hazardous weather events of great concern to society. In recent years the understanding of these phenomena has grown due to increased and improved surveillance by satellites and aircraft as well as by numerical modelling and simulation, theoretical studies and laboratory experiments.

The symposium on "Intense Atmospheric Vortices" was held at the European Centre for Medium Range Weather Forecasts (ECMWF), Reading, England, July 14-17, 1981. The subject area of the Symposium was concerned with observational work, experimental models, theoretical and numerical studies involving hurricanes, typhoons, tornadoes and related phenomena. The aim was to bring together experts on these meteorological processes and on the fundamental fluid-dynamic mechanisms for vorticity intensification from all parts of the world. Thirtyfour scientists participated in the Symposium, including more than half of those leading world experts in the field whom the organizers had invited.

The Symposium examined the different mechanisms for vorticity intensification that operate in two diverse types of meteorological phenomena of great importance, namely the tropical cyclone and the tornado. At the same time, for both of the two main classes of atmospheric phenomena, the Symposium allowed a close discussion between those expert in obtaining and analysing observations (from radiosonde data, aircraft, satellites and radars) and those expert in probing physical mechanisms by means of theoretical, numerical or laboratory models.

Both for the intense tropical cyclones (hurricanes and typhoons) and for the still more intense tornadoes, efforts were made to relate the highly concentrated local development of extremely strong winds to a definition of those types of meteorological patterns on a very much larger scale which are essential for the generation of such phenomena. Studies of this kind are of great importance to forecasters who are seeking methods for giving reliable predictions of dangerous concentrated storms. The Symposium demonstrated, and further developed, a valuable increase of convergence between the observationalist point of view and that of the modeller concerning the nature of the large-scale conditions favourable to severe-storm development.

Data from atmospheric soundings, as summarized in the paper by W.M. Gray, give an overview of the large-scale conditions favourable to hurricane formation in the Inter-Tropical Convergence Zone which is similar to that provided by numerical modelling work, for example as described in the paper by Kurihara. The argument was put forward that two large-scale features must be simultaneously present in the 'outer core' region (extending over more than 1000 km) within which the hurricane develops. One is a large-scale dynamic pattern which, typically, involves cyclonic shear at low levels combined with anticyclonic shear at high levels. The other is a thermal pattern, mobilising convection action to allow a very high rate of transfer of heat and moisture to the atmosphere by bringing relatively cold air forcefully into contact with a warm sea surface. As the aircraft data summarized by R.C. Sheets show, this energy input can act very effectively in positions far from the centre of circulation, allowing an asymmetric initial development of strong winds before they ultimately wrap around themselves to form a symmetric 'mature hurricane' with a well developed 'eye' wall.

The tropical cyclone develops over a time-scale of several hours or a few days. Accordingly, Coriolis forces have time to act, and it is mainly the Coriolis vorticity that intensifies through stretching in the updraughts associated with low-level convergence and high-level divergence.

With phenomena on a much shorter time-scale, evolving over a period of many minutes or a few hours, such as tornadoes, or the waterspouts analysed in the paper by J. Simpson, Coriolis forces are relatively insignificant. Thus, while updraughts are again essential to the stretching of vortex lines, they do not act on Coriolis vorticity but on the far bigger local vorticity in the atmospheric boundary layer. The relatively rare boundary layer with uni-directional flow, when subjected to a local lifting-up by a powerful updraught, may as shown by the paper by D.K. Lilly generate a pair of rotating storms. These storms are cyclonic on the right and anticyclonic on the left (Northern Hemisphere).

The more typical Ekman-spiral boundary layer, however, in which we observe a wind veering to the right (Northern Hemisphere) with increasing altitude, produces a strong cyclonically rotating storm after vorticity stretching by a local updraught.

The large-scale environment essential for the development of a tornado is defined with considerable precision by observational studies, as outlined in the paper by J. Klemp. These, again, are supported by parallel numerical-model studies, for example in the paper by R.K. Smith and L.M. Leslie. It is the massive type of rotating thunderstorm known as the super-cell storm, with its huge anvil cloud at the top, that is needed to generate the enormous updraught velocities (of the order of 10 ms^{-1}) which can stretch the vorticity sufficiently in the atmospheric boundary layer to initiate the tornadic phase.

Once again, a convergence line plays a vital role. This time, it is the line of convergence between the main updraught that soars up to form the anvil cloud far above and the corresponding downdraught which had long ago been shown by Ludlam and Browning to be so tortuously and intricately linked around the updraught. This convergence line or 'gust front' develops a curled edge which is where rapidly rotating high-speed updraughts predominantly tend to occur. They are indicated, both in observations and numerical experiments, by funnel clouds peeping down from the main thundercloud. In a manner dependent on local thermal conditions, these may or may not develop into a tornado that reaches the ground.

Improved understanding of at least two important features of tornadoes has been derived from a variety of laboratory-model experiments. Thus, those described in papers such as that by R. Rotunno give a valuable indication of mechanisms facilitating the development of multiple tornadoes. Again, factors limiting tornado development and persistence are brought out in the experimental work described by M.P. Escudier on the vortex-breakdown phenomenon.

A general survey of laboratory-model phenomena with axisymmetric input is given in the paper by T. Maxworthy. They include at least one hurricane simulation; although certain limitations in present-day laboratory models of hurricane are made apparent by the need to fix the 'eye' position and dimensions geometrically. More generally, this authoritative survey makes clear both the value of laboratory experiments and the danger that details of boundary geometry may exercise a disproportionate influence on their results.

Finally, this last theme was taken up in the paper on a laboratory model with non-axisymmetric input by E.J. Hopfinger. It shows, in particular, how waves propagating along vortices help to extract energy from a region of three-dimensional turbulence. This paper also describes a model of vorticity intensification that has a very broad interest for meteorologists. It is one where vigorous three-dimensional turbulence inserted at the bottom of a rotating fluid system produces a most striking effect: it causes the vorticity in that system to be intensified by factors of well over 10 within a relatively small number of concentrated vortices. At the Symposium this paper led to interesting speculations about the mechansims by which prolonged random stretching and progressive contortion of vortex lines might, after the operation of various re-connection processes, form into an assemblage of intense cyclonic motions relatively free of three-dimensional turbulence.

We wish to extend our gratitude to the International Union of Theoretical and Applied Mechanics (IUGG) for sponsoring the Symposium. We also wish to extend our thanks to ECMWF, the venue for this Symposium, for providing its outstanding facilities. It is a pleasure to acknowledge all those who have been involved in typing the manuscript and drafting the figures which are included in this book. We are particularly grateful to Dr. Erland Källén at ECMWF, who has supervised the technical work and contributed to the editing of this book.

Organizers of the Symposium:
Sir James Lighthill
Dr. Lennart Bengtsson

Contents

TROPICAL CYCLONES

TORNADOES

List of Participants

Dr. L. Bengtsson, European Centre for Medium Range, Weather Forecasts, Reading, Berks. RG2 9AX/England

Dr. F.K. Browand, University of Southern California, Dept. Aerospace Engineering, Los Angeles, CA 90007/USA

Dr. M. Challa, Florida State University, Geophysical Fluid Dynamics Institute, Tallahassee, FL 32306/USA

Prof. D.R. Davies, Department of Mathematics, Exeter University, Exeter/England

Dr. R. Davies-Jones, National Severe Storms Laboratory, 1313 Halley Circle, Norman, OK 73069/USA

Dr. M.P. Escudier, BBC Research Centre, CH-5513 Baden Dättwil/Switzerland

Dr. A.E. Gill, University of Cambridge, Department of Applied Mathematics and Theoretical Physics, Silver Street, Cambridge CB3 9EW/England

Dr. W.M. Gray, Colorado State University, Department of Atmospheric Science, Fort Collins, CO 80523/USA

Dr. Raymond Hide, Meteorological Office (Met O.21), Bracknell, Berks. RG12 2SZ/England

Dr. G. Holland, Colorado State University, Department of Atmospheric Science, Fort Collins, CO 80523/USA

Dr. E.J. Hopfinger, Institut Mechanique, University de Grenoble, B.P. 53X, F-38041 Grenoble CEDEX/France

Dr. J. Klemp, National Centre for Atmospheric Research, Boulder, CO 80307/USA

Dr. Y. Kurihara, NOAA Geophysical Fluid Dynamics Laboratory, P.O. Box 308, Princeton, NJ 08540/USA

Dr. L.M. Leslie, Australian Numerical Meteorology Research Center, P.O. Box 5089 AA, Melbourne, Victoria 3001/ Australia

Dr. W.S. Lewellen, ARAP (Aeronautical Research Associates of Princeton Inc.), P.O. Box 2229, Princeton, NJ 08540/ USA

Sir James Lighthill, University College London, Gower Street, London WC1E 6BT/England

Dr. D.K. Lilly, National Center for Atmospheric Research, P.O. Box 1470, Boulder, CO 80302/USA

Dr. B.A. Lugovtsov, Institute for Hydrodynamics, Siberian Branch of USSR Academy of Sciences, 630090 Akademogorodok, Novosibirsk/USSR

Professor T. Maxworthy, Department of Aerospace and Mechanical Engineering, University of Southern California, Los Angeles, CA 90007/USA

Dr. A. McEwan, CSIRO, Division of Atmospheric Physics, P.O. Box 77, Mordialloc, Victoria 3195/Australia

Dr. M.E. McIntyre, DAMTP, University of Cambridge, Cambridge CB3 9EW/England

Professor B.R. Morton, Monash University, Department of Mathematics, Clayton, Victoria 3168/Australia

Dr. K.V. Ooyama, Atlantic Oceanographic and Meteorological Laboratories, National Hurricane Research Laboratory, 1320 South Dixie Highway, Coral Gables, FL 33146/USA

Dr. R.E. Peterson, Atmospheric Science Group, Texas Technical University, Lubbock, TX 79409/USA

Professor R.L. Pfeffer, Florida State University, Geophysical Fluid Dynamics Institute, Tallahassee, FL 32306/USA

Professor R.P. Pearce, University of Reading, Department of Meteorology, Reading, Berks/England

Dr. R. Rotunno, National Center for Atmospheric Research, P.O. Box 1470, Boulder, CO 80302/USA

Dr. R.C. Sheets, National Hurricane Center, 1320 South Dixie Highway, Coral Gables, FL 33146/USA

Dr. D.R. Sikka, Indian Institute of Tropical Meteorology, Pune/India

Professor Joanne Simpson, NASA - Code 914, Goddard Space Flight Center, Greenbelt, MD 20771/USA

Dr. R.K. Smith, Monash University, Department of Mathematics, Clayton, Victoria 3168/Australia

Dr. J.T. Snow, Purdue University, Department of Mechanical Engineering, West Lafayette, IN 47907/USA

Professor J.S. Turner, ANU Research School of Earth Sciences, Canberra CT 2600/Australia

Dr. Zhou Zi-dong, Central Meteorological Service, Baishiqiaolu No. 46, Western Suburb, Peking/People's Republic of China

Tropical Cyclones

Tropical Cyclone Genesis and Intensification

W.M. Gray

ABSTRACT

This paper discusses the likely physical processes which are responsible for distinguishing between tropical disturbances which develop into tropical cyclones and disturbances which do not.

The pre-typhoon cloud cluster disturbance is differentiated from the prominent non-developing cloud cluster disturbance primarily by its surrounding 2-8° radius tropospheric wind and vertical wind shear patterns. Other parameters such as mean vertical motion, moisture, convergence, vertical stability, surface wind speed, etc. are not significantly different. The favorable wind and vertical wind shear patterns are initially established by general circulation features.

A paradox to the tropical cyclone development problem is the dual role of the tropical disturbance's transverse circulation. The in-up-out mean radial circulation needed in a mechanical sense to spin-up the tangential flow also acts in an opposite sense to export moist static energy from the disturbance. Outflow air has higher energy than inflow air. Weather system maintenance and development can occur only if the disturbance's surface energy flux is large enough to overcome the energy losses from the mean radial circulation and radiation. This requires extra energy flux from the ocean beyond that given by the bulk formula from space and time averaged wind speed and moisture values. Such extra surface energy fluxes have been calculated from budget analysis of our rawinsonde data as a residual and are observed to be related to the product of the weather system's surrounding vertical wind shear and upward vertical motion.

It is hypothesized that the pre-storm weather system's larger surrounding region tropospheric vertical wind shear patterns lead to the development of more frequent and more intense rainband convection. The downdrafts developed by these rainbands tap the ocean for more energy by simultaneous increasing surface wind speed and decreasing moisture and temperature. This results in a nonlinear increase of surface energy flux along the rainbands. These extra rainband energy gains cause a significantly larger surface energy flux than that specified by the broadscale bulk formula using composited data. This surface energy increase is diffused throughout the vortex and allows the energy depleting mean radial circulation to be balanced. Smaller residual energy amounts are often available to intensify the vortex. Recent numerical modeling experiments by W. Fingerhut (1981) are verifying that such enhanced surface energy fluxes and vertical wind shear relationships can differentiate developing weather systems from those which do not develop.

Topics in Atmospheric and Oceanographic Sciences
Intense Atmospheric Vortices, ed. by Bengtsson/Lighthill

1 OBSERVATIONAL INFERENCES

At locations and during seasons in which tropical cyclone development regularly occurs, as around Guam during the summer, favorable thermodynamic factors known to be necessary conditions for genesis such as deep warm ocean water, large vertical gradients of θ_e in the lower troposphere, high middle-level humidity, etc. are nearly always present. Explaining storm genesis therefore requires consideration of the daily variations of dynamic factors. Hundreds of individual case data sets of developing and non-developing disturbances in the western Pacific and western Atlantic have been assembled at Colorado State University (CSU) to analyze such dynamic factors which are associated with tropical cyclone development and those factors which inhibit development. These new data sets have been extensively analyzed by the rawinsonde data compositing techniques that have been developed at CSU and discussed in papers by Gray (1968, 1979a), Williams and Gray (1973), Frank (1977a, 1977b), McBride (1981a, 1981b), McBride and Zehr (1981), Nunez (1981) and Holland (1981).

To understand cyclone genesis one must be able to explain not just the processes which bring about cyclone genesis, but also the conditions which, in the majority of cases, inhibit genesis. Although the cloud patterns of the pretyphoon disturbance only extend to radii of from 2 to 8°, the wind field associated with the developing weather system typically extends outward to very large radii of 10-12° or so. As cyclone development occurs, surface pressures typically fall over a broad area extending to 4-6° radius (McBride, 1981a). Initial pressure falls typically do not begin at inner radii and spread outward as some researchers have envisaged but, instead, they appear to occur simultaneously at all radii out to 6-8°. The 50-100 mb inner core pressure falls of the mature typhoon appears to be primarily a response to outer region wind changes. One must better understand how the outer region of the weak tropical system spins itself up before the inner-core intensification can be properly explained.

The frequent clustering with time of tropical cyclone activity on a 10-15 day scale has been previously noted by the author (Gray, 1979). This clustering is believed to be the result of general circulation changes occurring on such a 10-15 day period. About 80-85% of all tropical cyclones form within specially strong doldrum or monsoon troughs which have substantial north to south gradients of tropospheric vertical wind shear. The other 15-20% of tropical cyclones form within trade-wind currents whose associated upper tropospheric anticyclonic flow patterns are significantly stronger than normal. Such conditions often occur on the southward margin of a Tropical Upper Tropospheric Trough (TUTT).

Comparison analysis of the composite moisture, vertical stability, and cloud area fields of developing and non-developing cloud cluster systems in both the west Pacific and the west Atlantic do not show major differences (McBride, 1981a; McBride and Zehr, 1981, McBride, 1981b). Primary differences occur only in the surrounding tangential wind, vertical wind shears and cloud banding fields. Typical vertical profiles of the tangential wind for developing and non-developing disturbances are shown in Fig. 1. Developing systems are characterized by significantly larger values of positive tangential wind at radii between 2-8° in the lower and middle troposphere than the non-developing cluster systems. Significantly more negative tangential wind is found at upper tropospheric levels. It is the 150 mb to 500 mb vertical wind shear which appears to be most crucial to specifying weather system development or non-development. Such vertical wind shears are often accomplished by general circulation and cross-hemispheric flow interactions which act to strengthen the ITCZ or which act to strengthen the trade winds and bring about TUTT enhancement.

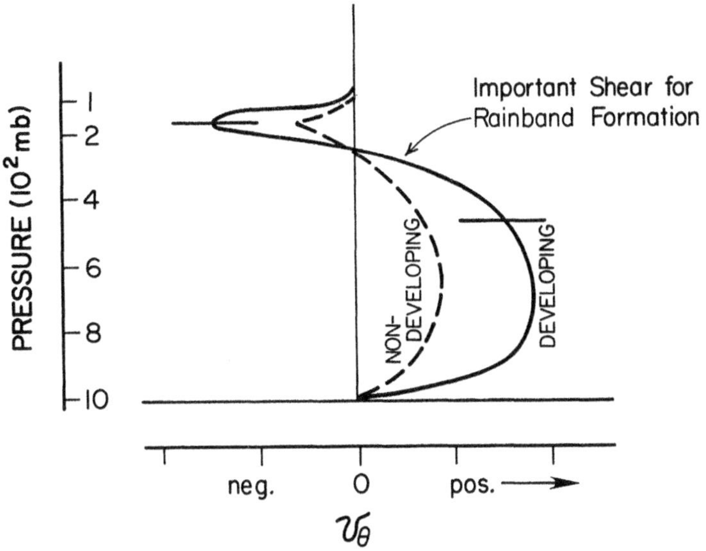

Fig. 1. Typical vertical distribution of the tangential wind surrounding developing and non-developing tropical weather systems at radii of about 4-8°. Note the large upper tropospheric wind shear of the developing systems

It is the author's belief that these favorable surrounding system vertical wind shear patterns are fundamentally important for rainband formation and enhanced cloud downdraft action. Such cloud downdraft action produces extra surface energy flux by virtue of the strong negative correlations between downdraft winds and the humidities and temperatures along the rainbands. This extra surface energy flux is necessary for the maintenance and/or growth of the disturbance because horizontal energy fluxes and radiation act to continuously deplete the western ocean weather system of its energy.

2 TROPICAL DISTURBANCE ENERGY BUDGET

Our extensive radial wind measurements surrounding the centers of oceanic tropical weather systems almost always indicate a deep surface to 400 mb inflow layer as portrayed in Fig. 2. Such deep-layer inflow causes an inward advection of low values of h (or moist static energy-$gz + C_pT + Lq$) at middle tropospheric levels. This leads to a net export of h-energy throughout the tropospheric vertical column.

Moist static energy (h) can increase or decrease in the tropical weather system through transfer from the sea by latent and sensible exchange (E_o), by radiation (R), or through horizontal transports on the boundaries ($\nabla \cdot \mathbb{V}h$). The energy balance of the tropical weather system may be expressed as:

$$\frac{\partial h}{\partial t} = \underset{(a)}{E_o} - \underset{(b)}{(\nabla \cdot \mathbb{V}h)} + \underset{(c)}{R} \qquad [1]$$

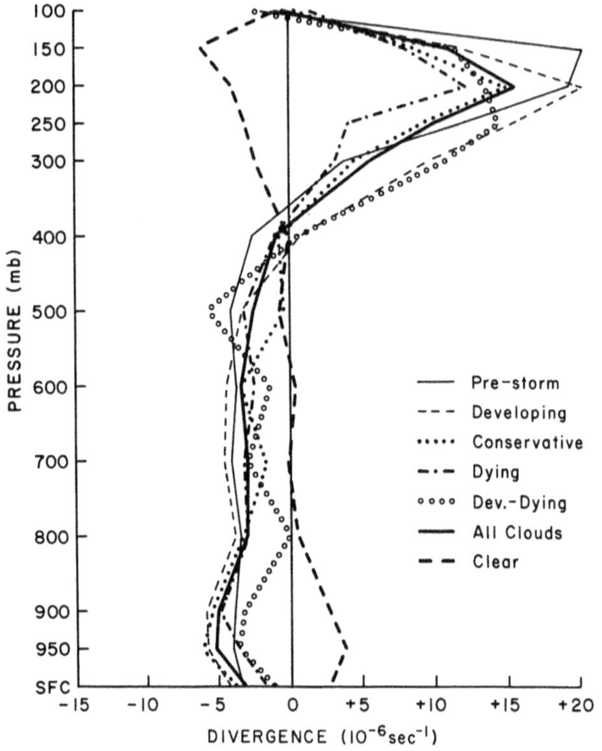

Fig. 2. Vertical profiles of 4°-square area-average divergence at cluster centers (Williams and Gray, 1973)

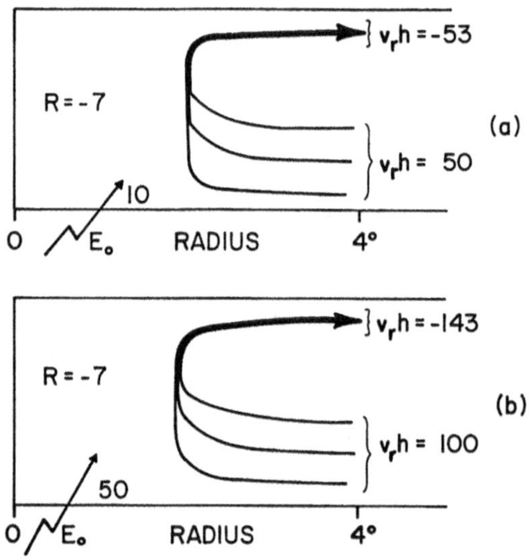

Fig. 3. Schematic representation of the typical h budget within 4° radius of the steady-state cloud cluster, (a), and of the steady-state hurricane, (b). Units are arbitrary. R = net tropospheric radiation, E_o = surface energy flux, and h is moist-static energy. A positive sign indicates disturbance energy gain

Each term has been integrated through the depth of the troposphere. Calculations of the $\nabla \cdot \mathbf{V}h$ term for western ocean tropical systems (the ones which can produce tropical storms) substantiate that there is appreciable amounts of horizontal advection of energy out of these systems; E_o must then be positive to balance both $\nabla \cdot \mathbf{V}h$ and R. An idealized steady state ($\partial h/\partial t = 0$) h-budget for the summertime pre-cyclone disturbance (diagram a) and the hurricane (diagram b) is shown in Fig. 3.

It is seen that the usual vertical circulation through the organized tropical weather system or cyclone acts to reduce the energy of the system. The tropical system can maintain itself only through substantial energy flux from the ocean.

Cumulus parameterization schemes which show disturbance energy accumulation depending on the magnitude of the mean upward circulation or of the inward horizontal water vapor convergence appear not to be valid. Energy received from the low-level inflow goes to enhance deep Cb convection. These Cb cloud elements rise to great heights, overshoot their level of zero buoyancy and induce outflow at levels (∿ 12-14 km) where h is larger than that of the inflow air. Figure 4 shows how this high level energy export occurs. Table 1 shows that the net import of water vapor energy (Lq) does not equal

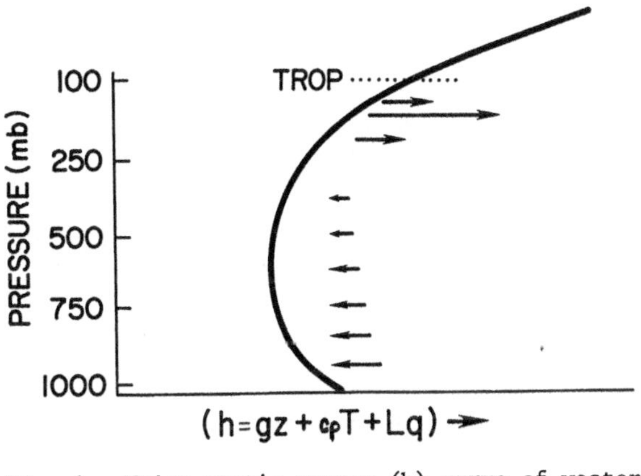

Fig. 4. Moist-static energy (h) curve of western ocean tropical weather systems and how the storm outflow occurs at a higher value of h than the inflow

TABLE 1

Typical 0-4° radius pre-cyclone cloud cluster energy gains and losses as a result of energy flux across the 4° radius - expressed in terms of °C/d mean surface to 100 mb tropospheric warming

Import of water vapor energy (Lq)	+ 4.5°C/d
Exports of dry static energy ($gZ + C_pT$)	- 5.5°C/d
Export of radiation energy	- 1.1°C/d
Required surface energy flux	+ 2.1°C/d
(Equivalent to 0.7 cm/d evaporation for a Bowen ratio of 0.1)	

the export of dry static energy (or $gZ + C_pT$). Thus, the paradox of the western ocean weather system's radial circulation; the in-up-out mean radial circulation needed in a mechanical sense to spin-up the weather system also acts in an opposite sense to export energy from the system and weaken it. This paradox has also been emphasized in a recent paper by Pearce (1981). Weather system maintenance and development require that the weather system's surface energy flux (E_o) be as large as the combined energy losses from the radial circulation and radiational divergence. Increase of h cannot occur unless term (a) of Eq. 1 is larger than terms (b) + (c). As the tropical storm becomes more intense the ratio of E_o/R becomes larger and larger. In the interior of the developed storm this ratio can reach values of 5-10 or greater.

Such large surface energy fluxes are quite consistent with the large vertical mass recycling which has been diagnosed for weather systems and typhoons by Gray (1973), Frank (1977c) and others.

2.1 Vertical mass recycling

In tropical weather systems and hurricanes (except in the inner core region, $r < 1^o$) there is a substantially greater amount of mass going up-and-down than that specified by the mean vertical circulation. The hurricane's vertical circulation is not just "in-up-out" as usually envisaged but rather is more typical of an "in,up-down-up, down-up, down-up, out" transverse circulation. Recycling is largest at cloud base and decreases to zero in the upper troposphere (see Fig. 5). A large amount of rainfall can thus occur within the tropical weather system beyond that specified by the magnitude of moisture convergence by the transverse circulation. Such a vertical recycling mass balancing circulation is indispensable for:

1) The maintenance of the weather system's convection. Cloud models show that, for the observed lapse rates of the typical summertime weather system, Cb convection requires large downdraft induced boundary layer mass forcing (Lopez, 1973),

2) Raising water vapor out of the planetary boundary layer, and

3) Obtaining the required surface ocean energy flux.

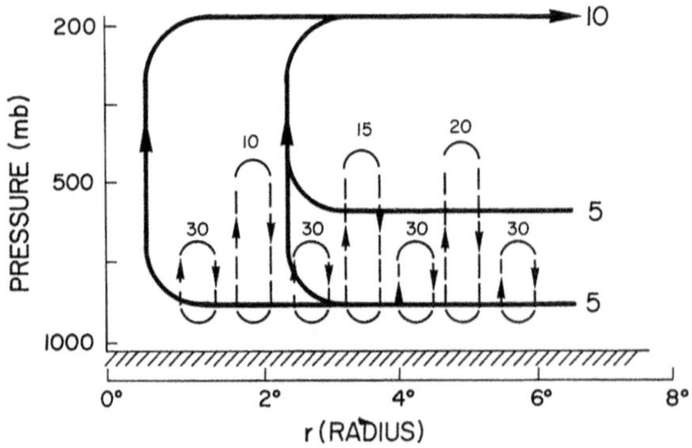

Fig. 5. Idealized x-section view of the hurricane's mean radial circulation with superimposed mass recycling. Units are arbitrary and represent mass transport

About two-thirds of the water vapor for the rainfall which occurs within 10° radius of the prehurricane and pretyphoon comes from local evaporation and not from inward horizontal advection (see Table 2). About a third of the rainfall within 0-4° radius comes from water vapor which is evaporated within that radius. Such evaporation rates are dependent upon the magnitude of the weather system's mass balancing vertical recycling.

TABLE 2

Percentage of pre-storm cloud cluster precipitation from evaporation by radial band from use of s-budget and horizontal vapor flux

Radius (^{o}Latitude)	Pretyphoon Cluster	Prehurricane Depression
0-2	24	37
0-4	30	39
0-6	36	46
0-8	52	54
0-10	65	65

2.2 Comparison of h-budget and bulk-formula estimated sea to air energy flux

It is observed that the surface energy fluxes calculated as residuals in our h-budgets are significantly larger inside 4-5° radius of developing weather systems than the surface energy fluxes we are able to calculate using mean composite quantities of the bulk formula $[C_E \rho_o U_o (q_s - q) + C_p \rho_o k U_o (T_o - T_a)]$. We observe that the ratio of these fluxes can vary from one at outer radii up to two at inner weather system radius.

We might think of the total sea to air energy flux (E_o or E_{TOTAL}) as being divided into a bulk formula component (E_{BULK}) and an extra or enhanced energy flux component (E_{EX}), thus,

$$E_o \text{ (or } E_{TOTAL}) = E_{BULK} + E_{EX} \qquad [2]$$

The extra surface flux appears related to the ability of the tropical weather system to develop rainbands. Such rainbands produce downdrafts inject faster, dryer, and colder air onto the ocean surface as indicated in Fig. 6.

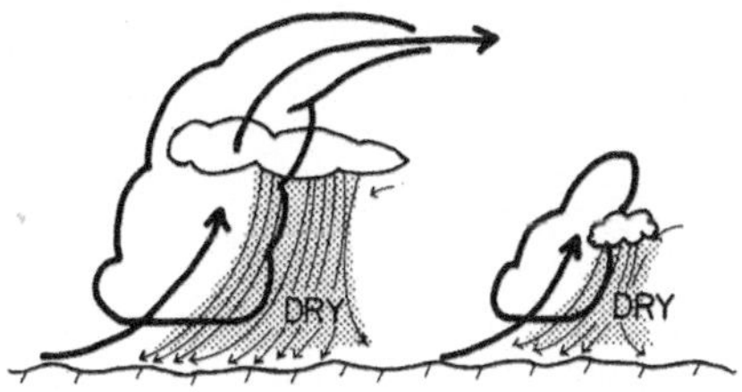

Fig. 6. Idealized vertical cross-section of downdrafts from line convection which brings high momentum and dry, cool air into the boundary layer. This enhances ocean energy extraction

It is hypothesized that the negative correlation of downdraft U_o with (q_s-q) and (T_s-T_o) on the rainband scale can cause a nonlinear local increase of surface energy flux on the rainband scale which can be 5-10 times larger than the surface energy flux specified by the bulk formula for observed mean conditions in the disturbance. The net effect on these downdrafts on the whole weather system is possibly an additional surface energy flux which is 50-100% larger than that specified by the bulk formula applied to mean conditions.

For instance, strong rainband downdraft action might cause the local surface wind (U_o) to increase over ambient conditions by a factor of 2-4 and at the same time through dry downdraft action bring about a doubling or a tripling of the (q_s-q) coefficient. Such simultaneous U_o and (q_s-q) alterations would cause a local and temporary increase of evaporation over that specified by the broadscale bulk formula by a factor of 5 to 10. A greater percentage increase would occur in the sensible heat flux because the normal air-sea temperature difference (or T_o-T_a) of 1° would likely be increased in downdrafts to a value of 5-6°C. Such (T_o-T_a) changes combined with local wind increases of a factor of 2-4 would cause sensible heat flux around a rainband to increase by as much as 10-20 times that of broadscale bulk formula values.

If downdraft rainband action were to be continuously active over 5-10 percent of the area of the typical weather system, then a net weather system increase of surface energy flux of 50-100 percent larger than that of the bulk formula might be expected.

We have made calculations of the ratio of total surface energy flux (E_{TOTAL}) to the surface energy flux specified by the bulk formula or E_{TOTAL}/E_{BULK}. Figure 7 shows this ratio as measured between 0-4° and 4-8° radius for our average west Pacific and west Atlantic developing and non-developing weather systems. Note that at the 0-4° radii the developing system data sets show a significantly higher value of total to bulk formula specified surface flux. Table 3 gives this ratio for all of our developing and non-developing data sets in each ocean basin. Data accuracy restricts calculations to a 4° average radial belt.

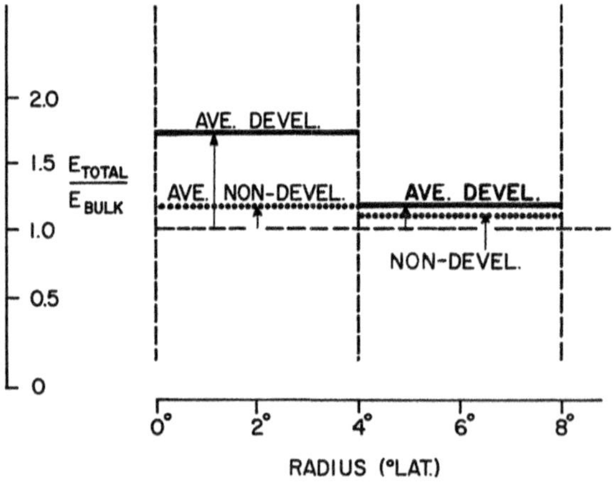

Fig. 7. Observed ratio of total surface energy flux (E_{TOTAL}) to broadscale bulk formula (E_{BULK}) surface flux for the average of our developing and non-developing weather systems

TABLE 3

Observed ratio of total surface energy flux (E_{TOTAL}) to broadscale bulk formula specified surface flux (E_{BULK}) between various radial belts

Weather Systems	0-4° Radius	4-8° Radius
West Pacific Cases		
Average Non-developing Systems	1.3	1.3
Average Developing Systems	2.0	1.3
West Atlantic Cases		
Average Non-developing Systems	1.1	1.0
Average Developing Systems	1.6	1.2

To alleviate much of the inherent observational noise in our cal-calculations, W. Fingerhut of our project has performed various smoothing analyses on our data sets. Figure 8 shows the ratio of ETOTAL to EBULK obtained by such data smoothing for the average of all of our early stage developing and all of our non-developing weather systems in both the Atlantic and Pacific. A distinct difference in this ratio is observed at radii inside 4-5°.

This ratio of total to bulk energy flux is closely related to (and appears parameterizable in terms of) the square root of the product of vertical motion at 300 mb and the tropospheric vertical wind shear between 150 and 500 mb or $[-\omega_{300\ mb}\ \partial V_\theta/\partial p]^{1/2}$. Thus,

$$\frac{E_{TOTAL}}{E_{BULK}} \simeq 1.0 + \frac{[-\omega_{300\ mb}(V_{\theta 150\ mb} - V_{\theta 500\ mb}) - 40]^{1/2}}{25} \qquad [3]$$

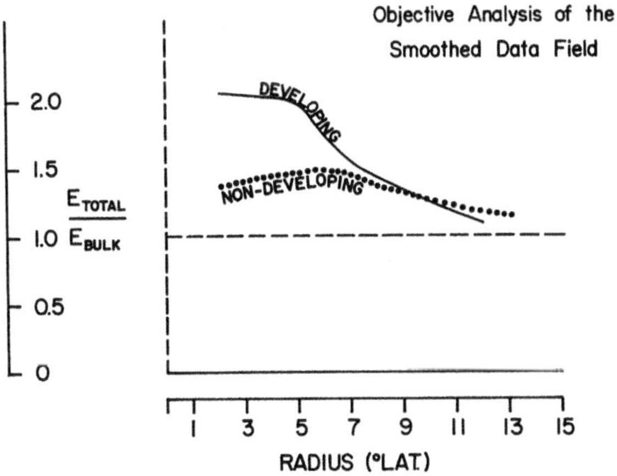

Fig. 8. Ratio of observed total surface energy flux to the bulk formula determined surface flux for the average of our observed developing and non-developing tropical weather systems

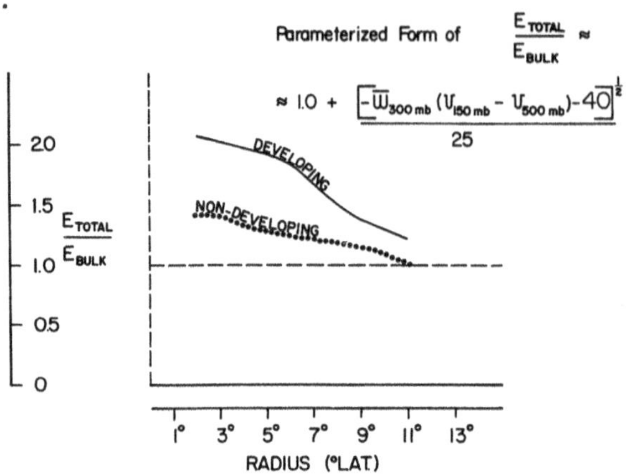

Fig. 9. Ratio of parameterized total surface energy flux to the bulk formula determined surface flux for the average of developing and non-developing weather systems

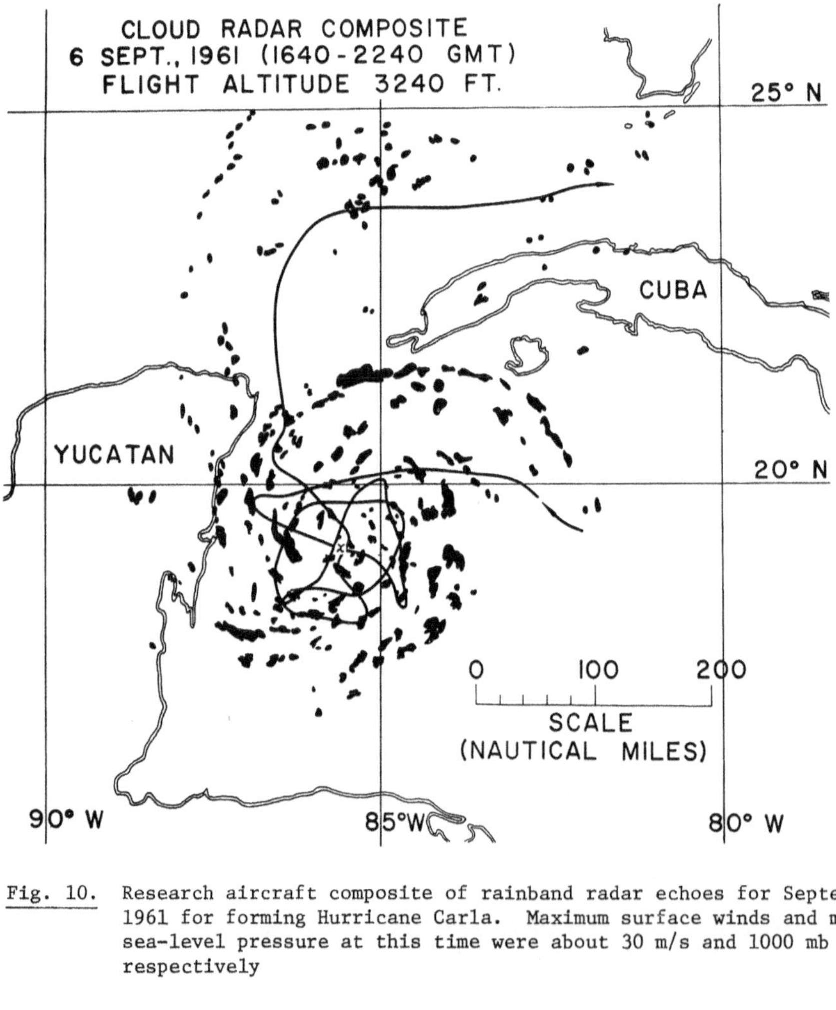

Fig. 10. Research aircraft composite of rainband radar echoes for September 6, 1961 for forming Hurricane Carla. Maximum surface winds and minimum sea-level pressure at this time were about 30 m/s and 1000 mb respectively

Figure 9 shows the parameterized form of this ratio as a function of radius for our smoother data sets. This parameterized form of E_{TOTAL}/E_{BULK} has a reasonable comparison with the observed values.

The author believes that the coefficient $(-\omega_{300\ mb}(\partial V_\theta/\partial p))^{1/2}$ is related to the tendency of deep convection to align itself into rainbands which are parallel with the upper tropospheric shear vector. William Fingerhut (1981) has recently tested such enhanced surface energy fluxes in his symmetrical PE-equation numerical model. He finds that these enhanced surface energy fluxes beyond that specified by the bulk formula are a crucial factor in distinguishing intensifying from non-intensifying weather systems. Our observational data clearly shows that the $(-\bar{\omega}_{300\ mb}(\partial V_\theta/\partial p))^{1/2}$ parameter is distinctly larger for the developing than for non-developing tropical systems.

An observed favorable arrangement of rainbands about a developing weather system is shown in Fig. 10. An idealized flow diagram summarizing the author's view of the fundamental processes of tropical cyclone genesis is shown in Fig. 11. Intensification requires continuous cycling around the bottom four boxes of this figure.

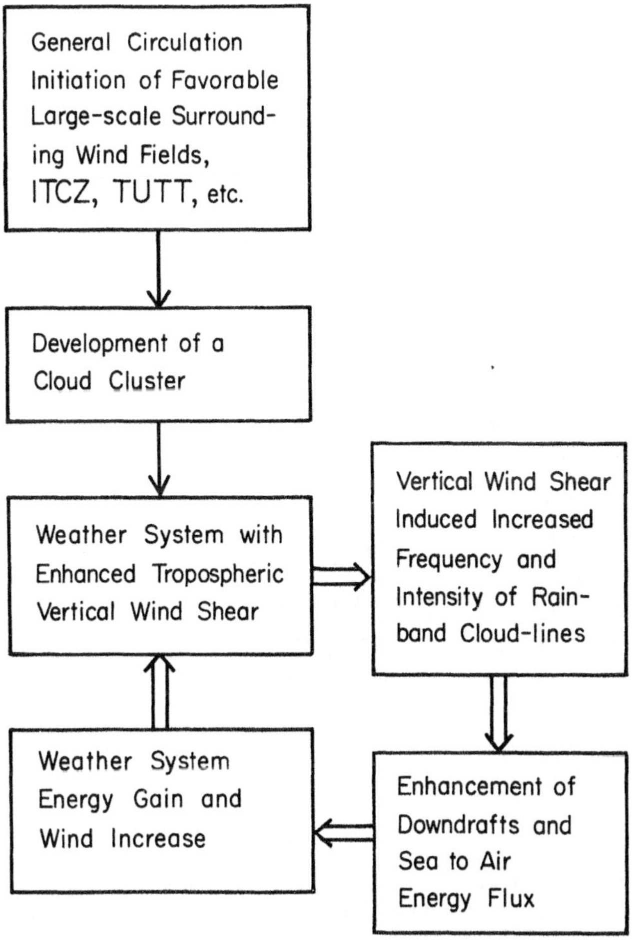

Fig. 11. Summary of hypothesized primary physical processes occurring to bring about tropical storm formation. Intensification requires a continuous cycling about the bottom four boxes

3 ROLE OF VERTICAL MIXING MOMENTUM PROCESSES IN STORM DEVELOPMENT

The author has earlier proposed (Gray, 1979a,1980) that the deep convective activity embedded in the larger vertical wind shear patterns of the developing weather systems acted directly to intensify the system by increasing the vertical wind shear. The squall line model of Imperial College (Green and Pearce, 1962; Moncrieff and Green, 1972; Moncrieff and Miller, 1976) which brings about vertical wind shear increase through cloud induced pressure forces was suggestive of such a direct convective induced intensification influence.

More detailed analysis of our observational data by Nunez (1981) and Lee (1981) show that this is not the case, however. The overall influence of cumulus convection in tropical weather systems apparently is to reduce and not increase the tropospheric vertical wind shear.

Nunez (1981) and Lee (1981) have solved for friction as a residual in the equations of motion and find that the net influence of friction and the mean vertical motion is to degrade the tropical weather system's vertical wind shear (see the left diagram of Fig. 12).

The tangential equation of motion may be written as:

$$\underset{(a)}{\frac{\partial \overline{V}_\theta}{\partial t}} = \underset{(b)}{-\overline{V_r \zeta_a}} + \underset{(c)}{\left(-\overline{\omega}\frac{\partial \overline{V}_\theta}{\partial p} + F_\theta\right)} \qquad [4]$$

where V_r and V_θ are the radial and tangential wind respectively

ω is the vertical p-velocity

ζ_a is the absolute vorticity, and

F_θ is tangential acceleration due to all subgrid scale processes.

The horizontal over bar $\overline{\quad}$ indicates spatial averaging over a horizontal grid size of 100-200 km.

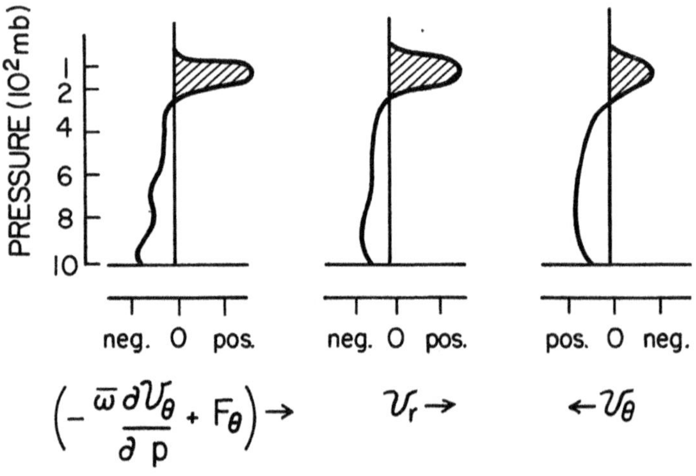

Fig. 12. Comparison of the typical vertical profile between 1-5° radius of the pre-cyclone weather system's tangential wind alteration due to vertical motion processes (left diagram), and the weather system's typical radial (center diagram) and tangential (right diagram) wind profiles. Note how similar these profiles are

It is observed that terms (b) and (c) are considerably larger than term (a). Terms (b) and (c) largely balance each other. Horizontal momentum rearrangements by the vertical motion processes thus bring about a radial motion response which acts to balance such vertical motion influences. In this sense the weather system's radial circulation may be considered to be primarily maintained by the horizontal acceleration action caused by the vertical motion processes. Figure 12 shows how very similar are the vertical profiles of $(-\bar{\omega}\, \partial V_\theta/\partial p + F_\theta)$, V_r and $-V_\theta$. The net influence of the vertical motion processes and the almost identical radial wind response (that is the balance between terms (b) and (c)) appears to be a fundamental process to the maintenance of the shape of the tangential wind profile in the vertical. This allows for continued rainband formation. Although vertical motion processes directly act to weaken the system's tropospheric vertical wind shear and apparently cause a weakening of the disturbance, they also (and perhaps more importantly) acts to bring about an alteration of the V_r wind such that the tangential wind and upper tropospheric vertical wind shears can be maintained. This is necessary for continued rainband formation and extra surface energy flux. Thus, the overall influence of the vertical motion processes might be interpreted (in an indirect sense) as being a necessary component to the weather system's development.

Fingerhut's (1981) modeling results show that the horizontal momentum alteration by the vertical motion processes (term c), are crucially important for the weather system's inner region wind concentration. Although the influence of the vertical motion processes is to inhibit vertical wind shear, such vertical motion mixing has a radial dependence. It inhibits inner weather system intensification less than outer system intensification. Fingerhut finds that this is a fundamental process to the system's later stage inner-region wind concentration.

4 THE CISK HYPOTHESIS AND TROPICAL CYCLONE DEVELOPMENT

Nothing appears to have more captured the imagination of tropical cyclone research meteorologists over the last 15 years than the loosely defined general theory which became known as Conditional Instability of the Second Kind (CISK). This theory has its origin in the paper by Charney and Eliassen (1964) and has been interpreted in different ways by a variety of meteorologists. Although a number of modifications to this theory were advanced over the years, such as wave-CISK, most of these new modifications retain a basic semblance of the original ideas.

The theory appears to have reasonable applicability to the inner-core and later stage development of the tropical cyclone. There appear to be basic problems with the general CISK concept (at least as it was generally perceived by many meteorologists) for explaining early stage tropical weather system intensification for the following reasons:

1) Only a small portion of the disturbance's inflow is driven by boundary layer convergence (Gray, 1979). The majority of the disturbance's inflow occurs above the boundary layer and cannot be well explained by Ekman convergence or wave convergence arguments. Only about a third of the rainfall in the pre-cyclone disturbance comes from vapor convergence between the surface and 900 mb.

2) The assumption of a balance between the wind and the pressure field of the early CISK models produced a substantial wind increase due to condensational heating. This effect suggests that the temperature increase due to condensation is over-estimated.

3) Increase of the transverse circulation and moisture convergence (other factors being equal) leads, as previously discussed, to a reduction of the disturbance net energy; not a gain as CISK theory implies. Thus, the sign of the process appears incorrect. The real energy gain of the tropical weather system can come only from a mass balancing vertical up-moist and down-dry recycling circulation resulting from surface energy flux, not from horizontal energy convergence.

4) The CISK theory does not explicitly deal with the more prevalent cases of weather system non-intensification. Many non-developing weather systems have massive amounts of inward moisture convergence and rainfall with no intensification.

5) The theory does not deal with the fundamental need for intensification of having large amounts of enhanced energy flux.

The most important missing links in the CISK ideas for explaining early stage disturbance intensification (as perceived by the author) is the lack of realization that the tropical disturbance's mean radial circulation does not act to import but rather exports energy from the disturbance (shortcoming 3) and the lack of realization as to the basic need of the developing disturbance to tap the ocean for large amounts of energy (shortcoming 5).

The type of wind increase and weather system intensification which is envisaged is illustrated in Fig. 13. This figure outlines the

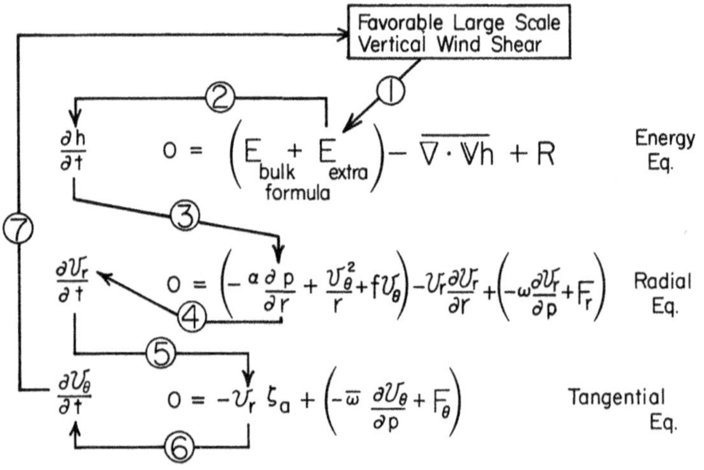

Fig. 13. Hypothesized progressive intensification steps (circled) whereby a weather system's favorable large scale wind shear brings about extra surface energy flux (step 1) and subsequent increases of h, $-V_r$, and V_θ. Intensification involves a continuous cycling through of these seven processes

various progressive intensification steps (circled) from enhanced surface energy gains to radial, and tangential wind changes. Intensification requires positive changes in $\partial h/\partial t$, $\partial\text{-}V_r/\partial t$, and $\partial V_\theta/\partial t$. Storm development requires the continuous cycling of this process or

$$\begin{array}{l}\text{Favorable} \\ V_\theta,\ \text{and}\ \partial V_\theta/\partial p\end{array} \longrightarrow E_{EX} \longrightarrow \frac{\partial h}{\partial t} \longrightarrow \frac{\partial}{\partial t}\left(\frac{\partial p}{\partial r}\right) \longrightarrow \partial\text{-}V_r/\partial t \longrightarrow \partial V_\theta/\partial t$$

(C: cycle from $\partial V_\theta/\partial t$ back to favorable V_θ, and $\partial V_\theta/\partial p$)

It appears that tropical weather system intensification can occur only if rainbands can be formed such that enhanced surface energy flux (E_{EX}) can occur.

5 DISCUSSION

It is hypothesized that once the troposphere favorably arranges its surrounding broadscale circulation patterns such that the tropical weather system's angular momentum budgets can be maintained (as discussed by Holland, 1981), it is the relative frequency and intensity of rainband formation which is the most crucial distinguishing feature between weather system development or non-development. Without the rainband's induced extra surface energy flux, the cloud cluster's transverse circulation would export more energy than it inputs and the system must die. It is observed that the majority of tropical cloud clusters which do not have such surrounding rainband features have typical lifetimes of only a day or less.

Much observational information is available to support the existence of greater amounts of line type convection around the prehurricane disturbance than the non-developing disturbance . V. Dvorak (private communication) has often stated that cyclone development tip-off occurs with the beginning formation of curved cloud lines. Observational evidence also indicates the likely association of rainband formation and intensity with large upper tropospheric vertical wind shear.

It is important that one distinguish between the rainband cloud line of the tropical weather system and the typical middle latitude severe weather squall line. The tropical cyclone rainband is typically oriented parallel with the vertical wind shear vector while the middle latitude squall line tends to arrange itself more perpendicular to the vertical shear vector as indicated in Fig. 14. This is likely a consequence of the quite different vertical wind shear profiles in which these two types of line convection exist - see Fig. 15. The tropical cyclone squall line downdraft acts to increase the low level wind in the direction in which the wind is already blowing, while the opposite occurs with the downdraft from the middle latitude squall line. Many other differences exist. One must be clear to differentiate between these two classes of cloud lines. Both types of line convection do, however, act to increase the wind and bring about dryness and a cooling of the boundary layer.

The weakest link in this argument is the lack of sufficient verifying evidence on the surface energy flux around line convection. The US National Hurricane Laboratory currently is attempting to measure such fluxes in its hurricane's boundary layer flights. We are also planning an extensive analysis of all available surface ship reports in and around the center of tropical cyclones to see the extent to which strong wind speeds and temperature and moisture are negatively correlated.

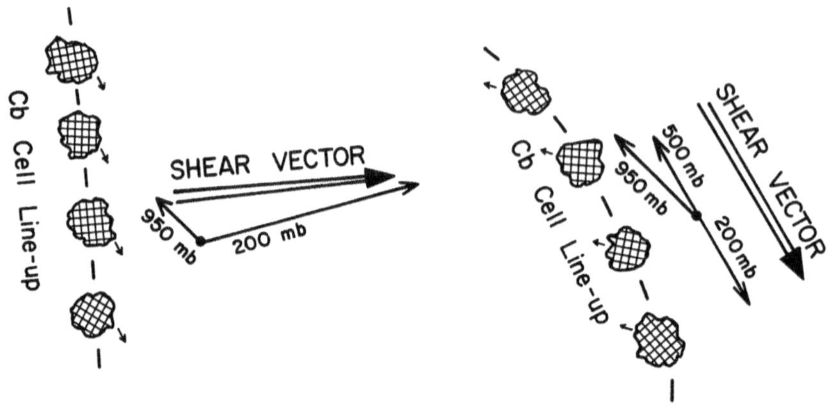

Fig. 14. Portrayal of how the usual US Great Plain squall line is oriented perpendicular to the 950 to 200 mb vertical wind shear vector (left diagram) while the usual tropical weather system rainband cloud line is typically oriented parallel to the vertical wind shear vector

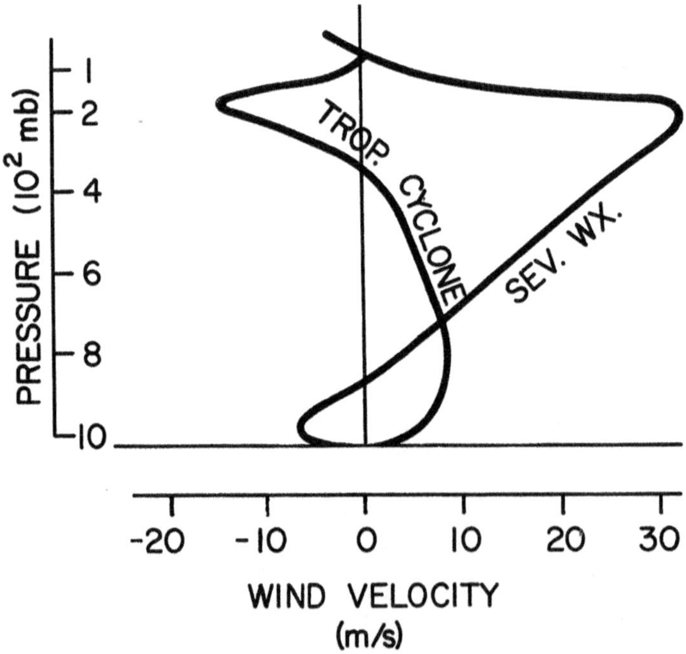

Fig. 15. Portrayal of the typical difference in the vertical profiles of wind velocity associated with the usual severe weather (SEV. WX.) event in the US Great Plains vs. the usual vertical wind profile associated with a tropical cyclone's rainband line convection

The second weakest link in this hypothesis is the lack of enough quantitative evidence on the line-up and intensity of deep convection downdrafts as a function of tropospheric vertical wind shear. We are planning a program to try to combine our rawinsonde observations with satellite observed cloud features to better document this relationship.

6 ACKNOWLEDGEMENTS

The paper is dedicated to Professor Herbert Riehl on the occasion of his 65th birthday. In the late 1950's while the author was a graduate student at the University of Chicago, Professor Riehl was instrumental in stimulating the author's curiosity concerning the nature of the tropical cyclone formation problem through repeated asking the question "What makes a hurricane?", "What makes a hurricane?, etc."

The author would like to acknowledge the many beneficial discussions he has had on tropical cyclone genesis over a number of years with Dr. William A. Fingerhut, Dr. John L. McBride (formerly of CSU), Geoff Love, and Professor Robert Pearce of the University of Reading, UK with whom the author has had a continuous and profitable dialogue on this subject for a number of years.

The author is indebted to Mr. Edwin Buzzell who has made most of the background rawinsonde calculations necessary for this discussion. He is also very appreciative of the overall research assistance provided by Mrs. Barbara Brumit. This research has been financially supported by a continuing grant from the US National Science Foundation.

REFERENCES

Charney, J. G. and A. Eliassen, 1964: On the growth of the hurricane depression. *J. Atmos. Sci.*, 21, 68-75

Fingerhut, W. A., 1981: Tropical cyclone genesis - numerical modeling inferences. Dept. of Atmos. Sci. Ph.D. Thesis, Colo. State Univ., Ft. Collins, CO, 192 pp

Frank, W. M., 1977a: The structure and energetics of the tropical cyclone, I: Storm structure. *Mon. Wea. Rev.*, 105, 1119-1135

Frank, W. M., 1977b: The structure and energetics of the tropical cyclone, II: Dynamics and energetics. *Mon. Wea. Rev.*, 105, 1136-1150

Frank, W. M., 1977c: Convective fluxes in tropical cyclones. *J. Atmos. Sci.*, 34, 1554-1568

Gray, W. M., 1968: Global view of the origin of tropical disturbances and storms. *Mon. Wea. Rev.*, 96, 669-700

Gray, W. M., 1973: Cumulus convection and large scale circulations, Part I: Broadscale and meso-scale interactions. *Mon. Wea. Rev.*, 101, 839-855

Gray, W. M., 1979a: Hurricanes/their formation, structure and likely role in the tropical circulation. *Quart. J. Roy. Meteor. Soc.*, 105, 155-218, supplement to *Meteorology Over the Tropical Oceans*

Gray, W. M., 1979b: Cbservational inferences concerning the occurrence structure, and dynamics of tropical cyclones. Australian Meteor. Mag., 27, 197-211

Gray, W. M., 1980: An individual view of the progress in hurricane research over the last 25 years. 13th AMS Tech. Conference on Hurricanes and Tropical Meteorology, Dec. 1-5, Miami Beach, FL, 13 pp

Green, J. S. A. and R. P. Pearce, 1962: Cumulonimbus convection in shear. Tech. Note 12, Dept. Meteorology, Imperial College, London

Holland, G., 1981: Angular momentum budgets in tropical cyclones. Paper prepared for the IUTAM/IUGG Symposium on Intense Atmospheric Vortices, July 14-17, Reading, England

Lee, C. S., 1981: Vertical rearrangement of tangential momentum in tropical cyclones. Dept. of Atmos. Sci. M.S. Thesis, Colo. State Univ., Ft. Collins, CO, 67 pp

Lopez, R. E., 1973: A parametric model of cumulus convection. J. Atmos. Sci., 30, 1354-1373

McBride, J. L., 1981a: Observational analysis of tropical cyclone formation, I: Basic description of data sets. J. Atmos. Sci., 38, June issue

McBride, J. L., 1981b: Observational analysis of tropical cyclone formation, III: Budget analysis. J. Atmos. Sci.,38, June issue

McBride, J. L. and R. Zehr, 1981: Observational analysis of tropical cyclone formation, II: Comparison of non-developing versus developing systems. J. Atmos. Sci., 38, June issue

Moncrieff, M. W. and J. S. A. Green, 1972: The propagation of steady convective overturning in shear. Quart. J. Roy. Meteor. Soc., 98, 336-352

Moncrieff, M. W. and B. Miller, 1976: The dynamics and simulation of tropical cumulonimbus and squall lines. Quart. J. Roy. Meteor. Soc., 102, 373-394

Núñez, E., 1981: Tropical cyclone structure and intensity change. Dept. of Atmos. Sci. Paper, Colo. State Univ., Ft. Collins, CO, (in preparation)

Pearce, R., 1981: A dynamical analysis of hurricane intensification. Archives for Meteorology, 54, 19-42

Williams, K. T. and W. M. Gray, 1973: Statistical analysis of satellite observed cloud clusters in the western North Pacific. Tellus, 28, 178-201

On Basic Problems in Theory and Modeling of the Tropical Cyclone

K. V. Ooyama

Abstract

The tropical cyclone is a mesoscale power plant with a cyclone-scale supportive system. While the cyclone can be handled basically as a quasi-balanced axisymmetric vortex, the energetic core of the vortex is in a fully developed nonlinear regime of three-dimensional convection. One of basic problems of tropical cyclone theory has been the handling of the two interacting scales as one system. The purpose of this paper is to examine the conceptual development in tropical cyclone modeling, which is now returning to explicit calculation of the convective scale in order to simulate the full spectrum of scale-dependent dynamics.

1 Introduction

The tropical cyclone is a complex system of interacting physical processes and multiscale motions. A complete description would have to cover nearly all the subjects in meteorology, from cloud physics within turbulent convection to general circulations of the tropics, and from interactions with the ocean to radiative heat transfer into outer space. One of the most difficult aspects to understand is the organized moist convection, which is also present in many other weather systems. The tropical cyclone, however, has certain unique characteristics, which, when considered according to their relative importance in the circulation, can enable us to deal effectively with this complex problem. The significance of understanding these characteristics can be clearly seen when one reviews the advances made in tropical cyclone research during the past two decades.

The progress has been neither smooth nor straightforward. At times, communication among investigators has seemed to have presented as difficult a problem as the tropical cyclone itself. Some views of the theoretical progress, such as those presented by GRAY (1980), are misleading. The confusion about the popular acronym CISK is so wide-spread that it has become a useless term in any sensible communication. Another difficulty is the wide latitude in the usage of words "formation" and "genesis". These words are probably as old as the literature of the tropical cyclone, but their precise meanings remain undefined. This paper is an attempt to reestablish a common basis of communication by summarizing tropical cyclone research and, thus, indicating where we stand in our theoretical view of the problem. For this purpose, mathematical equations are

Topics in Atmospheric and Oceanographic Sciences
Intense Atmospheric Vortices, ed. by Bengtsson/Lighthill

avoided intentionally and references to individual contributions are kept to an essential minimum. An extensive bibliography is available in ANTHES (1981).

2 The Steady State

In the fully developed stage, the tropical cyclone is a nearly circular, warm-cored vortex, occupying the entire height of the troposphere and extending radially many hundreds of kilometers. Continually active clouds surround the center of the storm and are organized as the eyewall and rain bands. Although small-scale details may change continuously and, sometimes, rapidly, the tropical cyclone, as a whole, is a stable system that may persist for many days over the warm tropical ocean as a recognizable unit. In comparison with severe weather systems elsewhere, the longevity of the tropical cyclone is one of its important characteristics.

In very simple terms, therefore, a mature tropical cyclone may be thought of as an axisymmetric, free-spinning vortex in a steady state. Such was the basis of both observational and theoretical analyses in early times and is still the basic premise in operational forecasts of the storm track. Physically, the simplification is recognition that the necessary rearrangements of the mass and angular momentum fields have already been achieved in the large volume of the vortex, and also that the kinetic and potential energies have been stored in a stable configuration to establish a large dynamic inertia. The low central pressure at sea level is in hydrostatic balance with the warm inner core aloft, and the resulting radial pressure gradient force is opposed by the inertial forces of the rotating wind.

The free-spinning vortex, however, does not account for other important characteristics of the mature tropical cyclone, such as the persistent updraft and precipitation in the eyewall clouds, the inward spiraling airflow in a relatively thin layer above the ocean surface, and the radial outflow in an upper layer of the vortex. The radial and vertical flows, called "the secondary circulation" in fluid dynamics, are as important as the primary (i.e., azimuthal) circulation is to the real tropical cyclone. The classical model of the secondary circulation, which consists of three legs of the in-up-out trajectory (RIEHL, 1954), is a distillation of those observed facts. It is this secondary circulation that allows a mathematical model of the primary vortex to make first contact with the physical world.

If the vortex together with the secondary circulation were to remain in a steady state, as the real storm does in good approximation, certain conditions would have to be satisfied along the trajectory of the secondary flow. It is obvious that the interior updraft must be in moist adiabatic ascent to maintain the warm core. However, to keep its temperature warm enough for the low surface pressure of a typical storm, the inflowing air must enrich its heat content by picking up sensible and latent heat from the ocean as the air moves toward the low pressure center. For the air to be able to move along the inward trajectory in the inertially stable, steady-state vortex, the angular momentum of the moving air must be partially diminished by friction against the ocean. The outflow leg of the secondary circulation may be physically inactive if the upper region of the primary circulation conforms to the angular momentum transported by the secondary circulation from below.

The above is the essence of the steady-state theory, presented by MALKUS and RIEHL (1960) and RIEHL (1963). Actually, their purpose was not to prove that a steady state is strictly possible, but, rather, to show that the exchange of both angular momentum and thermodynamic energy between the low-level inflow and the ocean is necessary, and empirically sufficient, to explain the coexistent primary and secondary circulations of a mature tropical cyclone in the inner area of, at least, a few hundred kilometers. Although numerical estimates of the boundary fluxes may vary with intensity of the storm or with different data sets (FRANK, 1977), the conclusion about the importance of the physically active secondary circulation does not become suddenly invalid if the storm is not yet fully developed, or if the idealized trajectory is modified. The qualification on the area of validity, though not of a primary significance, arises from the limit of observational data and also from theoretical considerations, such as inertial instability of the outflow.

The steady-state theory does not include any dynamic interaction between the secondary circulation and the primary circulation, although the former will not exist without the latter. In particular, the radial pressure gradient that drives the low-level inflow cannot be maintained without the deep layer of the primary circulation. To understand the dynamic relationship between the two circulations, we must consider a time-dependent problem of the combined system.

3 Genesis

The question at the other end of tropical cyclone problem, opposite the steady state, is that of tropical cyclogenesis. Speculations and suggestions abound, but we lack a clear understanding. Since it is still the most intriguing tropical cyclone question, we analyze the nature of the problem here.

When the characteristic rotational wind of an incipient tropical cyclone is detected, it appears to have developed in the area of a pre-existent disturbance with organized convective activity. However, although such disturbances frequently occur over the tropical oceans, most of them do not become tropical cyclones. There have been a number of studies on climatological and synoptic conditions (GRAY, 1978), but there are no hard data that might reveal the individual circumstances of tropical cyclone formation, or nonformation, in sufficient detail. Even if we were to carry out a field experiment to obtain such data, a diagnostic analysis of formative events would prove very difficult.

It is unrealistic to assume that the formation of an incipient vortex is triggered by a special mechanism or mechanisms, or that genesis is a discontinuous change in the normal course of atmospheric processes. For the reason that is discussed below, it is far more natural to assume that genesis is a series of events, arising by chance from quantitative fluctuations of the normal disturbances, with its probability of further evolution gradually increasing as it proceeds. According to this view, the climatological and synoptic conditions do not directly determine the process of genesis, but may certainly affect the odds of its happening. With a better understanding of the mesoscale dynamics of organized convection, the range of statistical uncertainty can be narrowed down. Nevertheless, the probabilistic nature of tropical cyclogenesis is not simply due to lack of adequate data, but is rooted in the scale-dependent dynamics of the atmosphere.

The schematic diagram in Fig. 1 summarizes the scale dependence. The abscissa represents a spectral decomposition of the motion in terms of horizontal scale ℓ. All the scales of motion on the abscissa may be present at any point in the geometrical space. The ordinate indicates Rossby's radius of deformation λ, which is a measure of the rotational constraint on the motion. In the general state of the atmosphere, in which disturbances are relatively weak, the rotational constraint arises mainly from the earth's rotation. Thus, the spectrum of motion at a given latitude may be represented on an appropriate horizontal line in the diagram. A typical line for the tropics (λ = 1000 km) and one for the extratropics (λ = 300 km) are shown. The third horizontal line is discussed later. Static stability of the atmosphere, except for the moist process, is assumed to be constant.

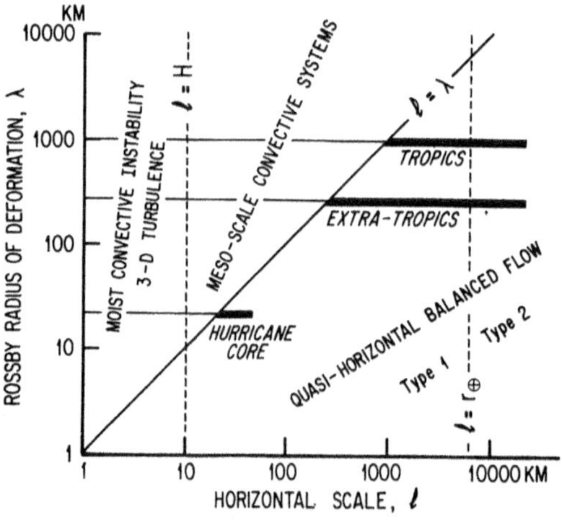

Fig. 1. Scale-dependent dynamics in the normal states of the atmosphere

For a given λ, the large-scale components of motion with ℓ greater than λ are quasi-horizontal and nearly in geostrophic balance with the pressure field, being strongly constrained by the earth's rotation. Vertical motion does occur, but only to the extent that is required by the adjustment of one balanced state of horizontal motion to another. PHILLIPS (1963) divides geostrophic motion into two types, but we shall refrain from discussing the detail. According to the theory of geostrophic turbulence (e.g., RHINES, 1979; see also a very illuminating review paper by TENNEKES, 1978), the spectral evolution of a two-dimensional flow is characterized by a slow rate of enstrophy cascade; for the atmosphere, the time scale is about a day. The kinetic energy cascade into smaller scales is essentially prohibited. Translated into more practical terms, the theory implies that the quasi-balanced synoptic-scale flow, shown by heavy-line segments in Fig. 1, is deterministically predictable for a period of, at least, a few days. It may be noted that the synoptic scale, here, is dynamically defined relative to λ, rather than by a fixed geometrical scale.

On the small-scale end of the spectrum is a different regime of motion, characterized as three-dimensional turbulence. Since the atmosphere is statically stable, very little energy should exist in this part of the spectrum, unless the air is heated from below, as in the boundary layer, or internally by latent heat of condensation, as in convective clouds. Over the warm tropical ocean, both conditions are regularly met; a well-mixed boundary layer of moist air and numerous convective clouds are ubiquitous phenomena. Horizontal scales of the energy-producing primary eddies (individual clouds) are limited by the vertical scale of a moist-adiabatically unstable layer. Thus, the area to the left of ℓ=H in Fig. 1 represents the regime of free convection, where H is either the scale height or the depth of the troposphere. According to the theory of three-dimensional turbulence, energy rapidly cascades into smaller and smaller scales, and no deterministic prediction in this regime is possible.

Since the atmosphere is not horizontally uniform, the small-scale free convection does not occur uniformly everywhere, but is modulated in patches of greater horizontal scales. Under certain conditions, the energy of the modulating scale also grows with the constituent free convection, and the modulation may eventually become a self-regulating mechanism. The result is the mesoscale system of organized convection. Unlike the synoptic-scale flow that can persist by itself, the mesoscale system cannot exist without active convection within. Thus, although it may last much longer than individual clouds, the time evolution of a mesoscale system is essentially probabilistic, unless it is strongly controlled by a more deterministic synoptic-scale environment.

Typical scales of the mesoscale convective systems observed during the GARP Atlantic Tropical Experiment were several hours long and a few hundred kilometers, at most. These scales were considerably smaller than those of the easterly wave, the prominent synoptic feature in that part of the tropics. Although the occurrence of mesoscale systems was statistically related to the wave phase, the association in terms of individual systems was highly variable. The composite average structure of the easterly wave by THOMPSON et al. (1979) completely smoothes out the mesoscale variability. In the extratropics, on the other hand, the scale separation between the mesoscale and the synoptic scale is small, suggesting a closer control of the former by the latter. In fact, a reasonable warning of severe weather can be issued from a synoptic forecast in mid-latitudes, even though individual tornadoes are not predictable.

Returning to the main topic, we now perceive the question of tropical cyclogenesis to be that of placing probabilistic mesoscale convective systems under the control of a deterministic environment. In the tropics, the mesoscale organization cannot grow to the normal limit of the synoptic scales. However, if the relative rotation, or vorticity, is increased in an area, the environment of mesoscale systems in that area is stiffened by the rotation. In other words, λ is locally decreased and, thus, the lower boundary of the dynamically defined synoptic scale reduces more closely to the mesoscale. If, by chance, this trend were continued, the deterministic dynamics of the balanced flow would begin to take over the control of the mesoscale convection. The final stage of this process, in which the control is complete, is indicated by the third horizontal line in Fig. 1. A full view of the scale diagram for a mature cyclone is shown in Fig. 2, in which λ is dependent on local inertial stability and, thus, a function of radius.

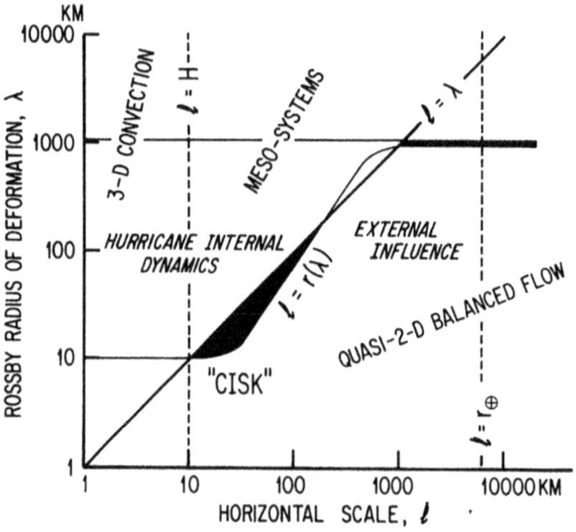

Fig. 2. Scale-dependent dynamics in a mature tropical cyclone

At the beginning of the genesis process, the moist convection is modulated and sustained by the three-dimensional dynamics of mesoscale systems, in which the vertical shear, the rain-induced downdraft, and the propagation of the system relative to the conditionally unstable mixed layer, are known to be important factors. The so-called Ekman pumping, which is a valid approximation for the frictional boundary layer of a deep balanced vortex, has no place in the self-regulating mechanism of the usual variety of mesoscale system. Currently, however, we do not have sufficient understanding of the control of organized convection during the transition from the probabilistic stage to the deterministic stage during tropical cyclogenesis.

This analysis prompts us to express a word of caution with respect to recent studies aimed at the genesis question. There is no doubt, in principle, that there exist large-scale influences on the genesis process. However, the composite studies of observational data that have been summarized in GRAY

(1978) do not necessarily reveal causal effects of large-scale conditions. Their findings in prehurricane and pretyphoon cases may be merely confirmation that the genesis process is already on its way. Numerical simulation studies, such as KURIHARA et al. (1981), are necessary and important means of exploring the multiscale three-dimensional problem of tropical cyclogenesis. In numerical models, however, the physics and the resolution of the dynamics are truncated. We must be very careful not to play the game with loaded dice.

4 Intensification and Maintenance

Until the early 1960's, attempts were made to theorize and numerically simulate the formation of a tropical cyclone, starting directly from moist convective instability (conditional instability). The intention was correct, in principle, but the execution was naive for lack of understanding that such a theory or model would have to go through many levels of nonlinear, multiscale, interactions before a tropical cyclone could emerge from individual convective clouds. Contrary to the premise of those earlier attempts, modern progress in theory and modeling of the tropical cyclone has descended from the steady-state theory and worked its way backwards, so to speak, to the question of genesis. (The physical time in a theory or model, of course, runs forward.) In the course of this progress, we have gained an invaluable insight into the complex dynamics of the tropical cyclone by being able to explain and simulate all the important aspects of the tropical cyclone in a dynamically coherent framework.

The basis of recent progress is an understanding of the intensification and maintenance mechanisms as a cooperative process between the organized moist convection and the cyclone-scale vortex, or, in the terminology of section 2, a cooperative process between the primary and secondary circulations. In the steady-state theory, the question is the compatibility of the two circulations. In the cooperative intensification theory, it is the dynamic interaction between the two. The importance of physical processes acting on the secondary circulation is unchanged, except that the geometry of the trajectory should not be fixed and the updraft cannot be strictly in moist neutral ascent. It is the presence of the primary circulation that was taken for granted earlier and that now must be explained.

The intensity of a tropical cyclone usually is measured in terms of either the minimum sea-level pressure at the center or the maximum speed of rotational wind that occurs in the zone of eyewall convection, usually within 50 km of the center. As we have mentioned, a low sea-level pressure requires a warm column of air aloft, which, in turn, requires the rotation of air around it to stay in place. If there were no rotation in a deep layer of the cyclone, it would be a matter of minutes, not hours, for the warm inner core to be completely dispersed in all directions. Therefore, regardless of the measure of intensity, the intensification of a tropical cyclone is possible if, and only if, the strength of the primary circulation increases. The frictionally induced inflow in the thin boundary layer cannot be a direct cause of intensification of the primary circulation in the main body of the cyclone, because the radial pressure gradient that drives the frictional inflow can increase only when the primary circulation in a deep layer increases. (For this reason, the sectionalized vortex in the theory by CARRIER (1971) would collapse if the artificial walls in his model were removed.) A simple and effective way to increase the rotation in the main body is to induce a radial inflow in the deep layer while conserving the absolute angular momentum. A layer of outflow at the top is needed to keep the inflowing air from accumulating in the interior.

The deep-layer inflow is, in essence, all that is needed for intensification of the cyclonic rotation. Its effect is cumulative and not influenced by surface friction. It may simply cease when the cyclone reaches a mature stage. There is one problem with this scenario, however. Energy is required for it to happen, especially an input of thermodynamic energy to lift the air at the end of its inflow leg to the higher level of the outflow. The kinetic energy necessary for increasing rotational winds is only a fraction of the required thermodynamic energy. As the supplier of the required thermodynamic energy, the secondary circulation in the frictional boundary layer, and in the moist updraft, becomes an indispensable partner of the cyclonic intensification.

The moist updraft must be convectively unstable; that is, the latent heat released by condensation of water vapor in the updraft must be greater than the amount of heat energy required by the moist air to rise from the boundary layer to the outflow level, so that the air from the middle layer can be entrained into the updraft and also lifted to the outflow level. In brief, the excess amount of latent heat released in the unstable convection does not directly raise the temperature of the inner core, but allows the mass flux of the updraft to increase with height by entrainment. This process further permits contraction of the air in the main body of the cyclone. If the cyclonic intensification continues, however, the inner core becomes so warm that the updraft is no longer unstable. Then, there is no more excess energy for entrainment, and the intensification comes to an end, although a slow change may continue in the outer area where the convection is still unstable. In this quasi-steady state, the frictionally induced inflow does not stop, and the heat input to the boundary layer from the ocean must continue in order to maintain the neutral moist ascent in the warm inner core. Thus, the intensity of a mature cyclone may rapidly respond to a change in the ocean surface temperature, and more drastically to a substantial reduction of latent heat transfer in the case of landfall.

The above is a very abbreviated description of the cooperative intensification theory. To formulate the theory, and numerical models, in a specific and quantitative form, one more point of great technical importance should be considered. Moist convective instability in numerous clouds is an undeniable fact of the tropical cyclone, and we have just shown that it is, indeed, necessary for intensification. As we have shown, convective clouds are in the energy-cascading regime of dynamics. Our problem, here, is how to handle the moist convection without letting the theory, itself, cascade into intractable turbulence. Obviously, we have to institute a closure assumption at some point in the spectrum of cascading scales. The closure, or the so-called parameterization, is an art of approximation by which implicit effects of the truncated smaller scales are put into an explicit consideration of the remaining larger scales. Actually, more than just geometrical scales are involved; that is, the dynamic characters of motion change with the scale. Therefore, we may attempt to bring a closure on organized convective systems, on individual clouds, or on in-cloud turbulence. The rationale for such a closure, as well as the complexity of the consequences, will vary accordingly.

Since the closure is not an exact science, there have been, and will be, different proposals and debates on specifics. Numerical techniques necessary in model simulation are not trivial problems, either. Nevertheless, it is important to recognize the achievements, as well as the limitations, of various model studies in general terms. For this purpose, the main direction of progress in tropical cyclone modeling may be divided into three phases as discussed below.

The models of the first phase, the so-called balanced models, capitalize on the extraordinary strength of local rotational constraint, which characterizes a mature tropical cyclone. As shown in Fig. 2, the motion in a mature cyclone is in a quasi-balanced state (gradient wind balance) on nearly all scales, except for that of individual clouds. Therefore, any organization of clouds in the inner core should be strongly controlled by the balanced flow of the vortex. In other words, the organized convection in the eyewall and in inner rain bands, although geometrically of the mesoscale, is not controlled by the ordinary mesoscale dynamics, but by the deterministic dynamics of the balanced flow. This is the fundamental rationale of the balanced model, with the closure assumption that puts the dynamics and thermodynamics of moist convection completely on the implicit side, and the primary and secondary circulations of the balanced vortex on the explicit side, of the scale division. This closure is invalid in the early stage of the tropical cyclone in which the scale gap between the convective clouds and the balanced flow is too wide, and never was intended for inquiry into the question of tropical cyclogenesis. On the other hand, the validity of the closure improves asympototically as the cyclone intensifies. Therefore, although a model simulation may be started from as weak a vortex as desired, the results are not interpretable physically until the model cyclone develops into a fully non-linear stage. For the later stage of intensification and also in the mature stage, however, the model results are not only realistic, but provide us with a clear insight into causal relationships among observed characteristics of the tropical cyclone. The most representative paper of the first phase of modeling is OOYAMA (1969).

The second phase of tropical cyclone modeling is marked by the use of primitive equations. Although the balanced vortex is conceptually helpful, it is overly restrictive and even inaccurate in areas of weak inertial stability and in the frictional boundary layer. However, the removal of the balance requirement from the explicit side of the closure brings a drastic change to the dynamic characteristics of the new closure. Even though the thermodynamic aspects of moist convection, which is mainly in the vertical, are still parameterized, the use of primitive equations of motion dictates that the cloud-scale dynamics, especially in terms of horizontal accelerations, be shifted from the implicit side to the explicit side of the closure. Thus, dynamically, the closure of the primitive equation model is on the vertically parameterized individual clouds. This is a welcome change, because the model is now, to a degree, able to generate the mesoscale organization of clouds, such as the eyewall and rain bands, explicitly. On the other hand, the computational burden is substantially increased to cope with cloud-scale instability, even though clouds are vertically parameterized. After considerable efforts to put numerical problems under control, many investigators have succeeded in simulating the tropical cyclone in the intensifying and mature stages. In fact, as documented by KURIHARA (1975), those results are a splendid confimation of the earlier results by the balanced model, as well as the present conceptual understanding of the intensification and maintenance processes.

With the use of primitive equations, there should be no restriction on model experiments starting from a very weak disturbance. Several experiments, including so-called sensitivity studies of initial growth rates, have been reported. The results, thus far, defy a definitive interpretation. There are wide variations among different models and a high dependence on various technical specifications within a model. One of many possible causes of uncertainty is the parameterization of convective clouds. Under a weak or nonexistent rotational constraint, the cloud-scale vertical circulation is a crucial factor in organizing mesoscale convective systems. The vertically parameterized cloud, mainly on the basis of thermodynamical considerations, does not produce the necessary effects, such as rain-driven downdraft. Thus, YAMASAKI (1977) and ROSENTHAL (1978) have opened the third phase of tropical cyclone modeling, by explicitly calculating cloud-scale circulations. The parameterization of clouds is completely removed and the closure is moved down in scale to in-cloud turbulence and cloud physics. The model results demonstrate the formation of squall lines in the early stage of a model run, and the cooperative mode of cyclone intensification in the later stage.

It is too early to say how these numerical models, with or without cloud parameterization, will contribute to our understanding of tropical cyclogenesis. The trend toward bigger and less restrictive models will continue, by allowing a larger domain to represent synoptic-scale conditions, and by improving geometrical and physical resolutions to replicate small-scale processes more faithfully. However, even if we could build a physically perfect model, the probabilistic nature of the genesis question would not change, as we have discussed in section 3. Besides the problem of logistics in running such a model, we must consider the problem of drawing conclusions from a few predictions of the event that has a very low probability of happening. Sensitivity studies with a model are easier to plan and execute, but conclusions may or may not be related to the real question.

5 CISK

Although we have intentionally avoided use of the acronym CISK, it deserves a few comments. Historically, the closure for the balanced model was introduced by OOYAMA (1963) and, in a somewhat different form, by CHARNEY and ELIASSEN (1964), as part of their respective linear theories of tropical cyclone intensification. The newly discovered instability was called conditional instability of the second kind (CISK) by CHARNEY and ELIASSEN. Linear mathematics necessitated that the theoretical demonstration be confined to the growth of small-amplitude disturbances on the basic state of no-motion. This is techically inconsistent with the physical reasoning of the closure. As we have shown, the true concept of CISK, as the cooperative intensification process, is appreciated only through the nonlinear model.

It is unfortunate that needless criticisms of the CISK concept is still raised, picking on the technical inconsistency of the linear theory and ignoring later contributions that have cast a better light on the theory. Meanwhile, the acronym has taken on its own life. A certain scheme of parameterizing moist convection, which was only a component of the CISK theory, has become known as CISK. Now, every mathematical exercise that employs the particular scheme, or similar ones, is called CISK, regardless of the original reason for the closure (see below). Many studies of preferred scales, or short-wave cutoff, of the linear CISK were irrelevant to the validity of CISK as a concept. The size of the eye, for example, is determined by a balance among nonlinear terms. It is almost impossible to list all examples of the abused acronym, but two such instances may be given. One is the Wave-CISK by LINDZEN (1974), which is nothing but conditional instability (of the first kind) in a sophisticated disguise. The conclusion of LINDZEN's paper, when stated clearly, is that there should be no convective clouds in the tropics. Another example is a puzzling reference to EKMAN-CISK by PEARCE (1980). By that, the author seems to mean the theory by CARRIER (1971), which was then proposed as an antithesis of CISK and still is.

For the closure of the balanced model of the tropical cyclone, an assumption (OOYAMA, 1963 and 1969) is made that the release of latent heat by moist convection in a vertical column is proportional to the supply of unstable moist air at the bottom of the column (more specifically to the vertical motion at the top of the mixed boundary layer). This assumption is commonly, but confusingly, referred to as the CISK parameterization and attributed to CHARNEY and ELIASSEN (1964). As a linear theory or a concept, CISK does not require this particular assumption. In fact, CHARNEY and ELIASSEN use a different assumption, and many numerical models of the tropical cyclone operate on a variety of parameterization schemes. In other words, there are many acceptable ways to parameterize the effects of moist convection for the same purpose. However, this does not imply that a scheme acceptable for one purpose would be acceptable for another purpose. In different applications of a parameterization scheme, the total closure assumption should be examined for each case. In this respect, many so-called CISK theories in the literature are physically unjustified. Below, we attempt to shed a little more light on this confusion.

Fig. 3 illustrates three processes that are important to moist convection over the tropical oceans. The abscissa is a spectral representation of horizontal scales, which we are interested in for the closure. The ordinate indicates the time rates of the three processes. Specifically, moist convective instability of individual convective clouds is shown as the rate of consumption of unstable moist air in the mixed layer. The clouds are on the implicit side of the closure and are independent of the scale of our interest on the explicit side. A rate of (1/2 hour)$^{-1}$ is shown by line (c) on the diagram. To maintain convection over the tropical oceans, either continuously or continually, the mixed layer must be maintained by transfer of sensible and latent heat from the ocean. A typical rate of mixed-layer recovery (1/2 day)$^{-1}$ is shown by line (b). It is obvious that moist convection cannot continue just by tapping the ocean below. If the convection were to continue in an area for many hours, the required moist air would have to be imported from the mixed layer outside that area. In the case of the tropical cyclone (except for the incipient stage), the advective supply is accomplished by the frictionally induced radial flow, which is on the explicit side of the closure. The advective supply rate, corresponding to 10 m s^{-1} of the convergent component of wind, is shown by a slanted line (a). For this speed, the convection can be supported continuously over an area of 15 km in linear scale, which may well be the eyewall convection.

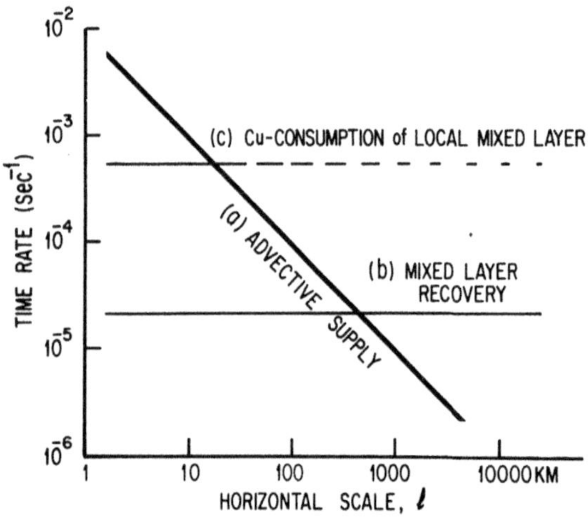

Fig. 3. Scale-dependent relation of three basic processes in organized moist convection

In the normal state of the tropical atmosphere, as well as in an incipient cyclone, moist convection can continue for hours in the form of organized systems. This is possible for such a system, when it generates mesoscale circulations to secure the supply of moist air. Squall lines, for example, achieve the supply by moving relative to the mixed layer air.

For the closure or parameterization, the problem is entirely different from that in a mature cyclone, since the mesoscale circulation is part of the dynamics to be parameterized, and cannot be put on the explicit side of the closure.

An attempt to apply the closure of CISK to synoptic-scale disturbances, such as the easterly wave, is also dubious, but for a different reason. Since the motion of scales greater than 1000 km may be assumed to be in a quasi-balanced state, the approximation of the EKMAN boundary layer is valid. However, as shown in Fig. 3, the boundary-layer convergence in such large scales has practically no effect on moist convection. The evaporative heat supply is the dominant factor that determines the overall activity of moist convection over the large-scale area. Of course, clouds are not necessarily distributed uniformly, but will be organized in mesoscale systems, over which the synoptic-scale field does not have direct control.

6 Conclusion

Progress in tropical cyclone theory and modeling during the past two decades has been reviewed, with emphasis on conceptual understanding. It is hoped that this paper will help to clarify the certain confusion in the current literature about basic problems of the tropical cyclone.

Acknowledgments

The author sincerely appreciates the encouragement given by Dr. S.L. Rosenthal and the helpful discussions with Dr. L.J. Shapiro in preparation of this manuscript.

References

Anthes, R.A., 1981: Tropical cyclones - their evolution, structure and effects. (To be published in Meteor. Mono.)

Carrier, G.F., 1971: The intensification of hurricanes. J. Fluid Mech., 49, 145-158

Charney, J.G., and A. Eliassen, 1964: On the growth of the hurricane depression. J. Atmos. Sci., 21, 68-75

Gray, W.M., 1978: Hurricanes: Their formation, structure and likely role in the tropical circulation. Meteorology over the tropical oceans. Roy. Meteor. Soc., 155-218

______, 1980: An individual view of the progress in hurricane research over the last 25 years. Selected Papers of the 13th Technical Conference on Hurricanes and Tropical Meteorology, Amer. Meteor. Soc., Miami Beach, Fla., 17-29

Frank, W.M., 1977: The structure and energetics of the tropical cyclone II. Dynamics and energetics. Mon. Wea. Rev., 105, 1136-1150

Kurihara, Y., 1975: Budget analysis of a tropical cyclone simulated in an axisymmetric numerical model. J. Atmos. Sci., 32, 25-59

______, and R.E. Tuleya, 1981: A numerical simulation study on the genesis of a tropical cyclone. (To be pulished)

Lindzen, R.S., 1974: Wave-CISK in the tropics. J. Atmos. Sci., 31, 156-179

Malkus, J.S., and H. Riehl, 1960: On the dynamics and energy transformations in a steady-state hurricane. Tellus, 12, 1-20

Ooyama, K., 1963: A dynamical model for the study of tropical cyclone development. New York Univ., 26 pp. (Unpublished manuscript)

________, 1969: Numerical simulation of the life cycle of tropical cyclones. J. Atmos. Sci., 26, 3-40

Pearce, R.P., 1980: A dynamical analysis of hurricane intensification. Contributions to Atmos. Phy., 54, 19-42

Phillips, Norman A., 1963: Geostrophic motion. Rev. of Geophysics, 1, 123-176

Rhines, P.B., 1979: Geostrophic Turbulence. Ann. Rev. Fluid Mech., 11, 401-441

Riehl, H., 1954: Tropical Meteorology. McGraw-Hill, New York, 342 pp

________, 1963: Some relations between wind and thermal structure of steady state hurricanes, J. Atmos. Sci., 20, 276-287

Rosenthal, S.L., 1978: Numerical simulation of tropical cyclone development with latent heat release by the resolvable scales I: Model description and preliminary results. J. Atmos. Sci., 35, 258-271

Tennekes, H., 1978: Turbulent flow in two and three dimensions. Bull. Amer. Meteor. Soc., 59, 22-28

Thompson, R.T., S.W. Payne, E.E. Recker and R.J. Reed, 1979: Structure and properties of synoptic-scale wave disturbances in the intertropical convergence zone of the eastern Atlantic. J. Atmos. Sci., 36, 53-72

Yamasaki, M. 1977: A preliminary experiment of the tropical cyclone without parameterizing the efects of cumulus convection. J. Meteor. Soc. Japan, 55, 11-30

On the Structure of Hurricanes as Revealed by Research Aircraft Data

R. C. Sheets

I INTRODUCTION

Research aircraft have been probing the inner core of tropical cyclones on a regular basis for more than two decades. Most of what we know today about the detailed internal structure and dynamics of these storms has resulted from studies based upon data collected from these airborne instrumented platforms. Noteworthy among earlier studies are those by Riehl and Malkus (1961) and LaSeur and Hawkins (1963). Riehl and Malkus used data collected in Hurricane Daisy (1958) to try to deduce the thermal and dynamical characteristics of a developing and mature storm. LaSeur and Hawkins used multiple level data collected in Hurricane Cleo (1958) to study the three-dimensional structure of a mature hurricane. Later, Hawkins and Rubsam (1968) conducted a detailed diagnostic study of Hurricane Hilda (1964), computing many of the same quantities as those computed by Riehl and Malkus in their 1961 study. Although, in general, findings were similar, significant differences between the two studies appeared, such as the depth of the inflow layer and the role of local generation as compared to advection for increases in the kinetic energy in the inner core of the hurricane. Similarly, several other case studies (Colon, 1961, 1964; Sheets, 1968, 1973; Hawkins and Imbembo, 1976) have shown gross features similar to those of the early Hurricane Cleo study, but have also shown wide variations from storm to storm and even from day to day within a given storm.

Shea and Gray (1973) composited or averaged data collected from more than 500 radial passes into and out of hurricanes to study the structure, dynamics and variability of the hurricane's inner core. Some of their principal results were the large variability of individual storms from the mean, supergradient winds just inside the radius of maximum winds in the lower half of the troposphere, and unstable lapse rates in the lower portion of the storm even in the inner core.

Comparisons of individual case studies with results of the composited analyses show significant differences. Gray used the results of the composite studies to speculate about the role of the supergradient wind in the formation and maintenance of the eyewall, and the role of the eyewall in the warming of the inner core. Disagreement continues over the existence as well as role of these supergradient winds.

Earlier studies including those by Shea and Gray relied upon doppler wind measurements which are affected by water motion. Shea and Gray as well as others tried to remove these possible errors, but questions continued concerning how accurate the wind measurements were in the inner core of the hurricane. In addition, most of the earlier work relied upon more qualitative than quantitative radar depictions of the convective activity to relate to other measurements. Fortunately, recent advancements in the development and acquisition of quantitative Plan Position Indicator (PPI) and Range Height Indicator (RHI) radar systems coupled with inertial navigation systems have permitted much more detailed documentation of the hurricane's

Topics in Atmospheric and Oceanographic Sciences
Intense Atmospheric Vortices, ed. by Bengtsson/Lighthill

structure. Special data sets were collected in Hurricanes Anita (1977), Frederic (1979) and Allen (1980) as well as Australian area severe Tropical Cyclone Kerry (1979) by utilizing these new systems.

Special emphasis in this paper is placed upon the Hurricane Anita case; this data set exemplifies state of the art capabilities. Also, these data were collected during the immature, developing and mature stages which permit intercomparisons between the various stages as well as some comparisons with earlier studies and speculations.

2 HURRICANE ANITA

Hurricane Anita developed from a highly asymmetric convective distribution and vortex into a highly symmetric "classical" hurricane over a period of three days. This storm progressed from one with no eyewall mechanism but maximum wind speeds in excess of 30 m/sec on August 30 to one with a highly organized eyewall and maximum wind speeds near 80 m/sec by September 1.

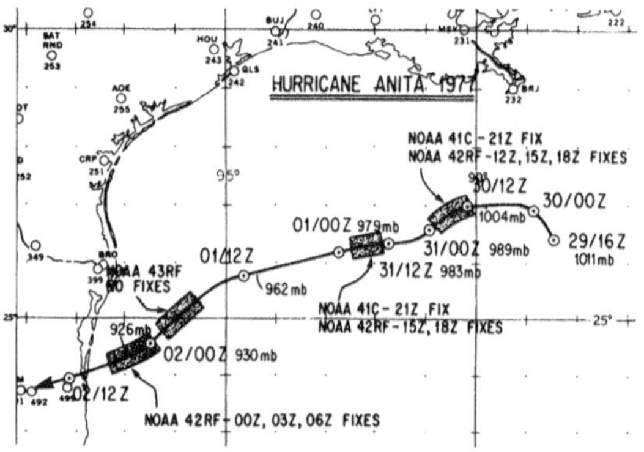

Fig. 1. Track of Hurricane Anita from August 21 to September 2, 1977, with superimposed locations of research aircraft data collection

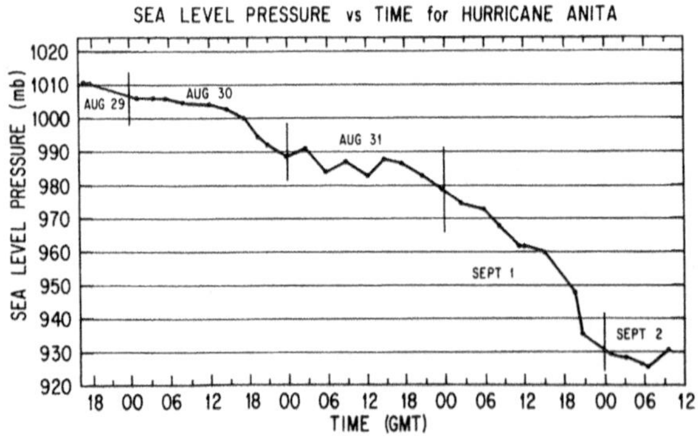

Fig. 2. Minimum sea level pressures recorded by reconnaissance aircraft during penetrations of Hurricane Anita

Fig. 1 shows the track of Anita as it formed, intensified and moved west-southwestward across the Gulf of Mexico. Also shown are the locations along the track where the data were collected that are analyzed and discussed in this report. Fig. 2 shows a graph of the minimum sea level pressure (MSLP)

versus time. It should be noted that the storm was slowly deepening on August 30 and August 31 (with some short-term fluctuations), deepened rapidly during most of the day on September 1 and finally leveled off just prior to going inland early on September 2.

2.1 August 30

Fig. 3 shows a visible spectra satellite depiction of Hurricane Anita for 1800 GMT August 30, 1977. Note the highly asymmetric cloud distribution with deep convection located to the east and southeast of the center of the vortex. Only low level cloud banding is shown over the western half of the storm. Fig. 4 shows the digitized PPI radar as recorded onboard the research aircraft at 1645 GMT on August 30. This radar depiction shows that the convection was poorly organized in terms of banding but was intense with large areas of greater than 46 DBZ level reflectivities which correspond to rain rates of approximately 37 mm/hr.

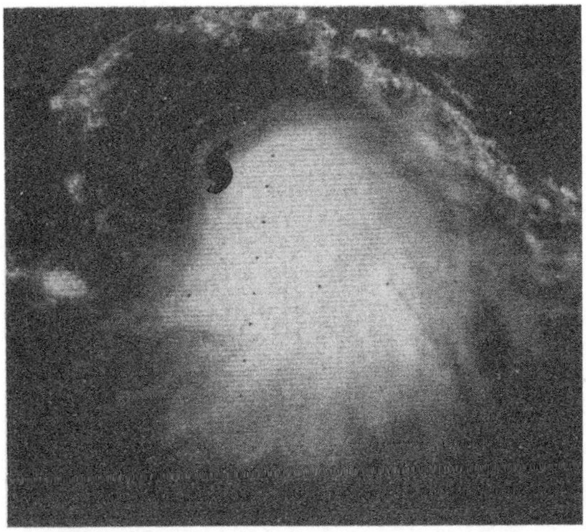

Fig. 3. Visible spectra satellite depiction of Hurricane Anita for 1800 GMT August 30, 1977

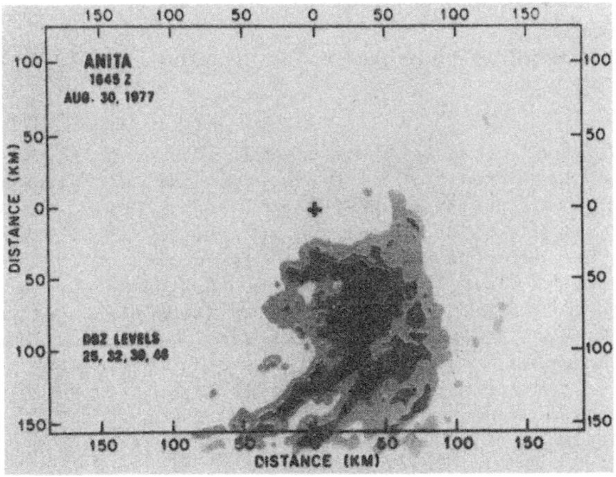

Fig. 4. Digitized Plan Position Indicator radar depiction of Hurricane Anita recorded onboard NOAA research aircraft at 1645 GMT August 30, 1977

Fig. 5 shows a low level streamline and isotach analysis of the relative winds (translational component removed) of Anita for 1130-1453 GMT August 30. Prominent features in this analysis are the extreme asymmetries in the wind field with dominance by the convection on the east side of the storm. Maximum sustained winds of more than 30 m/sec at a radius of near 60 km on the east side were recorded as compared to less than 10 m/sec at the same radial distance on the west side. The MSLP was only 1004 mb at this time. Pressure wind relationships used operationally (Holliday, 1969) for tropical storms and hurricanes give about a maximum wind of 20 m/sec for this pressure. Another analysis (not shown) for this same level utilizing data recorded over the two-hour period following the time of the data used in the construction of Fig. 5 indicates that these dominant asymmetric features were maintained over a period of several hours.

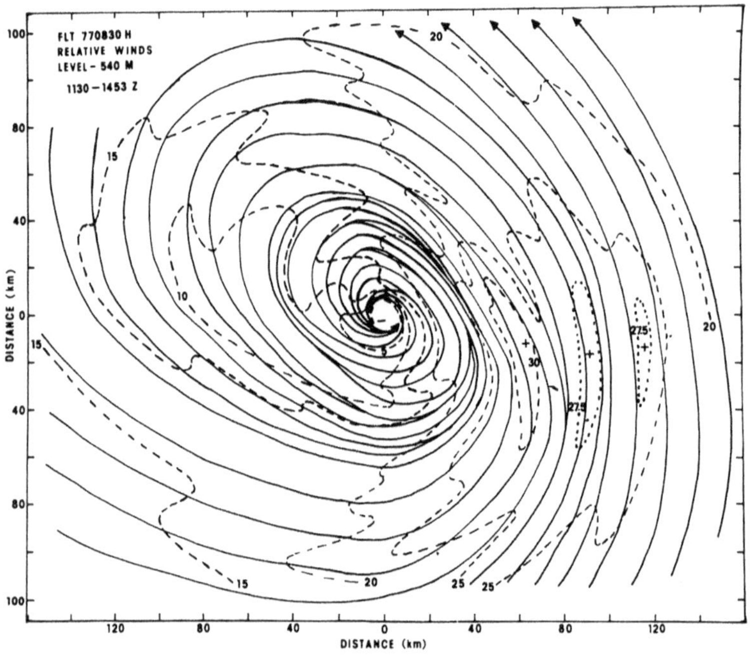

Fig. 5. Relative wind streamline and isotach analysis of data recorded for 1130-1453 GMT at the 540 m level in Hurricane Anita on August 30, 1977

Figs. 6 and 7 show vertical radar cross sections coupled with profiles of data recorded on 540 m level traverses from north to south and east to west through the storm. The RHI radar depictions were obtained by compositing the 3 CM radar reflectivities in the vertical above the aircraft during these radial passes. Therefore, the data shown in the profiles correspond directly in time to the radar depictions. These RHI radar depictions again show areas of intense convection on the south and east sides of the storm with a near total absence of precipitation on the north and west sides. In neither case is there any semblance of an eyewall mechanism. Even though convection existed on the south side of the storm, there were large regions of relatively dry air at this level (temperature dew point differences of more than 3°C). Note the relative warmth of the west and south sides of the system as compared to the east and north sides. Vertical winds were all less than

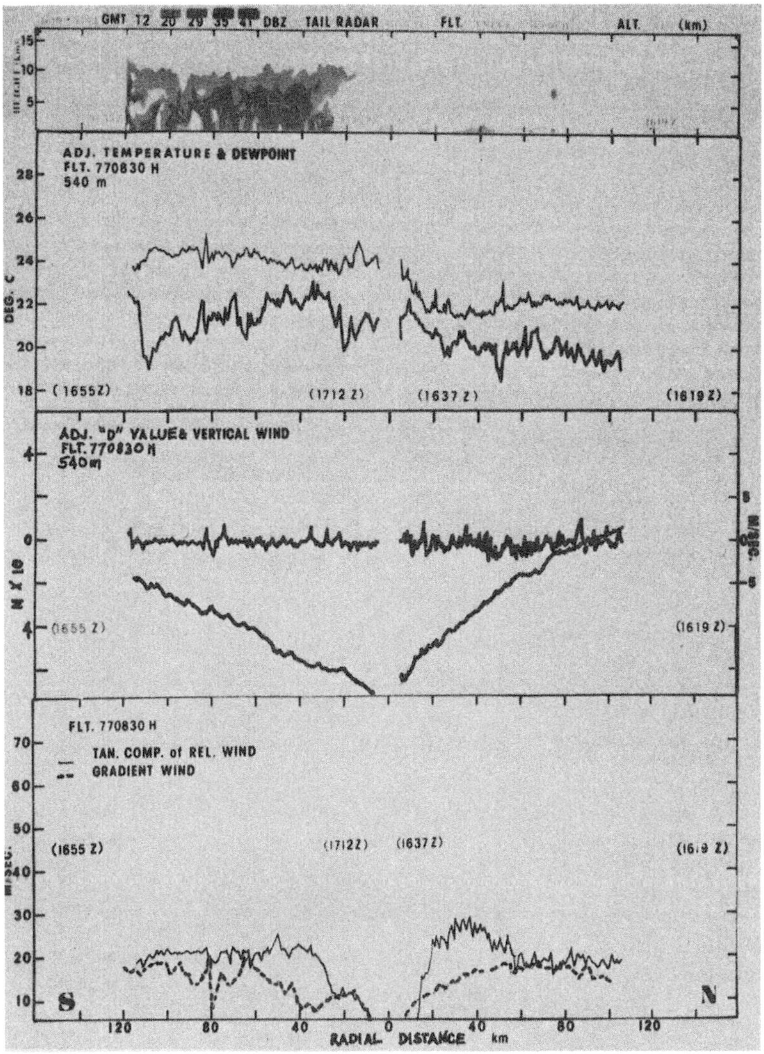

Fig. 6. Vertical radar cross section (top) with corresponding profiles of: a) temperature and dew point temperature, b) vertical wind, and c) tangential component of the relative wind and computed gradient wind for a north-south traverse through Anita at the 540 m level from 1619 to 1712 GMT August 30, 1977

5 m/sec with weak and variable pressure gradients within 150 km of the center of the vortex. Tangential wind speeds over most of the intense convective region as well as immediately downstream from the convection (north side) were generally supergradient while those over the convectively suppressed western portion of the vortex were near gradient or sometimes subgradient.

A streamline and isotach analysis for the 500 mb level (not shown) based upon data collected within two hours after the final analysis at the low level showed the center of the vortex displaced nearly 40 km south of the surface

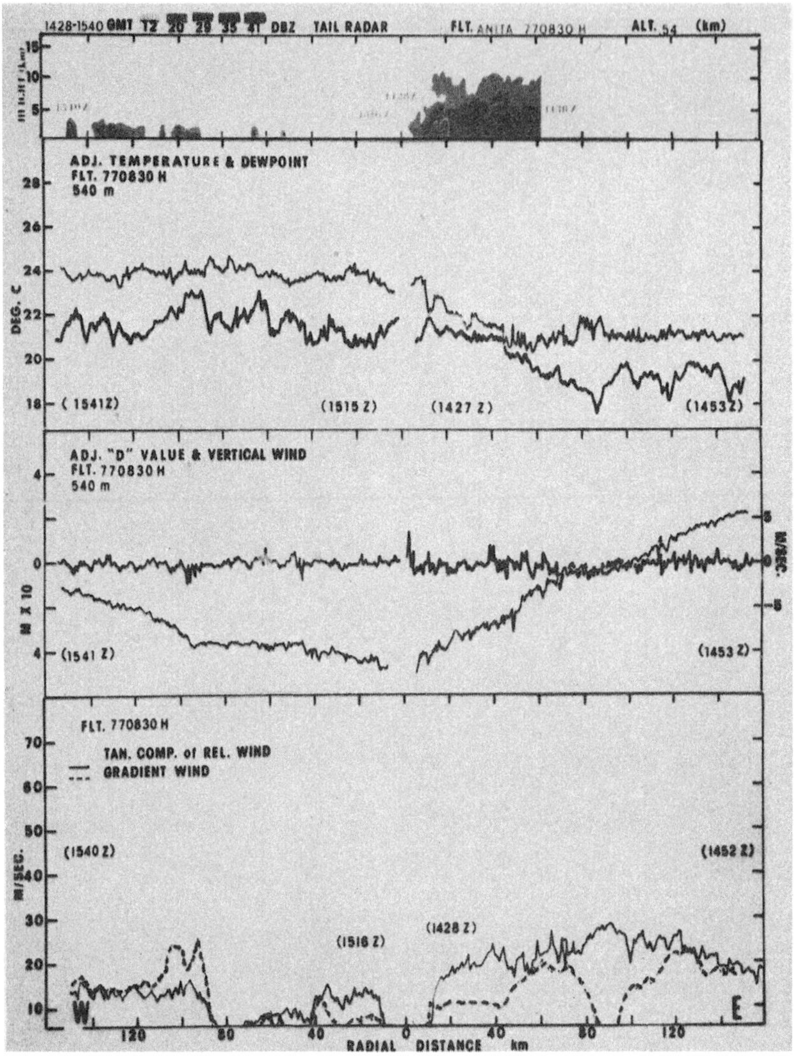

Fig. 7. As in Fig. 6 except for east to west pass for 1428-1540 GMT August 30, 1977

vortex center. However, winds over the central portion of the vortex at this level and extending northward to beyond the low level center were 5 m/sec or less. Again, the vortex flow was highly distorted at this level as was the case at the low level in the vicinity of the deep convection on the south and east sides. Maximum wind speeds were near 25 m/sec in the convective region with the widespread area of stronger winds primarily over the northern half of the storm at the 500 mb level as compared to the eastern half at the low level.

2.2 August 31

Fig. 8 shows a satellite picture of Hurricane Anita at 1930 GMT August 31, 1977. The storm appeared much better organized than on the previous day. The digitized PPI radar presentation for 1845 GMT on this day (Fig. 9) showed an eyewall. The intensity of the convection near the center of the storm was considerably reduced from that of the previous day. (Part of this apparent reduction may be due to the fact that these data were collected at a higher level (500 mb) in the storm than for the previous day (540 m). This possibility is partially supported by the RHI composites presented later.) A streamline and isotach analysis of the low level relative winds for this day (Fig. 10) showed that the region of strongest winds had rotated to the northern sector of the storm and exceeded 45 m/sec. The wind field was still highly asymmetric with maximum wind speeds of less than 25 m/sec in the southwest portion of the storm. Minimum sea level pressures were near 985 mb at this time which would correspond to surface winds of 35 m/sec.

Fig. 8. Visible spectra satellite depiction of Hurricane Anita for 1930 GMT August 31, 1977

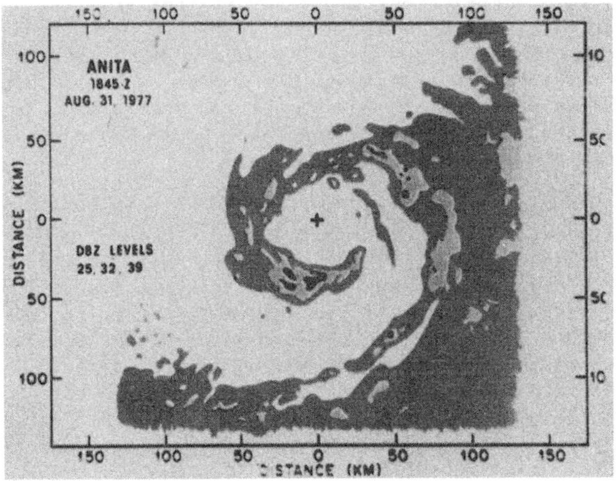

Fig. 9. As in Fig. 4 except for 1845 GMT on August 31, 1977

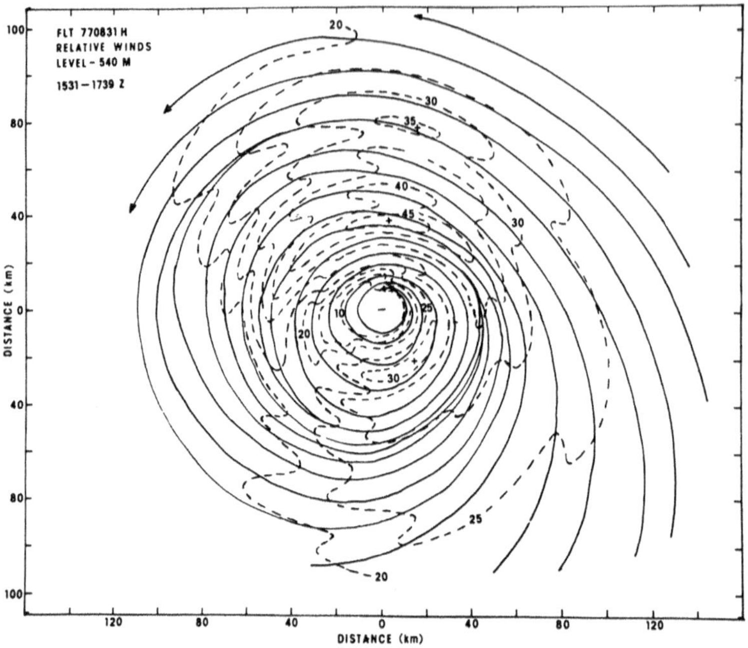

Fig. 10. As in Fig. 5 except for 1531-1739 GMT on August 31, 1977

Figs. 11 and 12 show the north-south and east-west RHI radar cross sections and corresponding data profiles for August 31. Here we see some central warming had occurred from that of the previous day but the west and south portions remained relatively warm and dry while the east and north were cooler and more moist. Also, vertical winds at these low levels had increased with some values exceeding 5 m/sec. Wind profiles started to take on a more characteristic decrease with increased distances outside of the radius of maximum winds. A large region of nearly constant supergradient relative tangential winds was present on the north side in the area where an eyewall was apparently trying to form. Winds on the interior edge of this maximum wind band were supergradient but were subgradient in the edge of the forming eye. Relative tangential winds on the nearly nonconvective west side were almost all subgradient while those on the east side where large radial inflow was taking place were in near gradient balance.

Analysis of the 500 mb level relative winds for this day (not illustrated) showed that the strongest winds were primarily confined to the north and west portions of the storm and exceeded 25 m/sec. This was nearly the same maximum speed observed on the previous day but stronger winds were more widespread and were further displaced cyclonically around the storm to the northwest sector. An analysis of the same variables for the 350 mb level for this day (not shown) was quite similar to the 500 mb level. A strong confluent zone was centered on the region of the forming eyewall on the western side of the storm at a radius of 40 km. On this day, the mid-level center was not displaced from the low level center as had been the case for the previous day.

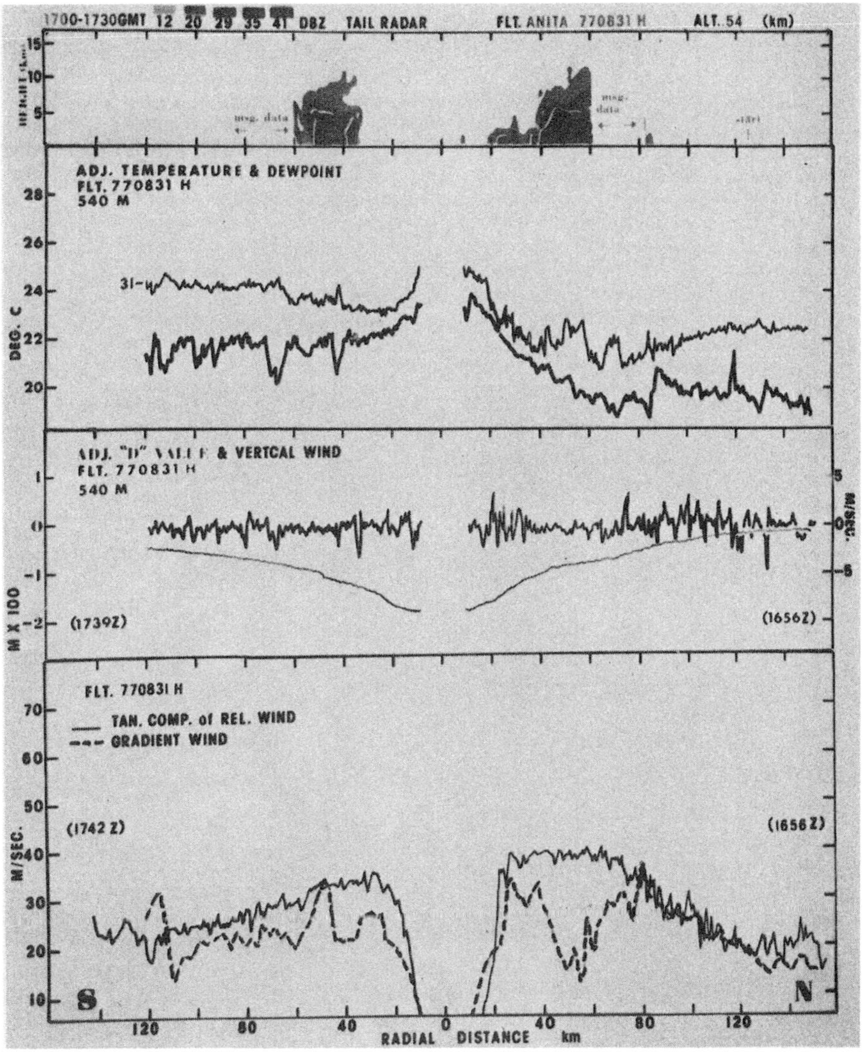

Fig. 11. As in Fig. 6 except for 1656-1742 GMT on August 31, 1977.

2.3 September 1-2

Hurricane Anita deepened rapidly as it moved across the western Gulf of Mexico. It had strengthened at a rate of 16 mb in 24 hours on August 30, 10 mb on August 31 and 48 mb on September 1 with 30 mb of the 48 mb deepening occurring during the last eight hours of the day! The storm then reached its minimum pressure near 0600 GMT on September 2 before moving inland over northeastern Mexico.

A satellite view of Hurricane Anita at 2300 GMT September 1, 1977 (Fig. 13) shows a "classical" hurricane with a well formed eye. Streamline and isotach

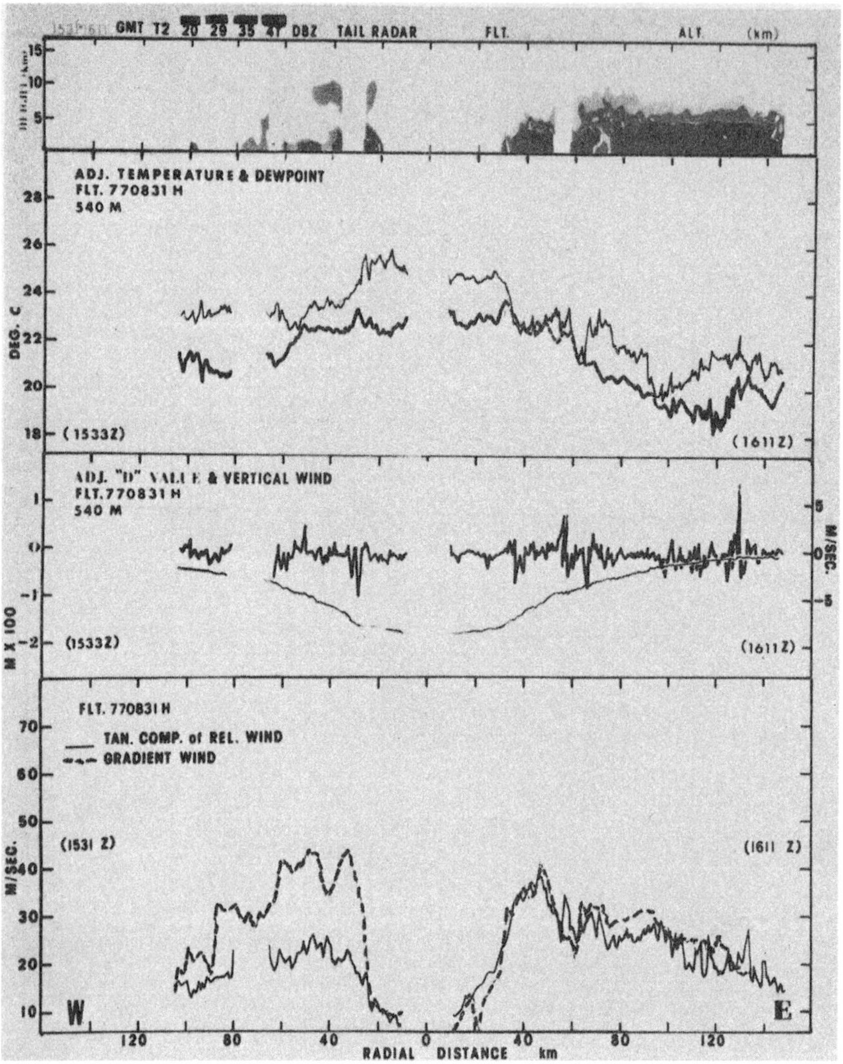

Fig. 12. As in Fig. 7 except for 1533-1611 GMT on August 31, 1977

analyses (not shown) of data collected at the 700 mb level early on September 2 (data are labeled as Flight 770901H since the flight originated on September 1) showed a highly symmetric storm with maximum winds exceeding 60 m/sec completely encircling the eye. Fig. 14 shows a PPI radar depiction of Anita. The aircraft was located near the first strong rain band 40 km north of the center at the time this recording was made. Some attenuation of the 5 CM radar return resulted when looking south through both the north and south eyewalls. More radar echoes were present on the south and east sides of the storm than are depicted here as shown on the RHI cross sections in Figs. 15 and 16. These cross sections show a well organized eyewall

Fig. 13. Visible spectra satellite depiction of Hurricane Anita for 2300 GMT September 1, 1977

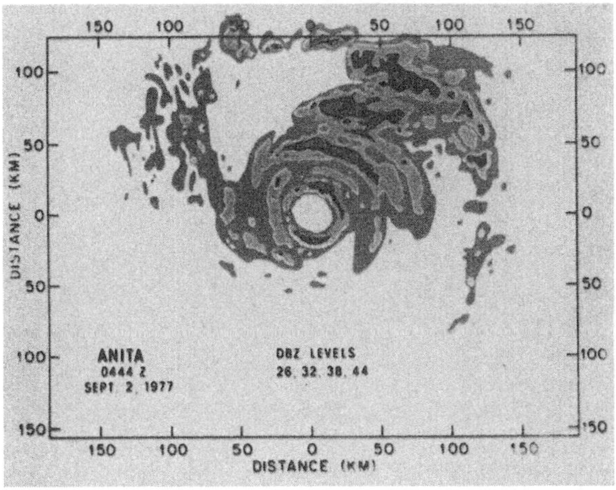

Fig. 14. As in Fig. 4 except for 0444 GMT on September 2, 1977

mechanism with varying degrees of outward tilt with height. Several instantaneous as well as composite RHI cross sections through the eye have been examined for the six-hour period monitored by this flight. Some profiles show a nearly vertical inner edge of the eyewall while most show a considerable tilt especially above the 5 km level. Due to an apparent stadium or bowl type structure, the eyewall tilt is sometimes exaggerated by passes which were not directly through the center (for composites) or instantaneous "slices" not taken directly at the center.

Many interesting features particularly of the eye and eyewall structure are shown in Figs. 15 and 16. Wind speeds are in near gradient balance in all quadrants except for some deviations in the immediate eyewall. Willoughby (1979) averaged Hurricane Anita's winds azimuthally for several radial passes and found the tangential wind to be in gradient balance. Also, as

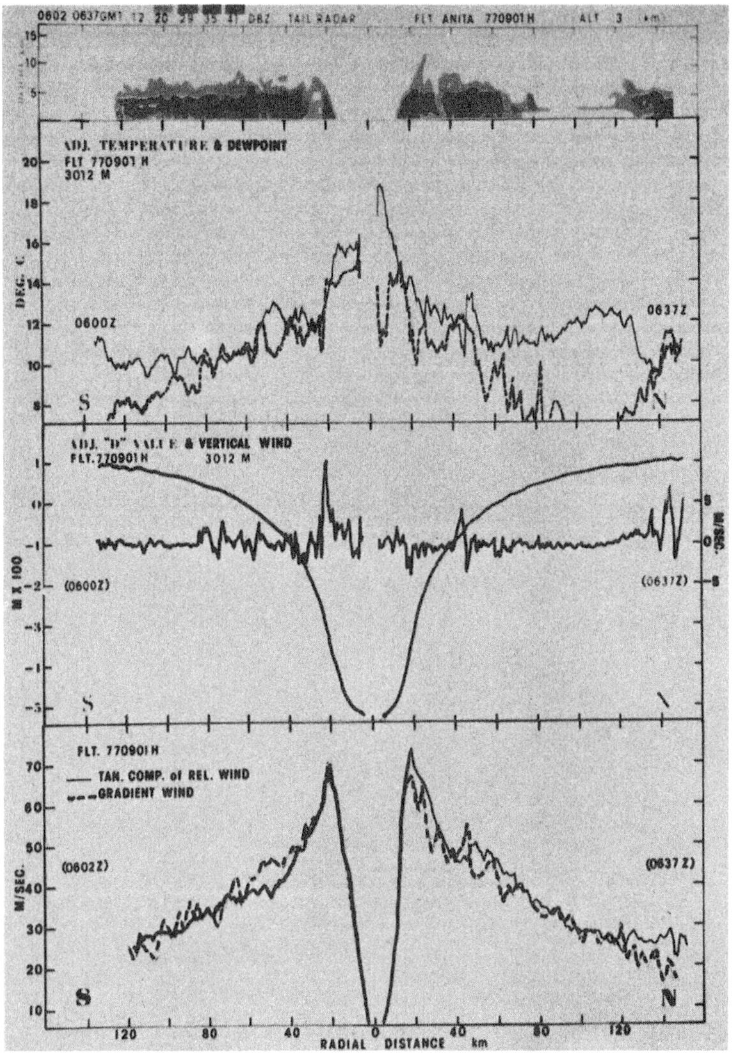

Fig. 15. As in Fig. 6 except for 0602-0637 GMT on September 2, 1977, at the 700 mb level

expected, winds are nearly tangential at this level for a mature hurricane. No comparable low level data were obtained on this day due to flight safety considerations. Although short-term fluctuations in the maximum wind occurred as maxima areas rotated through the region being sampled, the relative wind field was quite symmetrical. Secondary wind maximas were noted associated with most of the active rain bands.

Updrafts of 10 m/sec were recorded in the south and east eyewalls during these two passes. Other vertical winds in excess of 5 m/sec were frequently observed with one updraft in excess of 15 m/sec in a band near a radius of 140 km on the east side. Evidence of small downdrafts with extreme warming and drying on the inner edge of the eyewall similar to mean conditions found

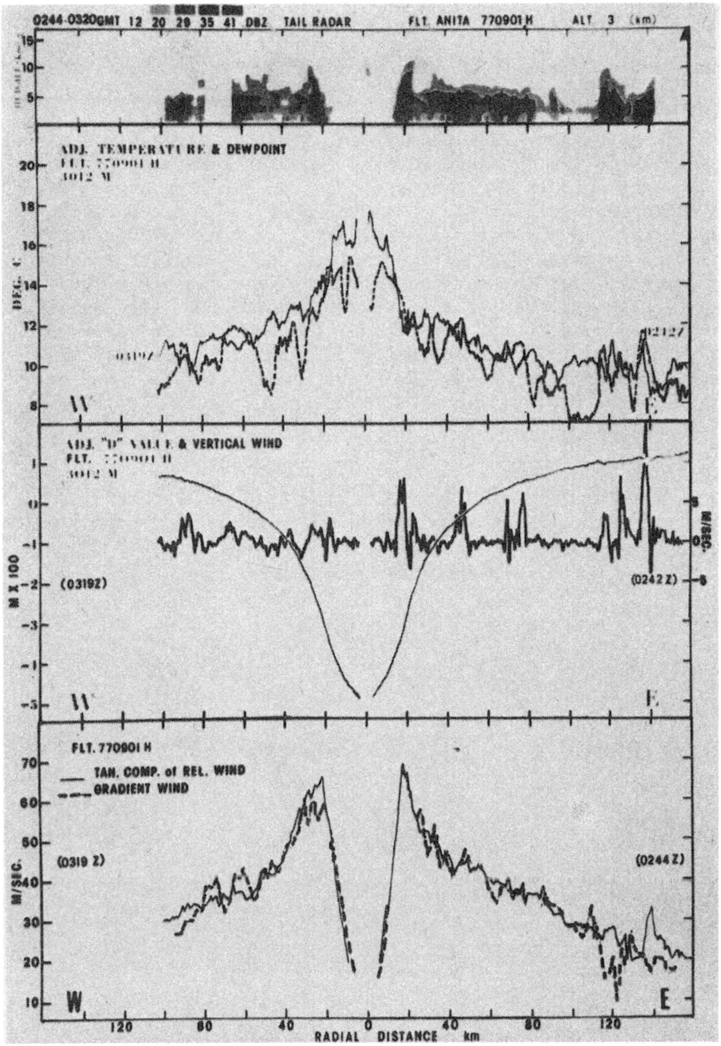

Fig. 16. As in Fig. 7 except for 0244-0319 GMT on September 2, 1977, at the 700 mb level

by Shea and Gray are shown on the west and north edges of the eye. In addition, even though conditions were near saturation over much of these profiles, high correlations are seen between many moist updrafts and dry downdrafts on each pass. The determination as to whether or not the moist updraft is warmer or cooler than its environment at this level is not easily made. Some cases show slight positive temperature anomalies in the updrafts, while others clearly show negative anomalies. However, in most cases where the updraft exceeds 5 m/sec there appears to be generally small positive anomalies.

3 SUMMARY AND CONCLUSIONS

Hurricane Anita was monitored over a three-day period by highly instrumented research aircraft during which it evolved from a weak vortex dominated by a highly asymmetric convective field into an intense symmetrical vortex and inner core convective field. Fig. 17 shows relative wind profiles for east-west and north-south traverses through the storm on each of the three days. These profiles can be characterized by the wind relationship (not shown) of $V_{\theta} r^{x}$ = const where x $\approx$ .2 on August 30 and early on August 31, to x $\approx$ .5 late on September 1 and early on September 2. By comparison, Tropical Cyclone Kerry was monitored on February 21 and 22, 1979. This storm had reached its maximum intensity one or two days earlier, appeared to decrease in intensity (as interpreted from satellite pictures), and then became nearly steady state for the two days monitored. Like Anita during its formative stages, Kerry had a highly asymmetric convective field. Also like Anita, the wind profiles could best be characterized by x $\approx$ .2 in the wind relationship cited above. Relative winds were supergradient on the inner edge of the maximum winds on at least two of the low level passes and in near gradient balance on another pass.

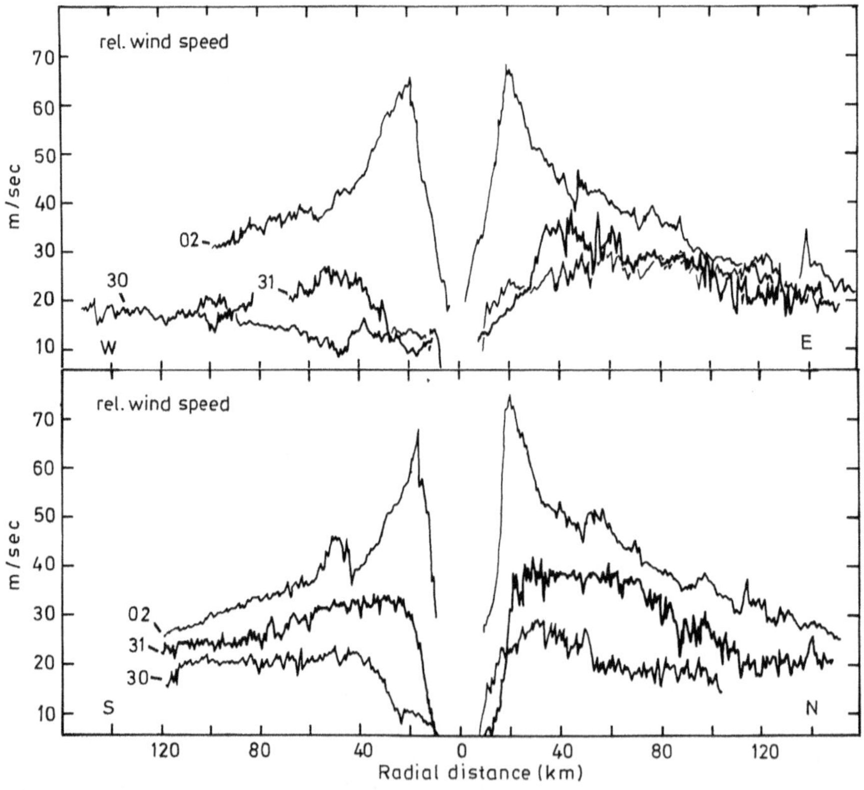

Fig. 17. Relative wind profiles for Hurricane Anita on August 30, August 31, and September 2 as recorded on east-west and north-south traverses through the hurricane

In summary, no clear consistent depiction of where super- or subgradient winds occur and their role in the location and formation of the eyewall can be determined from these individual cases. Winds were nearly in gradient balance at the mid-levels in intense Hurricane Anita but were generally out of balance at low levels during the formative (Anita) and dissipative (Kerry) stages when asymmetries were largest. Strong saturated updrafts at the middle level in mature Hurricane Anita showed small positive temperature anomalies while strong downdrafts generally exhibited dry and warm conditions. Weaker vertical wind speed areas were more difficult to characterize. Secondary wind maximums were often clearly evident in the mature hurricane as has been noted in other case studies cited earlier. This particular data set as well as previous studies cited did not cover extended periods where contraction or expansion of the eyewall and radius of maximum winds were observed. Such a data set, however, was obtained in Hurricane Allen (1980). This storm went through at least three large oscillations in the radius of maximum winds accompanied by large changes in the magnitude of the wind. This case is being studied by Willoughby of the National Hurricane Research Laboratory (NHRL). The data sets discussed here as well as others recently collected indicate that much is still to be learned about the internal dynamics of the tropical cyclone. Equipment now exists to make definitive studies of convection and its role or interaction with the tropical cyclone vortex. Jorgensen and Marks of NHRL, in particular, have developed radar data processing schemes which, coupled with planned airborne doppler radar being developed by Trotter, should provide new insight in the near future to convective processes in tropical cyclones.

4 BIBLIOGRAPHY

Colon, J. and Staff, NHRP 1961 On the structure of hurricane Daisy (1958). National Hurricane Research Project No. 48, pp.102 (available from NOAA, NHRL, Coral Gables, FL)

Colon, J. 1964 On the structure of hurricane Helene (1958). National Hurricane Research Project Report No. 72, pp.56 (available from NOAA, NHRL, Coral Gables, FL)

Hawkins, H.F. and D.T. Rubsam 1968 Hurricane Hilda, 1964. II. Structure and budgets of the hurricane on October 1, 1964. Mon.Wea.Rev., 96, 617-636

Hawkins, H.F. and S.M. Imbembo 1976 The structure of a small intense hurricane--Inez 1966. Mon.Wea.Rev., 104, 418-442

Holliday, C. 1969 On the maximum sustained winds occurring in Atlantic hurricanes. ESSA Technical Memorandum WBTM-SR-45, pp.6 (available from U.S. Dept. of Commerce, Weather Bureau-Southern Region, Ft. Worth, TX)

LaSeur, N.E. and H.F. Hawkins 1963 An analysis of hurricane Cleo (1958) based on data from research reconnaissance aircraft. Mon.Wea.Rev., 91, 694-709

Riehl, H. and J. Malkus 1961 Some aspects of hurricane Daisy, 1958. Tellus, 13, 181-213

Shea, D.J. and W.M. Gray 1973 The hurricane's inner core region: I, Symmetric and asymmetric structure; II, Thermal stability and dynamic characteristics. J.Atmos.Sci., 30, 1544-1576

Sheets, R.C. 1968 The structure of hurricane Dora (1964). National Hurricane Research Laboratory Report No. 83, pp.64 (available from NOAA, NHRL, Coral Gables, FL)

Sheets, R.C. 1973 Analysis of hurricane Debbie modification results using the variational optimization approach. Mon.Wea.Rev., 101, 663-684

Willoughby, H.E. 1979 Some aspects of the dynamics in hurricane Anita of 1977. NOAA Technical Memorandum ERL NHEML-5, pp.30 (available from NOAA, NHRL, Coral Gables, FL)

Lagrangian Angular Momentum Transports in Tropical Cyclones

G. J. Holland

ABSTRACT

Angular momentum budget equations for a moving tropical cyclone are presented and the physical role and relative magnitude of each component is described.

1 INTRODUCTION

Absolute angular momentum, defined here relative to a vertical axis at the tropical cyclone centre, is a quasi-conservative parameter and thus is a valuable tool in the analysis and interpretation of tropical cyclone dynamics. A number of authors have presented angular momentum budgets in Eulerian coordinates (cf. the survey in HOLLAND, 1981). However, tropical cyclones are typically not stationary relative to the earth and an analysis in moving, or Lagrangian coordinates should provide a more complete description and a better insight into their dynamical processes.

In this paper we present some results from the development and application of Lagrangian budget equations by HOLLAND (1981). We present the equations in section 2, describe the physical role and relative magnitude of each component in section 3, and synthesize the results in section 4.

2 LAGRANGIAN BUDGET EQUATIONS

The complete angular momentum budget equations in Lagrangian, cylindrical coordinates have been derived in a boundary flux form by HOLLAND (1981). They are:

$$\frac{\partial}{\partial t}\int_m M_r \, dm = F_H + F_V + T_m + F_\theta \qquad [1]$$

where M_r is the relative angular momentum (RAM) given by rV_θ, m is the mass of the volume under consideration, F_H and F_V are the horizontal and vertical boundary fluxes into the cyclone, T_m is a torque which the cyclone experiences when moving meridionally, and F_θ is the frictional dissipation at the earth's surface. Since the vertical motion is independent of the choice of coordinate system we will not discuss the vertical fluxes.

The horizontal fluxes, F_H, are given by:

$$F_H = -\frac{2\pi r^3}{g}\int_{P_2}^{P_1} \left(\frac{\overline{V}_r\overline{V}_\theta}{r} + \frac{\overline{V_r{}'V_\theta{}'}}{r} + \frac{f_o\overline{V}_r}{2} + \frac{\overline{f'V_r}}{4}\right) dp \qquad [2]$$

Topics in Atmospheric and Oceanographic Sciences
Intense Atmospheric Vortices, ed. by Bengtsson/Lighthill

where V_r, V_θ are radial and tangential components of wind speed, $V_{c\theta}$ is the tangential component of the cyclone speed and P_2, P_1 are the pressures of the upper and lower boundaries. The mean and perturbation quantities are defined by:

$$\overline{A} = \frac{1}{2\pi}\int_o^{2\pi} A d\theta$$

$$A' = A - \overline{A}$$

$$f_o = f \text{ at cyclone centre} \qquad [3]$$

$$f' = f - f_o$$

For ease in later discussion, we shall refer to Eq. 2 symbolically as follows:

$$F_H = F_H\,(\overline{V}_r\overline{V}_\theta,\ \overline{V_r'V_\theta'},\ f_o\overline{V}_r,\ \overline{f'V_r}) \qquad [4]$$

Hence, for example

$$F_H\,(\overline{V}_r\overline{V}_\theta) = \frac{-2\pi r^3}{g}\int_{P_2}^{P_1} \frac{\overline{V}_r\overline{V}_\theta}{r}\,dp$$

is the flux of RAM into the cyclone by the mean radial circulation. This and the other terms will be discussed fully in section 3.

The meridional motion torque, T_m is given by

$$T_m = \frac{-V_{cn}\beta_o}{4}\,(r_1^2 + r_2^2)\int_m dm \qquad [5]$$

where V_{cn} is the northward component of cyclone motion, β_o $(=\partial f_o/\partial y)$ is evaluated at the cyclone centre, and r_1, r_2 are the inner and outer boundaries ($r_1 = 0$ for the whole cyclone).

Following standard bulk aerodynamic methods, the surface frictional dissipation is given by

$$F_\theta = -0.64\int_o^{2\pi}\int_{r_1}^{r_2} r^2\rho C_D|V_9^*|V_{\theta 9}^*\,dr d\theta \qquad [6]$$

where $V_9^* = V_9 + V_c$ and $V_{\theta 9}^* = V_{\theta 9} + V_{c\theta}$ are the 900 mb Eulerian total and tangential wind speeds; 0.64 is a conversion to surface wind speeds; and C_D is a drag coefficient.

In deriving these equations HOLLAND (1981) started with the tangential equation of motion in cylindrical, Eulerian coordinates, changed to Lagrangian coordinates and converted to a momentum equation.

The major assumptions were that the cyclone is moving at a constant speed, V_c; and that the effects of earth curvature on radius can be neglected (introduces at most a 5% error).

3 DISSECTION: THE MOMENTUM BUDGET COMPONENTS

In this section we consider each component separately. We discuss its physical significance and present its typical magnitude for each of the complete set of developing northwest Pacific cyclones listed in Table 1. Full details on these data, and on the compositing philosophy, technique and statistical justification may be found in GRAY (1981).

TABLE 1

The northwest Pacific composite data sets used in this study

NAME	DESCRIPTION	MAXIMUM SUSTAINED WINDS ($m\ s^{-1}$)	LAT.	LONG.	MOVEMENT DIRECTION TOWARDS	MOVEMENT SPEED ($m\ s^{-1}$)
WPD2	Pretyphoon Cloud Cluster	<10	10^oN	153^oE	286	5.0
WPD3	Intensifying Cyclone	20	13^oN	144^oE	298	4.4
WPD4(D)	Intensifying Typhoon	40	19^oN	138^oE	302	4.0
WPD5	Supertyphoon	>50	22^oN	137^oE	323	4.0

In calculating the budgets we used Eulerian data on a cylinder with an eight (azimuthal) by six (radial every 2° latitude) grid with eighteen levels between the surface and 80 mb. The data were transferred to Lagrangian coordinates by subtracting the mean cyclone motion from each grid point. Standard techniques were used to compute all but the eddy RAM transport components. If these latter are computed from the above grid, only smooth eddies up to a maximum wavenumber of four can be resolved. Hence, we calculated the rV_rV_θ term for each Lagrangian observation in the original data and composited it separately to attempt to obtain a more realistic estimate of the eddy contribution. The resulting values are typically twice those obtained from the grid point data alone.

The results are presented in terms of tangential accelerations ($m\ s^{-1}d^{-1}$) resulting from the integrated angular momentum vergence into 2° latitude radial bands centered on 2, 4, 6, 8 and 10° latitude radius. This provides a simpler discussion framework than the more common method of presenting the actual momentum fluxes.

(a) The Horizontal Mean RAM flux, $F_H(\overline{V}_r\overline{V}_\theta)$ and Surface Frictional Dissipation F_θ

The F_H $(\overline{V}_r\overline{V}_\theta)$ term in Eq. 2 is the RAM flux by the mean radial, or Hadley type, circulation and is independent of cyclone motion. A positive transport, or cyclone spinup, will occur when this circulation imports high cyclonic momentum, and exports lower cyclonic momentum air. Air flowing inward in the lower levels imports cyclonic angular momentum, but much of this momentum is lost to surface friction and lower angular momentum air is removed in the upper level outflow. Hence, as implied by ROBOCK (1975), for a steady state the frictional dissipation and mean circulation enjoy a symbiotic relationship; the friction enables the mean circulation to import angular momentum and the mean circulation imports the angular momentum necessary to balance the frictional dissipation.

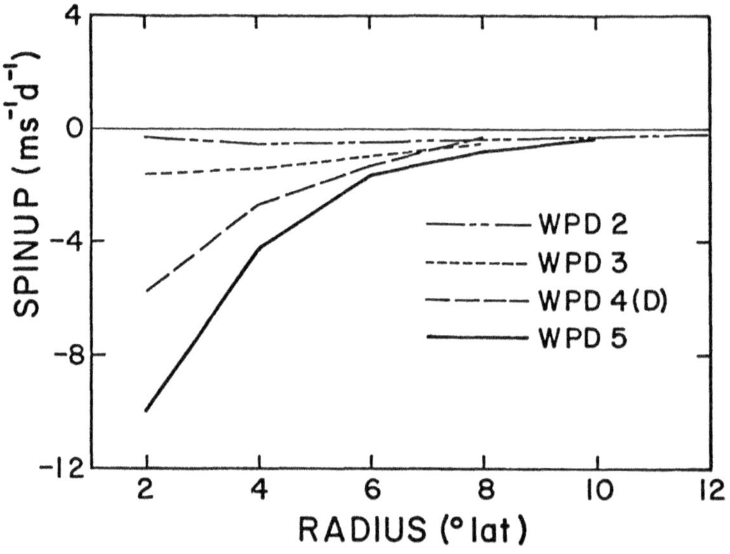

Fig. 1. Spinup by surface frictional dissipation within developing northwest Pacific tropical cyclones

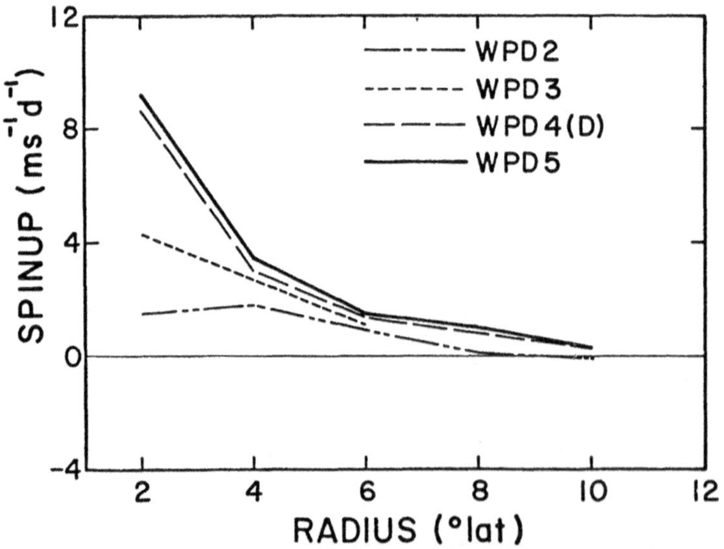

Fig. 2. Spinup by mean radial transports of relative angular momentum in developing northwest Pacific tropical cyclones

The surface dissipations and mean RAM flux accelerations for the cyclones listed in Table 1 are shown in Figs. 1 and 2. There is a substantial increase in friction with intensity in the inner region, as the dissipation increases with the square of the strengthening winds there, but negligible deceleration occurs in the outer region. As the surface friction and radial circulation increase, so also does the mean RAM transport in the symbiotic relation discussed above. However, in the inner region of the D2 through D4(D) cyclones the mean RAM transport is greater than the frictional dissipation. Thus implying that surfaces of constant AAM are contracting towards the centre, an expected result since these cyclones are intensifying. The opposite occurs for the D5 supertyphoon, which is on average weakening.

The surface friction also acts in an upgradient fashion to generate anticyclonic wind shear in the vertical. As originally noted by PALMEN and RIEHL (1957), the surface sink of cyclonic angular momentum inside the tropical cyclone generates an anticyclonic shear between the inflow and outflow layers; the stronger the radial circulation and frictional dissipation, the stronger the vertical shear. Since the maintenance or development of the warm core is dependent on such a shear configuration, this process is an important component of the overall cyclone dynamics.

(b) The Horizontal Eddy RAM Flux. $F_H(\overline{V_r'V_\theta'})$

The $F_H(\overline{V_r'V_\theta'})$ term in Eq. 2 is the RAM flux by azimuthal eddies. These eddies are not independent of cyclone motion because the cyclone motion adds a wavenumber one asymmetry to the Eulerian as compared to the Lagrangian winds. These azimuthal eddies may be visualized as Rossby type horizontal waves which contain no net mass transport but can import cyclonic angular momentum when the inflowing air has more cyclonic tangential wind speed than the outflowing air at the same level. The waves may vary from the cumulus, wavenumber one hundred scale to the full cyclone, wavenumber one scale.

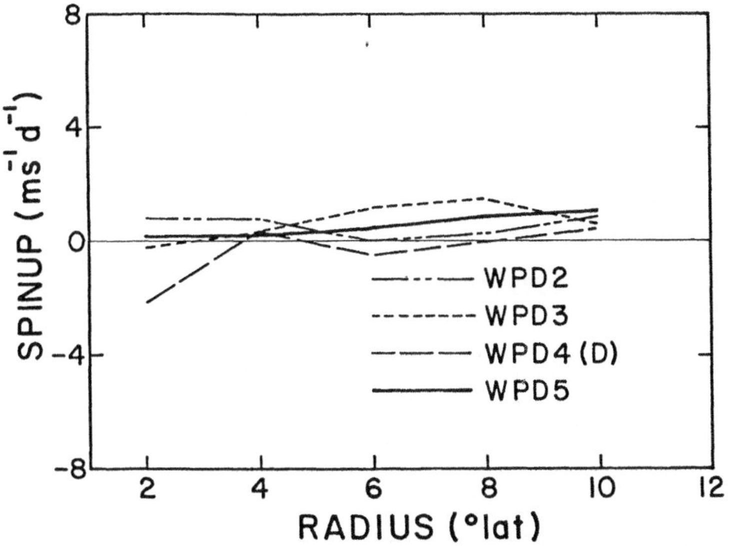

Fig. 3. Spinup by eddy radial transports of relative angular momentum in developing northwest Pacific cyclones

The eddy RAM transports for the northwest Pacific cyclones are shown in Fig. 3. By comparison with the mean RAM transports shown in Figs. 1 and 2, the eddy transports are smaller in the inner region and are comparable, or larger, in the outer region. They are also mainly confined to the outflow layer (a result of the large anticyclonic outward transports in the intense outflow jets). These observations are in agreement with earlier finding by ANTHES (1970), BLACK and ANTHES (1971), ROBOCK (1975) and FRANK (1977).

(c) The Horizontal Earth Angular Momentum Flux. $F_H(f_o\overline{V}_r)$

The $F_H(f_o\overline{V}_r)$ term in Eq. 2 represents the earth angular momentum (EAM) flux by the mean radial transport and is independent of cyclone movement. Unless there is a mass accumulation in, or loss from the cyclone this transport integrates to zero in the vertical. However, it provides a large cyclonic import in the inflow layer and export in the outflow layer. Thus, as originally noted by PALMEN and RIEHL (1957) and PFEFFER (1958b), it has the important role of rearranging angular momentum in the vertical and creating the vertical wind shears so necessary for maintaining the warm core.

(d) The Eddy Coriolis Torque. $F_H(\overline{f'V_r})$

The $F_H(\overline{f'V_r})$ term in Eq. 2 should not be mistaken as an eddy earth momentum flux. The earth component of absolute angular momentum is defined as the moment of earth rotation about the local vertical at the cyclone centre. It is thus $\frac{1}{2} f_o r^2$ and contains no azimuthal dependence. This azimuthal dependence arose from a transfer of an internal Coriolis torque (the fV_r term in the tangential equation of motion) to the boundary (cf. HOLLAND, 1981); hence we call the $F_H(\overline{f'V_r})$ term the eddy Coriolis torque to avoid any misunderstandings.

These eddy Coriolis torques have generally been neglected in previous studies (FRANK, 1977 is one exception). Yet, as shown in Fig. 4, they provide a large outer region acceleration in Lagrangian coordinates. Since there is no zonal gradient in f, the eddy Coriolis torque depends only on the meridional flow through the cyclone. Thus, the accelerations in Fig. 4 result from the observed fact that cyclones consistently move polewards faster than their environmental "steering current", and encounter an equatorward flow through and import of high Coriolis air.

(e) The Meridional Motion Torque. T_m

The T_m term in Eq. 1 arises from the meridional movement of the cyclone and the concomitant conversion between EAM and RAM. As the cyclone moves, say, poleward its axis tilts towards the earth's axis, the magnitude of f_o increases and the EAM becomes more cyclonic. If the absolute angular momentum is conserved the RAM must become more anticyclonic to compensate and thus the poleward movement is in effect applying an anticyclonic torque to the cyclone. The effect increases with radius and can be quite large; for example, moving a system with initially no tangential circulation from 10°N to 20°N will generate an anticyclonic circulation with tangential winds of 7.5 m s^{-1} at 600 km radius.

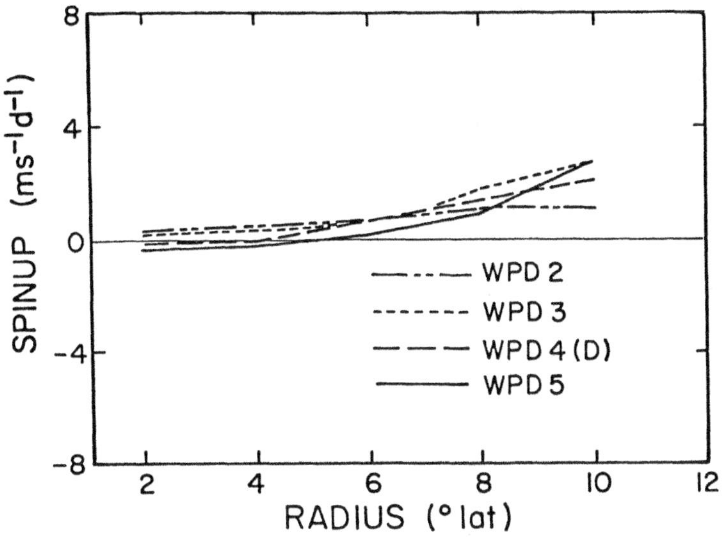

Fig. 4. Spinup by the eddy Coriolis torque in developing northwest Pacific cyclones

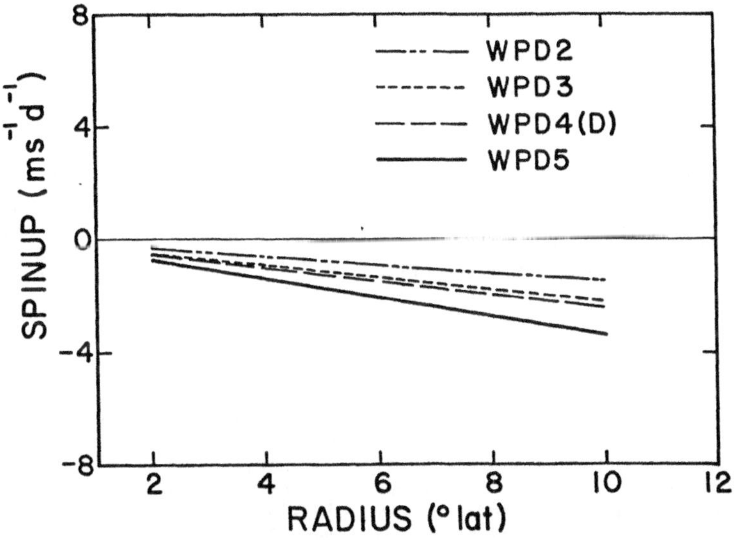

Fig. 5. Spinup by the meridional motion torque in developing northwest Pacific cyclones

From Eq. 1 the meridional motion torque is a function of the meridional speed of movement and β_o. As cyclones intensify their poleward motion component increases and β_o decreases (Table 1). But the motion increase dominates and, as shown in Fig. 5 this torque increases with cyclone intensity.

4 SYNTHESIS: THE TOTAL MOMENTUM BUDGET

Following our discussion and the observations for northwest Pacific cyclones in section 3, we summarize the relative importance of each component to the total momentum budget in Table 2. The inner regions of the cyclone are clearly dominated by the mean RAM transports, $F_H(\overline{V}_r\overline{V}_\theta)$, and frictional dissipation, F_θ. The mean transport and frictional dissipation become small at outer radii and are replace in importance by the eddy transports $F_H(\overline{V_r'V_\theta'})$ and $F_H(\overline{f'V_r})$ and the meridional motion torque, T_m.

TABLE 2

Relative role of each component of the angular momentum budget at different radii

Radius (°Lat.)	$F_H(\overline{V}_r\overline{V}_\theta$,	$\overline{V_r'V_\theta'}$,	$f_o\overline{V}_r$,	$\overline{f'V_r}$)	T_m	F_θ
2	***	0	0	0	0	***
2-6	**	*	0	*	*	**
6-10	*	**	0	***	***	0

Legend: *** Major and dominating component
** Major but not dominating component
* Small, component, may be large in some cases
0 Negligible component

No previous author has presented a complete Lagrangian budget equation, and only FRANK (1977) has considered the eddy Coriolis torque, $F_H(\overline{f'V_r})$. We have shown that, at least for northwest Pacific cyclones, the eddy Coriolis torque is the largest single contributor to the outer region spinup; the eddy RAM transports $F_H(\overline{V_r'V_\theta'})$ seem to be smaller. Further, even though the sink to surface friction decreases with radius, since most cyclones move poleward a large sink of RAM still occurs through the meridional motion torque, T_m.

The mean EAM transport term $F_H(f_o\overline{V}_r)$ has no direct contribution to the total momentum budget. However, it does have a very important secondary role, that of rearranging angular momentum in the vertical. Most of the eddy transports and a substantial part of the mean transports bring cyclonic RAM inwards in the upper atmosphere. In precisely the wrong place for our warm-cored-low-level-wind-maximum cyclone. However, as we have discussed in section 3(c), the mean EAM transport produces a concomitant large anticyclonic acceleration of the outflow layer and cyclonic acceleration of the inflow layer. It is thus in effect taking the incoming cyclonic momentum in the upper levels and depositing it in the lower levels where it can be used to maintain or intensify the cyclone.

5 CONCLUSION

Some results from a current investigation of angular momentum transports in a moving tropical cyclone have been presented.

We have briefly described the complete budget equations, together with the role and relative importance of each term in intensifying tropical cyclones.

References

Anthes, R. A., 1970: The role of large-scale asymmetries and internal mixing in computing meridional circulations associated with the steady state hurricane. Mon. Wea. Rev., 98, 521-528

Black, P. G. and R. A. Anthes, 1971: On the asymmetric structure of the tropical cyclone outflow layer. J. Atmos. Sci., 28, 1348-1366

Frank, W. M., 1977: The structure and energetics of the tropical cyclone. Dept. of Atmos. Sci. Paper No. 258, Colo. State Univ., Ft. Collins, CO, 180 pp

Gray, W. M., 1981: Recent advances in tropical cyclone research from rawinsonde composite analysis. World Meteorological Organization Programme on Research in Tropical Meteorology, 407 pp

Holland, G. J., 1981: Angular momentum budgets in hurricanes. Forthcoming Dept. of Atmos. Sci. Paper, Colo. State Univ., Ft. Collins, CO

Palmen, E. and H. Riehl, 1957: Budget of angular momentum and energy in tropical storms. J. Meteor., 14, 150-159

Pfeffer, R. L., 1958: Further examination of the balance of angular momentum in the mature hurricane. National Hurricane Research Project Report 19, US Weather Bureau, Washington, DC, 15 pp

Robock, A., 1975: On the eddy structure of hurricanes. Quart. J. Roy. Meteor. Soc., 101, 657-663

Numerical Analyses on the Activities of Intense Atmospheric Vortices (Typhoons) Over China

D. Xing-yuan and Z. Zi-dong

1 INTRODUCTORY REMARKS

China is one of the areas upon which the activities of large-scale intense atmospheric vortices have their influence most frequently and seriously. Take typhoons for example, more than one third of the tropical storms are generated in the North-Western Pacific. In the recent 30 years, from 1949 to 1978, the total number of typhoons generated over the North-Western Pacific was 858, among which 204 typhoons had landed on Chinese coasts. And more than one third of these landed typhoons had moved farther into our interior provinces and regions with a recorded maximum wind speed of 60 m/sec.

One of the main tasks of our meteorological service is to predict the tracks of typhoons as correctly as possible. For this purpose, a statistical-dynamical analyses of typhoon trajectories from the historical data are very useful. The results of analyses and the conclusion drawn from them are of much benefit to the operational forecasting and long-term economical planning as well.

The powerful mathematical tool of data-analyses developed in various applications in recent years is the empirical orthogonal expansion [1]. Up to now, the empirical orthogonal expansion has been mainly applied to scalar variable analyses. But, in meteorology, quite a few variables are of vector character. To some vector variables, the scalar empirical orthogonal expansion method can also be applied. But this is not true for other vector variables. For example, wind variables canbe written as the sum of a non-divergent vector and an irrotational vector, both of them, in turn, can be derived from a scalar field by differentiation. Therefore, at the beginning, we may apply the empirical orthogonal expansion to these scalar fields, and then by differentiation the empirical orthogonal expansion for the wind fields is obtained.

But as far as typhoon trajectories are concerned, it is better to generalize the scalar empirical orthogonal expansion to that of two-dimensional variables. How to expand a vector field by empirical function is discussed in the next paragraph.

Topics in Atmospheric and Oceanographic Sciences
Intense Atmospheric Vortices, ed. by Bengtsson/Lighthill

According to the coefficients of expansion, the typhoon trajectories are classified. For this purpose, the fuzzy cluster analyses are used. The classification is separately made for the following five periods:
(1) from January to June (2) July (3) August (4) September and (5) from October to December.

Only the first main components in the expansion are retained. For each of the abovementioned periods, a similarity index is constructed. The low similarity indexes are neglected and only high similarity indexes are used to form the preferable tracks. This problem is discussed in the third paragraph. In the last paragraph, the obtained preferable tracks for each period are given and discussed.

2 EMPIRICAL ORTHOGONAL EXPANSION FOR TWO-DIMENSIONAL VARIABLES

Assume there are k tracks of intense vortices, k = 1,2,....,T. For each track, j positions are chosen, j = 1,....,N. Each position is a vector and can be represented by a two-dimensional variable

$$W_{jk} = \psi_{jk} + i\phi_{jk} \tag{1}$$

ψ: longitude, ϕ: latitude, i is imaginary unit. ψ_{jk} and ϕ_{jk} have been nondimensionalized by choosing 100 degrees as the characteristic unit. In totality we have NxT two-dimensional variables which may be used to form a datum-matrix W:

$$W = \begin{bmatrix} W_{11} & W_{12} & \cdots\cdots & W_{1T} \\ W_{21} & W_{22} & \cdots\cdots & W_{2T} \\ \cdots & \cdots & \cdots & \cdots \\ W_{N1} & W_{N2} & \cdots\cdots & W_{NT} \end{bmatrix}$$

We are going to find out a N-dimensional vector by which is most close to each column-vector of matrix. That is, under such a vector b the following quantity L reaches its maximum value:

$$L = b^* WW^* b \tag{2}$$

$$b^* b = 1 \tag{3}$$

Where * means conjugate-transpose and L is an inner product of $b^* w$ and $\overline{b^* w}$, and the over-bar means conjugate.

The N-dimensional vector b, which satisfies (2)-(3) is our empirical orthogonal function. Each entry of this N-dimensional vector b, in turn, is also a two-dimensional vector.

A correlation matrix G is introduced as follows:

$$G=WW^{*} \tag{4}$$

and (2) may be written as $L = b^{*}Gb$.

A Lagrange multiplier ν may be introduced to solve the above variational problem. And it can be deduced to the eigenvalue problem of the NxN matrix G: $Gb = b\nu$. Since the element of G is

$$g_{jm} = \sum_{k=1}^{I} (\psi_{jk}+i\phi_{jk})\ (\psi_{mk}-i\phi_{mk}),$$

then it is easily seen that $g_{jm}=\bar{g}_{mj}$. Therefore $G=G^{*}$, i.e. G is a matrix of Hermitian type. Its eigenvalues are all real, and eigenvectors are mutual orthogonal. Using these eigenvectors as the column vectors, we may construct an unitary matrix B so that

$$GB = BD \tag{5}$$

$$B^{*}B = E \tag{6}$$

D is a diagonal matrix which entries are the eigenvalues $\nu_1 > \nu_2 >\ldots> \nu_N$, E is an unit matrix. Each column vector of the matrix B is an empirical orthogonal function.

Left-multiplying (5) by B and using (4) and (6), we have $B^*WW^*B=D$.

Define a coefficient matrix $C=B^*W$. Then

$$W = BC \tag{7}$$

$$W_{jk} = \sum_{m=1}^{I} B_{jm}C_{mk} \quad (j=1,2,\ldots,M,k=1,2,\ldots,T) \tag{8}$$

The solution to the complex eigenproblem of equations (5) - (6) was obtained by a norm reducing Jacobi type method [2]. That is, the matrix G is first diagonalized and all its eigenvalues and eigenvectors are then obtained simultaneously. The admissible error is less than 10^{-13}.

It has been shown that 9 positions for each trajectory are sufficient to approximate the true typhoon trajectory, that is, N=9. Thus for each period, there are altogether 9 eigenvalues and 9 eigenvectors. The obtained 9 eigenvalues for each period are given in Table 1.

From this table one can see that the first few eigenvalues are much larger than the others. It is assumed that the eigenvectors corresponding to the first 5 eigenvalues reflect the main physical factors influencing the typhoon movements. These eigenvectors are stable and have a dominating role. Therefore, in order to have more stable results, the last 4 eigenvectors corresponding to the small eigenvalues are neglected in equations (7) - (8), that is $M=5<N$.

Table 1. The 9 eigenvalues for each period

period		Jan to Jun	Jul	Aug	Sep	Oct to Dec
	1	17.52	13.06	27.09	23.73	43.37
e i	2	4.484	2.564	5.160	4.179	9.595
g	3	0.725	0.459	1.228	0.638	4.195
e n	4	0.526	0.097	0.179	0.210	0.962
v	5	0.150	0.053	0.075	0.072	0.179
a l	6	0.068	0.021	0.038	0.045	0.064
u	7	0.021	0.015	0.019	0.030	0.040
e	8	0.015	0.009	0.015	0.022	0.023
	9	0.009	0.007	0.010	0.011	0.019

The elements of the coefficient matrix C have the following form

$$C_{jk} = C^R_{jk} + iC^I_{jk} \qquad (j = 1,2,\dots,5; \quad k=1,\dots,T).$$

3 THE FUZZY CLUSTER ANALYSES

For the purpose of classification of the typhoon tracks, the similarity indexes are calculated by using the expansion coefficients C_{jk} obtained in the last paragraph. As we have mentioned, j is restricted up to 5. Introduce the following symbols:

$$\gamma_{tk} = \sum_{j=1}^{5} (C^R_{jt}\, C^R_{jk} + C^I_{jk}\, C^I_{jt});$$

$$\beta_{tk} = \sum_{j=1}^{5} (C^I_{jt}\, C^R_{jk} - C^R_{jt}\, C^I_{jk});$$

$$d_l = \sum_{j=1}^{5} [(C_{jl}^R)^2 + (C_{jl}^I)^2] \quad ; \quad l=t,k$$

$$\delta = 1, \text{ when } \gamma_{tk}/(d_t d_k) > 0; \quad \delta=0, \text{ otherwise.}$$

and the similarity indexes are defined as follows

$$S_{tk} = (1 - \delta|\beta_{tk}|/(d_t d_k))\ \gamma_{tk}/(d_t d_k);$$

From this definition, we can see that the degree of similarity is proportional to the scalar products of the position vectors, and inversely proportional to the absolute value of the sum of their vector products. In the case of $S_{tk} \leqslant 0$, there is little similarity.

The degree of similarity represented by S_{tk} is a fuzzy relationship. The set S_{tk} is possessed of reflexivity and symmetry, but not transitivity. In order to make it also be possessed of transitivity, we at first make the following linear transformation.

$$S_{tk}^{*} = 0.5\ S_{tk} + 0.5, \quad t,k = 1,2,\ldots, T. \quad \text{since}$$

$$-1 \leqslant S_{tk} \leqslant 1, \text{ then we have } 0 \leqslant S_{tk}^{*} \leqslant 1.$$

Let the set S_{tk}^{*} constitute a TxT matrix S. Then we raise it to the second, fourth power and so on. Suppose its 2^K-th power is equal to its 2^{k+1}-th power, $S^{2^K} = S^{2^{K+1}} = R$.

Let R_{tk} represent the elements of matrix R. Then the set of R_{tk} is also possessed of transitivity [3-4].

Therefore, we can make the trajectory classification based on some criterion λ. The criteria λ for the above-mentioned five periods are given equal to 0.97661, 0.97314, 0.98219, 0.98301 and 0.98936 respectively.

The value of each element R_{tk} of the matrix R is rounded to 1 or 0 according to the rule: $R_{tk}=1$, when $R_{tk} \geqslant \lambda$; $R_{tk}=0$, otherwise.

In the case when $R_{tk}=R_{tm},$ then the corresponding two typhoon trajectories are positioned into the same category. As only the first 5 main eigenvectors are used in (7) - (8), then equations (7) - (8) give us 9 filtered positions for each trajectory. A preferable track is considered as the average of the filtered trajectories in each category.

4 RESULTS AND DISCUSSION

The original typhoon trajectory data were used for our numerical analyses.

By using two-dimensional variable empirical orthogonal expansion and fuzzy cluster analyses, several preferable tracks are obtained for each period, and shown in Figs 4.1 to 4.5. The thickness of each line is proportional to the number of original trajectories included in each preferable track.

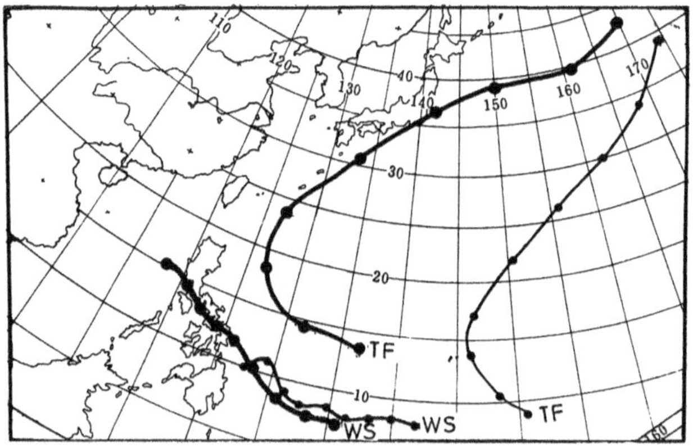

Fig. 4.1. The preferable tracks for Jan. to June

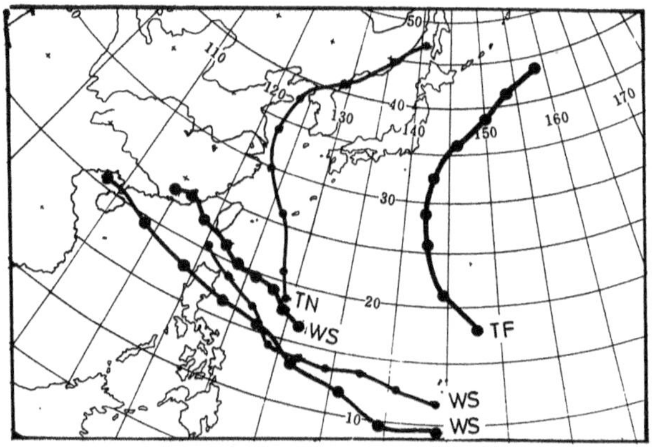

Fig. 4.2. The preferable tracks for July

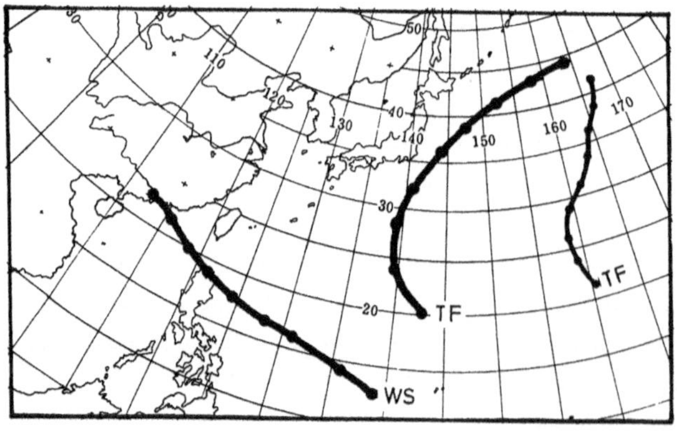

Fig. 4.3. The preferable tracks for Aug.

From these figures one can see the following:

(1) In the preferable track WS, typhoons move westward through the South China Sea. This preferable track exists for all five periods. It reflects the weather situation that the subtropic high over the Pacific is very stable. Under this circumstance typhoons move along the easterly stream in the southern periphery of the High. The typical weather situation for this category is given in Fig. 46. The 500 mb contours for 20 GMT 21 July 1980 are shown by solid thin lines; the real typhoon trajectory for 18 to 23 July 1980 by dashed lines and the preferable tracks by solid-dotted lines. In summer or autumn periods typhoons often pass through the South China Sea and then land in Southern China. In winter or spring periods typhoons often disappear over the South China Sea.

(2) In the preferable track TF typhoons recurve over the ocean. It reflects the weather situation that the subtropic high over the Pacific is unstable. Often the High is split into two parts, and typhoons pass through northwards between them. Sometimes, the High retracts eastwards and typhoons move northward along its western periphery.

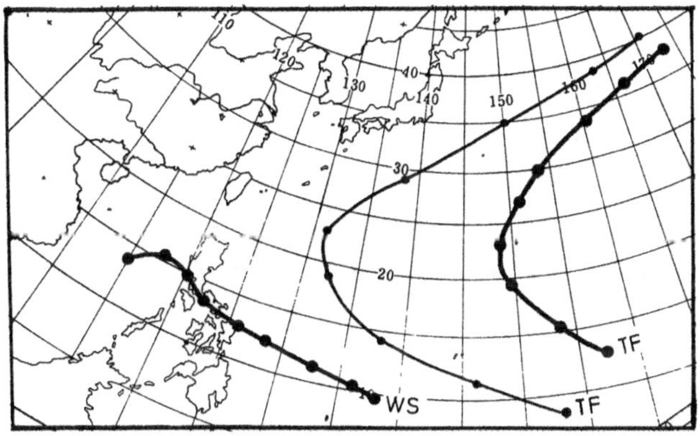

Fig. 4.4. The preferable tracks for Sept.

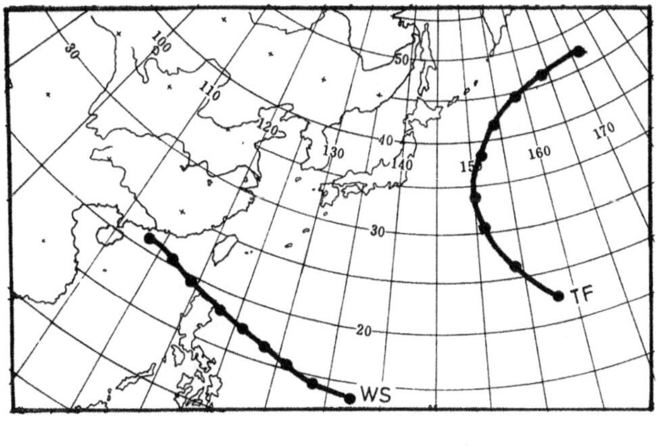

Fig. 4.5. The preferable tracks for Oct. to Dec.

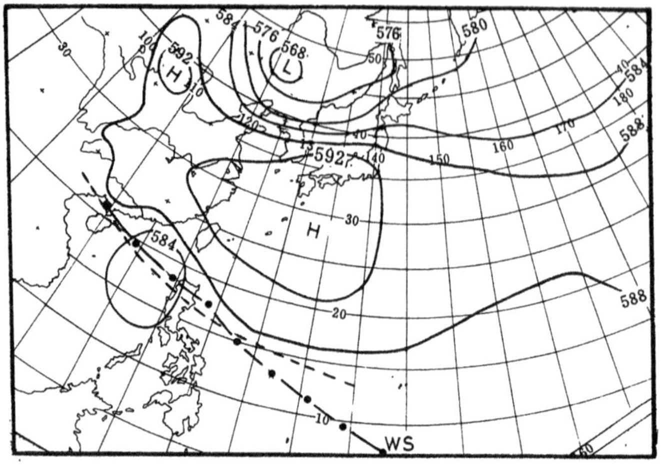

Fig. 4.6. The typical weather situation for WS

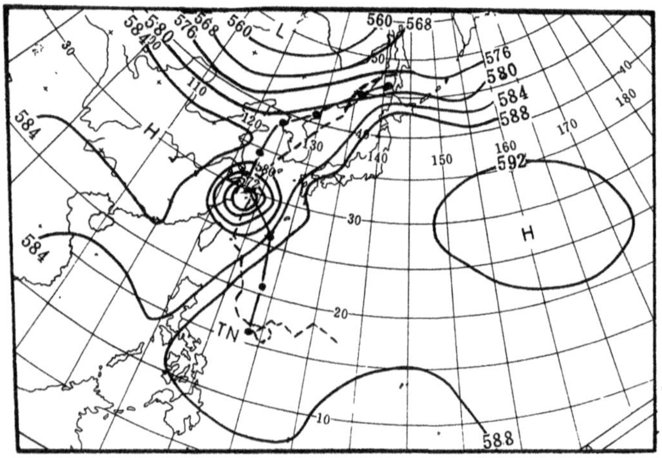

Fig. 4.7. The typical weather situation for TN

(3) In the preferable track TN, typhoon recurve over the East China Sea. All the coastal areas of East China are under the influence of typhoons. It reflects the weather situation that the meridional circulation is dominating over East Asia. A long-wave trough is often located over coastal areas, and typhoons recurve along the southerly stream in the fore part of the trough. The typical weather situation is shown in Fig. 4.7. The 500 mb contours for 16 GMT 16 August 1979 are given by solid thin lines; the real typhoon trajectory for 14 to 19 August 1979 by dashed lines and the preferable track TN by solid-dotted lines.

It should be mentioned that the two real typhoon trajectories shown in Fig. 4.6 and Fig.4.7 had not been used in our empirical analyses. From these two figures one can see that real typhoon trajectories are similar to our preferable tracks and this convinces us that the results in our numerical analyses, to some extent, have certain forecasting capacity.

ACKNOWLEDGEMENTS

The authors would like to thank Cao Hong-Xing, Chen Guo-Fan, Di Xin-Wu, Li Rui-Yun and Wang Wu-Zai for helpful suggestions and calculating assistance.

REFERENCES

1. ECMWF Workshop The use of empirical orthogonal functions in meteorology, ECMWF, 1977

2. Eberlein, P.J. Solution to the complex eigenproblem by a norm reducing Jacobi type method. Numer.Math. 14, 232-245, 1970

3. Zaden, L.A. Fuzzy sets, Information and Control, 8, 338-353, 1965

4. Dunn, J.G. A graph theoretic analysis of pattern classification via Tamur's fuzzy relation, IEEE Trans, on systems, man and cybernetics, Smc-4, 3110-313, 1974

Influence of Environmental Conditions on the Genesis of a Tropical Storm

Y. Kurihara and R. E. Tuleya

Abstract

Numerical experiments have been performed in search of the favorable conditions of environmental wind for the genesis of a tropical storm. The low level basic flow has an impact on the latent energy supply which is essential for genesis. The upper level basic flow has to be coupled with the phase speed of the low level incipient disturbance. The low level cyclonic shear of the basic flow is conducive to storm genesis.

1 Introduction

Over the tropical oceans, a weak atmospheric disturbance sometimes develops into a tropical depression, with winds up to $\sim 17\ ms^{-1}$, then perhaps into a tropical storm, and in certain cases into a hurricane or a typhoon, with winds greater than $\sim 33\ ms^{-1}$.

In this paper, we are concerned with the problem of genesis of a tropical storm. We treat this intriguing subject with the use of a numerical simulation model. A number of experiments have been integrated to investigate how different environmental flows affect the evolution of a shallow easterly wave. Extensive analysis of the numerical results has been done for both cases of development and non-development. We summarize in this paper those characteristics of the zonal mean flow which contribute to the formation of a tropical storm. Readers who are interested in the detailed results of each individual experiment are referred to the papers by KURIHARA and TULEYA (1981) and by TULEYA and KURIHARA (1981).

2 Brief Description of the model

The numerical model used in the present study is the eleven-level primitive equation model constructed at the Geophysical Fluid Dynamics Laboratory, National Oceanic and Atmospheric Administration. It is identical to that described by KURIHARA and TULEYA (1981). The model is governed by the equations of motion, the hydrostatic relation, the thermodynamical equation, the prognostic equation for the water vapor, the surface pressure tendency equation and the equation for the vertical velocity. The equations are expressed in the spherical coordinate system in which the σ-coordinate proposed by PHILLIPS (1957) is used in the vertical. The model covers a channel-type 25x25° domain centered at 18°N, with cyclic condition assumed for its

Topics in Atmospheric and Oceanographic Sciences
Intense Atmospheric Vortices, ed. by Bengtsson/Lighthill

25 degree longitudinal span and open conditions placed at 5.5 and 30.5°N. A longitude-latitude, non-staggered grid system with a resolution of 5/8 degree is used. In the vertical, eleven levels are specified at σ = .031, .120, .215, .335, .500, .665, .800, .895, .950, .977 and .992.

Subgrid scale processes are incorporated in the model by parameterization schemes. They include the non-linear viscosity scheme for horizontal diffusion (SMAGORINSKY, 1963), the turbulence closure scheme for vertical mixing (level 2 formula; MELLOR and YAMADA, 1974), the Monin-Obukhov framework for air-sea interaction and the adjustment scheme (KURIHARA, 1973) for free moist convection. The convective adjustment scheme in the present model does not necessarily eliminate conditional instability entirely. The effect of radiational heating or cooling is also included through a simple scheme.

3 Initial Condition

We specify the initial winds for various experiments by superposing the same shallow disturbance on different zonal mean flows.

The basic zonal mean flow at the initial time is defined as

$$\bar{u}(\sigma,\phi) = u_o(\sigma)\,\frac{\cos\phi}{\cos\phi_o} - u_1(\sigma)\,\tanh\left(\frac{\phi-\phi_o}{D}\right) \qquad (1)$$

where ϕ is the latitude, ϕ_o is the center latitude (18°N) of the channel and D = 2.5 degree latitude. The first term of the right side of (1) describes the zonal flow component at constant angular velocity. The second term yields the horizontal shear of the basic flow. Different environmental flow fields can be established through various combinations of u_o and u_1 profiles.

The initial disturbance, which is the same for all experiments, is a shallow sinusoidal wave of 25 degree wave length, without vertical or horizontal tilt of axis. Figure 3.1 shows the vertical and meridional profiles of the wave amplitude (for details, see KURIHARA and TULEYA, 1981). The maximum perturbation velocity is 5 ms^{-1}. The perturbation is initially confined below σ = .365 and to within 8.5° latitude of the channel center.

The temperature and the surface pressure fields at the initial time are obtained from the reverse balance equation for a given wind field. In obtaining the solution, we use the observed mean temperature and surface pressure for the GATE (GARP Atlantic Tropical Experiment) Phase III period (August 30 - September 19, 1974) at 80°W in the tropics to specify the zonal mean values at the center latitude of the channel. The initial relative humidity at each model level is determined using the observations during the abovementioned period. A dynamic initialization is performed to simulate the winds in the planetary boundary layer from the start of the integration experiments. Then, the time integrations are carried out to 96 hours with a time step of 90 seconds. During the integrations, the sea surface temperature is fixed at 302K (∿28.8°C).

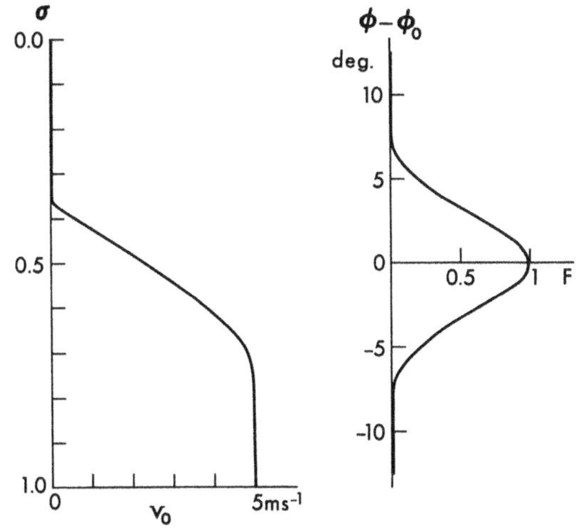

Fig. 3.1. Vertical variation of perturbation amplitude at the center latitude ϕ_o, 18°N, (left) and meridional profile of the ratio of perturbation amplitude to maximum amplitude (right)

4 Results and Discussions

The evolution of an incipient tropical disturbance is greatly influenced by the environmental wind. As an example, we choose two cases; one to represent a developed disturbance and the other to describe a non-developed disturbance. In specifying the initial basic flow, the two cases take common values for u_o at the surface and $u_1(\sigma)$ in equation (1). Specifically, $u_o(\sigma) = -5$ ms^{-1} and $u_1(\sigma) = 3.75$ ms^{-1}, which indicates surface easterly wind and cyclonic horizontal shear of the environmental flow. The two cases differ with each other in the u_o profiles. One case is characterized by easterly vertical shear (easterlies increasing with height) of -15 ms^{-1} between 150 and 850 mb levels, and the other case is featured by a westerly vertical shear of 15 ms^{-1}. The intensity of the disturbance after 96 hour integration in each of the above two cases is summarized in Table 4.1. It is clearly seen that the easterly shear of -15 ms^{-1} is contributive to the genesis of a tropical storm, while the westerly shear with the same magnitude is not propitious. Figure 4.1 shows the longitude-height distribution, through the disturbance center, of the wind relative to the moving system at 48 hour for the two cases. In the developed case, a deep radial-vertical circulation is already organized by 48 hour. No such circulation is found in the non-developed case.

The example presented above is a result from a series of experiments which have been performed to investigate the effects of the vertical and the horizontal shears of the basic wind as well as of the direction of a uniform flow on the growth of the disturbance. Based on the synoptic analysis of the integration results and the budget analyses of vorticity, heat and energy, we make the following summaries:

Table 4.1. Numerical results for the developed and non-developed disturbances

	developed disturbance	non-developed disturbance
at 0 hour		
central surface pressure (mb)	1008.4	
max. surface wind (ms^{-1})	9	
max. vorticity at σ = .95 ($10^{-6}s^{-1}$)	43	
at 96 hour		
central surface pressure (mb)	993.5	1006.9
max. surface wind (ms^{-1})	22.5	10.0
max. vorticity at σ = .95 ($10^{-6}s^{-1}$)	365	63

Two major processes which are associated with storm genesis at its early stage are (1) the latent energy supply and its release, and (2) the coupling between the upper level and the low level disturbance.

A sufficient supply of moisture to a disturbance system is essential for its growth. In fact, when either the model is made dry or the sea surface temperature is lowered below a certain value, no sign of development is noticed regardless of the initial environmental flow. The low level environmental flow has a great impact on the latent energy supply. Apparently, the evolution of a deep circulation as shown in the developed disturbance in Fig. 4.1 is generally accompanied by a large relative inflow. A simple measure of relative inflow may be the difference between the low level zonal mean wind $\bar{u}_L$, say $\bar{u}$ at σ = .895, and the zonal phase speed of the disturbance c, i.e. $|\bar{u}_L - c|$. The development potential may be increased at a large value of this parameter. It is also found that the wind speed maintenance in the lower part of the free atmosphere is likely to have a long-term effect on the genesis. Momentum transfer from aloft into the boundary layer over a large area surrounding the disturbance may be required in order to offset the frictional effect on the wind and, hence, sustain the evaporation at the sea surface.

The heat released by the condensation of the water vapor may induce and intensify the radial-vertical circulation mentioned before, only when the upper level solenoidal field is kept positioned over the moving low level disturbance. Since the temperature tendency at the upper levels at the early state is mainly due to the horizontal advection, the coupling between the upper and lower levels may be measured by the parameter $|\bar{u}_U - c|$, where $\bar{u}_U$ is the zonal mean wind at the level of warming, say $\bar{u}$ at σ = .335. (After a certain stage of development, the phase speed of the warm core becomes dependent not only on $\bar{u}_U$ but also on other factors.)

To support the above argument, we present Table 4.2, in which u_U - c, u_L - c and U_L after 12 hour integration and the intensity of the disturbance at 96 hour for various experiments are shown. The initial environmental flows for these experi-

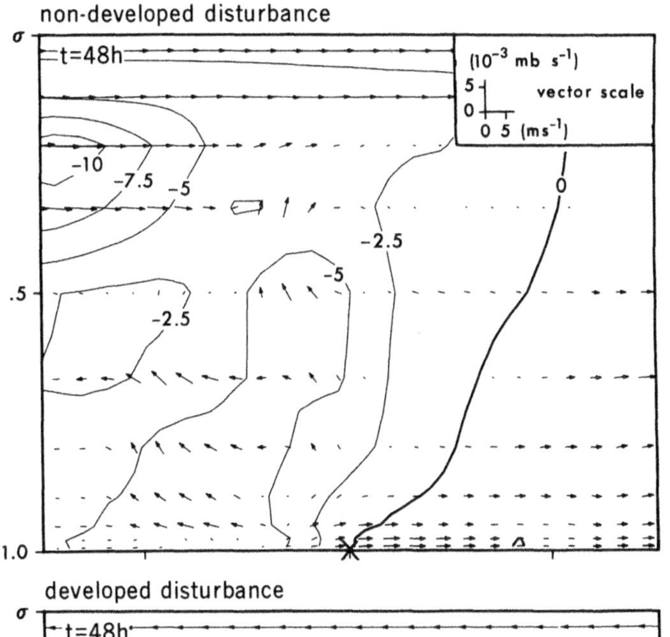

Fig. 4.1. Longitude-height distribution, through the disturbance center, of the wind relative to the moving system, at 48h, for the non-developed (top) and the developed (bottom) disturbance. Arrows represent the zonal and vertical components and isopleths show distribution of the meridional component (m s^{-1})

ments possess the same horizontal shear and the surface wind, i.e. $u_1(\sigma) = 3.75$ ms^{-1} and $u_o(1) = -5$ ms^{-1} in (1), but the different vertical shear as shown in the Table. The initial disturbance is the same for all experiments. Table 4.2 indicates that $\bar{u}_U - c$ at 12 hour is minimum at the easterly vertical

Table 4.2. Evolution of disturbances under the different vertical shears of the basic wind

$\bar{u}$ (150 mb) - $\bar{u}$ (850 mb)		easterly shear			westerly shear	
at 0 hour (ms^{-1})	-22.5	-15.0	-7.5	0	7.5	15.0
at 0 hour	central surface pressure = 1008.4 mb					
	maximum surface wind = 9 ms^{-1}					
	maximum vorticity at σ = .95 = 43 x $10^{-6}s^{-1}$					
at 12 hour						
phase speed c (ms^{-1})	-12.3	-11.0	-9.8	-8.6	-7.4	-5.9
$\bar{u}$ (σ = .335) - c (ms^{-1})	- 6.7	- 3.0	0.5	3.2	5.4	8.1
$\bar{u}$ (σ = .895) - c (ms^{-1})	6.3	5.5	4.8	4.0	2.8	1.4
$\bar{u}$ (σ = .895)	- 6.0	- 5.5	-5.0	-4.6	-4.6	-4.5
at 96 hour						
central surface pressure (mb)	1000.8	993.5	997.1	1002.1	1003.5	1006.9
max. surface wind (ms^{-1})	16.8	22.5	22.2	16.2	13.8	10.0
max. vorticity at σ = .95 ($10^{-6}s^{-1}$)	170	365	317	175	155	63

shear of $\sim$7.5 ms^{-1}. Such a bias in the storm genesis potential toward the easterly shear is further shifted because of the increase of $|\bar{u}_L - c|$ and $|\bar{u}_L|$ with easterly shear. As a result, the most intense vortex of all experiments with the surface easterly of -5 ms^{-1} develops under the easterly vertical shear of -15 ms^{-1}.

One of the other findings from the series of experiments is that the low level cyclonic shear of the basic flow and, to a lesser degree, the upper level anticyclonic shear of the basic flow provide a favorable condition for the growth of a disturbance. The low level anticyclonic shear is detrimental for storm genesis.

It is also found that the behavior of the disturbance in a uniform westerly flow differs considerably from that in a uniform easterly flow. When we specify the initial basic flow as uniform easterly ($u_o(\sigma) = -5\ ms^{-1}$, $u_1(\sigma) = 0$ in (1)) and as uniform westerly ($u_o(\sigma) = 5\ ms^{-1}$, $u_1(\sigma) = 0$) respectively in two integrations, we obtain vortices with different structure at 96 hour. Figure 4.2 shows the distribution of relative vorticity at $\sigma = .95$ at 96 hour for the two experiments. The vortex in the case of uniform westerly is more intense (maximum relative vorticity of $284 \times 10^{-6}s^{-1}$ compared to $175 \times 10^{-6}s^{-1}$ in the

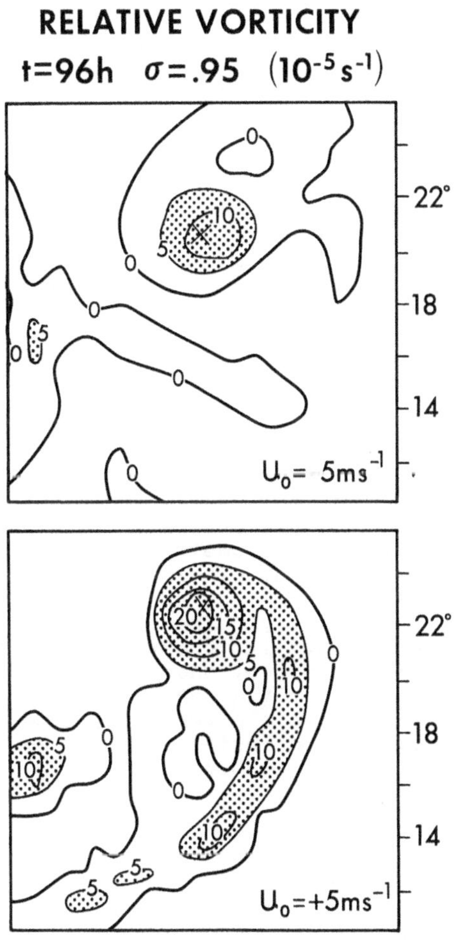

Fig. 4.2. Distribution of relative vorticity ($10^{-5}s^{-1}$) at $\sigma = .95$ at 96 hour for the case of uniform easterly flow, $u_o = -5\ ms^{-1}$, (top) and for the case of uniform westerly flow, $u_o = 5\ ms^{-1}$, (bottom). Shaded areas indicate values greater than $5\times10^{-5}s^{-1}$

case of uniform easterly), and its structure has similarity to that of extratropical cyclones. The difference of vortices in the two experiments may be attributed to the various factors including the beta effect, the location of the strong upward motion and the phase speed of the disturbance.

5 Critical Remarks

In the present study, favorable conditions of the environmental flow for tropical storm genesis are investigated with the use of a numerical simulation model. The simulation results suggest that the supply of latent energy and the coupling between the upper level and the low level disturbance are essential for the genesis of a tropical storm. The effect of condensation of water vapor are incorporated in the model through the method of parameterization of convective processes. Accordingly, the degree of validity of the parameterization scheme used may be questioned. Future research is needed on this point. It may involve the study concerning the effect of environmental flow on the behavior of individual clouds.

As mentioned in the preceding section, the dynamical parameters $\bar{u}_L - c$ and $\bar{u}_U - c$ may be used to measure storm genesis potential pertaining to an incipient disturbance. Although the phase speed of the low level disturbance c can be obtained from observations in practice, it is desirable to understand what factors determine c. Among other things, c is influenced by the vertical shear of the basic flow as shown in Table 4.2.

The problem of tropical storm genesis seems to be two-fold. On the one hand, favorable environmental flow condition should be present for a crucial period of a few days at least. In this respect, temporal and spacial variation of the basic flow can either enhance or suppress the developing disturbance. On the other hand, the evolution of the disturbance depends on its own structure and intensity at the incipient stage. When the maximum amplitude of the initial disturbance is specified at 5, 2.5 and 1.25 ms^{-1} in the model whose initial basic flow corresponds to the case of easterly vertical shear of -15 ms^{-1} in Table 4.2, the minimum surface pressure at 96 hour is found to be 993.5, 1000.1 and 1005.0 mb, respectively. The disturbance in each of the above three experiments attains the intensity of a tropical storm after 59, 93 and 113 hours of model integration, respectively. We speculate that, in the real atmosphere, a disturbance does not always develop within a realistic time range even if the environmental genesis potential exists. For this reason, a developing disturbance may be mistakenly categorized as a non-developing disturbance.

Acknowledgments - The authors are grateful to J. SMAGORINSKY and K. MIYAKODA for their encouragement. Thanks are due to J. KENNEDY, P. TUNISON and J. CONNER for their assistance.

REFERENCES

KURIHARA, Y., 1973: A scheme of moist convective adjustment. Mon. Wea. Rev. 101, 547-553

KURIHARA, Y., and R. E. TULEYA, 1981: A numerical simulation study on the genesis of a tropical storm. Mon. Wea. Rev., 109 (August issue)

MELLOR, G. L., and T. YAMADA, 1974: A hierarchy of turbulence closure models for planetary boundary layers. J. Atmos. Sci., 31, 1791-1806

PHILLIPS, N. A., 1957: A coordinate system having some special advantages for numerical forecasting. J. Meteor., 14, 184-185

SMAGORINSKY, J., 1963: General circulation experiments with the primitive equations: I The basic experiment. Mon. Wea. Rev., 91, 99-164

TULEYA, R. E., and Y. KURIHARA, 1981: A numerical study on the effects of environmental flow on tropical storm genesis. (submitted)

Model Development of Atlantic Hurricanes from Prehurricane Cloud Clusters, Depressions and Cyclones[1]

R. L. Pfeffer and M. Challa

Abstract - The role of large-scale eddy fluxes of momentum in the intensification of Atlantic hurricanes from prehurricane cloud clusters, depressions and cyclones is investigated. This is accomplished by numerically integrating SUNDQVIST's (1970) symmetric model for hurricane development with parameterized large-scale eddy fluxes of momentum. The initial wind and moisture distributions and the prescribed eddy fluxes are taken from atmospheric observations of Atlantic developing hurricanes as composited by MCBRIDE (1981a, b) and MCBRIDE and ZEHR (1981). With the use of the same initial wind, temperature and moisture distributions, numerical integrations were performed with and without the observed eddy fluxes of momentum.

Without eddy flux forcing, the prehurricane cloud cluster and the prehurricane depression do not develop into hurricanes, and the intensifying cyclone takes approximately 13 days to reach hurricane strength. With the observed eddy flux forcing, all three initial disturbances develop rapidly into hurricanes, the intensifying cyclone requiring less than 2 days to reach hurricane strength. The rate of hurricane development, and the size, intensity and eye structure at maturity are different, depending upon the details of the eddy momentum flux distribution. It is concluded that inward eddy fluxes of momentum associated with synoptic-scale wave asymmetries may be crucial to the development of Atlantic hurricanes from prehurricane cloud clusters, depressions and cyclones.

1 Introduction

The hurricane is a heat engine which is driven by the release of latent heat of condensation mainly in its eye wall. How moisture and latent heat are first concentrated to form such an intense heat source is a fundamental question relating to hurricane formation. In two recent papers, the authors (CHALLA and PFEFFER, 1980 and PFEFFER and CHALLA, 1981) presented evidence to the effect that inward eddy fluxes of momentum associated with large-scale wave asymmetries may be of fundamental importance in initiating and maintaining the radial circulation which draws moist air toward the hurricane center in the lower troposphere and pumps drier air out at higher elevations.

For the purpose of these studies, SUNDQVIST's (1970) symmetric numerical model for hurricane formation was modified to include parameterized eddy fluxes of momentum. In the first paper, the authors reported the results of a sensitivity study in which this model was integrated with different initial conditions and with various prescribed distributions of the eddy momentum flux. They found that model hurricanes developed during the course of every numerical

[1] Contribution No. 172 of the Geophysical Fluid Dynamics Institute, Florida State University.

Topics in Atmospheric and Oceanographic Sciences
Intense Atmospheric Vortices, ed. by Bengtsson/Lighthill

integration in which an eddy momentum flux convergence was prescribed in the domain, even when the sea surface temperature was taken to be slightly subcritical such that the purely symmetric model integration failed to develop a hurricane. The growth rates, sizes and final intensities of the model hurricanes differed from one another, depending upon the prescribed momentum flux distribution. Moreover, when an eddy momentum flux divergence was prescribed, model hurricane development was suppressed, even when the sea surface temperature was taken to be supercritical such that the purely symmetric model integration produced explosive growth.

In the second paper, the authors described the results of numerical experiments similar to the ones in the previous study, but with more realistic distributions of the eddy flux of momentum replacing the somewhat arbitrary distributions prescribed in the earlier study. The eddy flux data and the initial tangential velocity and moisture distributions were taken from the composite data analyses of Atlantic developing (prehurricane) and non-developing tropical disturbances prepared by MCBRIDE (1981a, b) and MCBRIDE and ZEHR (1981). For the purposes of that study, their data corresponding to each of several stages in the evolution of developing and non-developing disturbances were linearly averaged to form a single composite developing hurricane and a single composite non-developing disturbance. Numerical integrations were performed using the data for the composite developing hurricane, with and without the observed eddy flux of momentum, and for the composite non-developing disturbance with the observed eddy flux distribution corresponding to this disturbance. Without eddy flux forcing, the prehurricane developing vortex failed to intensify into a hurricane, even after 20 days of integration. With the observed eddy flux of momentum, the same initial vortex intensified rapidly, reaching hurricane strength within 4 days. Moreover, because of the weak and diffuse pattern of the eddy flux of momentum in the non-developing tropical disturbance, the initial vortex corresponding to this disturbance also failed to develop into a hurricane.

In the present paper, we shall investigate the role of the eddy flux of momentum in the intensification of Atlantic hurricanes from different stages in their evolution, namely prehurricane cloud clusters, depressions and cyclones. Once again we will use MCBRIDE's data and our modification of SUNDQVIST's numerical model.

2 Model Equations

The details of SUNDQVIST's hurricane model and numerical procedures are contained in his original paper (loc. cit.) and a review may be found in the paper by CHALLA and PFEFFER (loc. cit.). The following list of governing equations with the addition of the eddy fluxes of momentum (underlined terms) is provided for the convenience of the reader:

$$\frac{\partial \bar{u}}{\partial t} + \bar{v}\left(f + \frac{\partial \bar{u}}{\partial r} + \frac{\bar{u}}{r}\right) + \bar{\omega}\frac{\partial \bar{u}}{\partial p} = K_H\left(\nabla^2 - \frac{1}{r^2}\right)\bar{u} + g\frac{\partial \bar{\tau}}{\partial p} \underline{- \frac{1}{r^2}\frac{\partial r^2\overline{u'v'}}{\partial r}} \qquad [1]$$

$$\frac{\partial \bar{\theta}}{\partial r} = -\frac{p}{R\pi}\left(f + \frac{2\bar{u}}{r}\right)\frac{\partial \bar{u}}{\partial p} \qquad [2]$$

$$\frac{\partial \bar{q}}{\partial t} + \bar{v}\frac{\partial \bar{q}}{\partial r} + \bar{\omega}\frac{\partial \bar{q}}{\partial p} = K_H\nabla^2\bar{q} + g\frac{\partial \bar{m}}{\partial p} + \delta\bar{q} \qquad [3]$$

$$A \frac{\partial^2\psi}{\partial p^2} + 2B \frac{\partial^2\psi}{\partial r \partial p} + C \frac{\partial^2\psi}{\partial r^2} - \frac{4B}{r}\frac{\partial\psi}{\partial p} - \left[\frac{1-\kappa}{p} B + \frac{C}{r}\right]\frac{\partial\psi}{\partial r}$$

$$= r \frac{\partial}{\partial p}\left[\left(\frac{2\bar{u}}{r} + f\right)\left\{K_H \left(\nabla^2 - \frac{1}{r^2}\right)\bar{u} + g\frac{\partial\bar{\tau}}{\partial p} - \frac{1}{r^2}\frac{\partial r^2\overline{u'v'}}{\partial r}\right\}\right] \quad [4]$$

$$+ \frac{r\kappa}{p}\frac{\partial}{\partial r}\left(\bar{Q} + g\frac{\partial\bar{h}}{\partial p} + \pi C_p K_H \nabla^2\bar{\theta}\right)$$

where

$A = \left(f + \frac{2\bar{u}}{r}\right)\left(f + \frac{\partial\bar{u}}{\partial r} + \frac{\bar{u}}{r}\right)$ inertial stability parameter

$B = -\left(f + \frac{2\bar{u}}{r}\right)\frac{\partial\bar{u}}{\partial p} = \frac{R\pi}{p}\frac{\partial\bar{\theta}}{\partial r}$ baroclinic stability parameter

$C = -\frac{R\pi}{p}\frac{\partial\bar{\theta}}{\partial p}$ static stability parameter

$$\bar{\tau} = \begin{cases} \rho C_D |\bar{V}/| \bar{u} & \text{at the sea surface} \\ \frac{1}{g} K_V \frac{\partial\bar{u}}{\partial p} & \text{above the surface} \end{cases}$$

$$\bar{m} = \begin{cases} \rho C_D |\bar{V}/| (\bar{q}_w - \bar{q}) & \text{at the sea surface} \\ \frac{1}{g}\left(K_V \frac{\partial\bar{q}}{\partial p}\right) & \text{above the sea surface} \end{cases}$$

$$\bar{h} = \begin{cases} C_p\, \rho\, C_D |V/| (T_w - \bar{T}) & \text{at the sea surface} \\ 0 & \text{above the surface} \end{cases}$$

$\bar{Q} = \xi(\bar{T}_c - \bar{T})$ (rate at which the environment is heated by cloud)

$\delta\bar{q} = \xi(\bar{q}_c - \bar{q})$ (rate of addition of specific humidity to the environment from a model cloud in regions of condensation when the environmental air is unsaturated)

$\xi = \xi_o\, n(p)$

Here, t is time, r is radius, p is pressure, T is temperature, θ is potential temperature, u and v are the tangential (positive counterclockwise) and radial (positive outward) components of the wind velocity, V/, $\omega \equiv dp/dt$, ψ is the Stokes stream function in the r-p plane, q is the specific humidity, h and m are the vertical fluxes of sensible heat and moisture, respectively, f is the Coriolis parameter (considered constant at the value for 20° latitude), Q is the rate of diabatic heating per unit mass, g is the acceleration of gravity, R is the gas constant for dry air, C_p is the specific heat of air at constant pressure, $\kappa \equiv R/C_p$, $\pi \equiv (p/1000\text{ mb})^{\kappa}$, τ is the tangential component of the vertical stress, K_H and K_V are the horizontal and vertical components of the eddy viscosity, C_D is the surface drag coefficient, the overbar represents an azimuthal average and the prime a departure from this average; ξ_o is obtained by equating the rate of latent energy input into the cloud by the influx of specific humidity through the bottom of the cloud to the rate of transfer of latent plus sensible heat to the environment, K_V is specified as a linearly decreasing function of height ranging from 1.0×10^{-3} at 950 mb to 1.0×10^{-6} at 50 mb, and we take $K_H = 5 \times 10^3\ m^2\ s^{-1}$ and $C_D = 1.5 \times 10^{-3}$.

It may be noted that when the eddy momentum flux forcing function

$$\frac{1}{r}\frac{\partial}{\partial p}\left[\left(\frac{2\bar{u}}{r} + f\right)\frac{\partial r^2\overline{u'v'}}{\partial r}\right]$$

in equation (4) is positive at a point (for example, when the momentum flux convergence increases with altitude), it induces a clockwise circulation in the r-p plane; when this function is negative, it induces a counterclockwise circulation, as shown in Fig. 1.

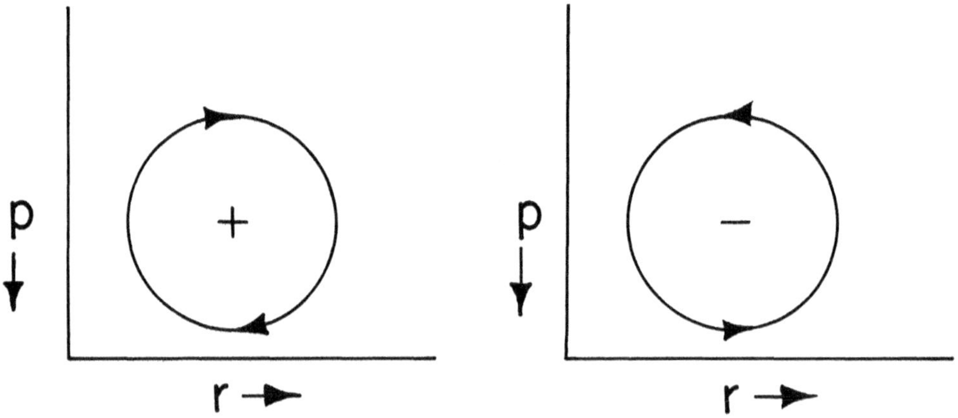

Fig. 1. Schematic drawing of the circulations produced by positive and negative sources of eddy momentum flux convergence

3 Distributions of Eddy Flux of Momentum in Hurricanes

The existence of significant large-scale inward eddy fluxes of momentum in hurricanes was first reported by PFEFFER (1955). More complete calculations, published later by PALMÉN and RIEHL (1957) and by PFEFFER (1958), confirmed the presence of such fluxes in mature hurricanes. The latter author showed how large-scale asymmetries in the hurricane flow pattern produce the observed fluxes. Additional evidence of large-scale inward eddy fluxes of momentum in hurricanes has been presented more recently by BLACK and ANTHES (1971), FRANK (1977) and MCBRIDE (loc. cit.). MCBRIDE classified seven categories of Atlantic tropical disturbances as follows: The non-developing cloud cluster (N1), the non-developing wave trough cluster (N2), the non-developing depression (N3), the prehurricane cloud cluster (D1), the prehurricane depression (D2), the intensifying cyclone (D3) and the hurricane (D4).

Fig. 2 shows the distributions of the eddy flux of angular momentum in the r-p plane based upon the values at 100 mb intervals from

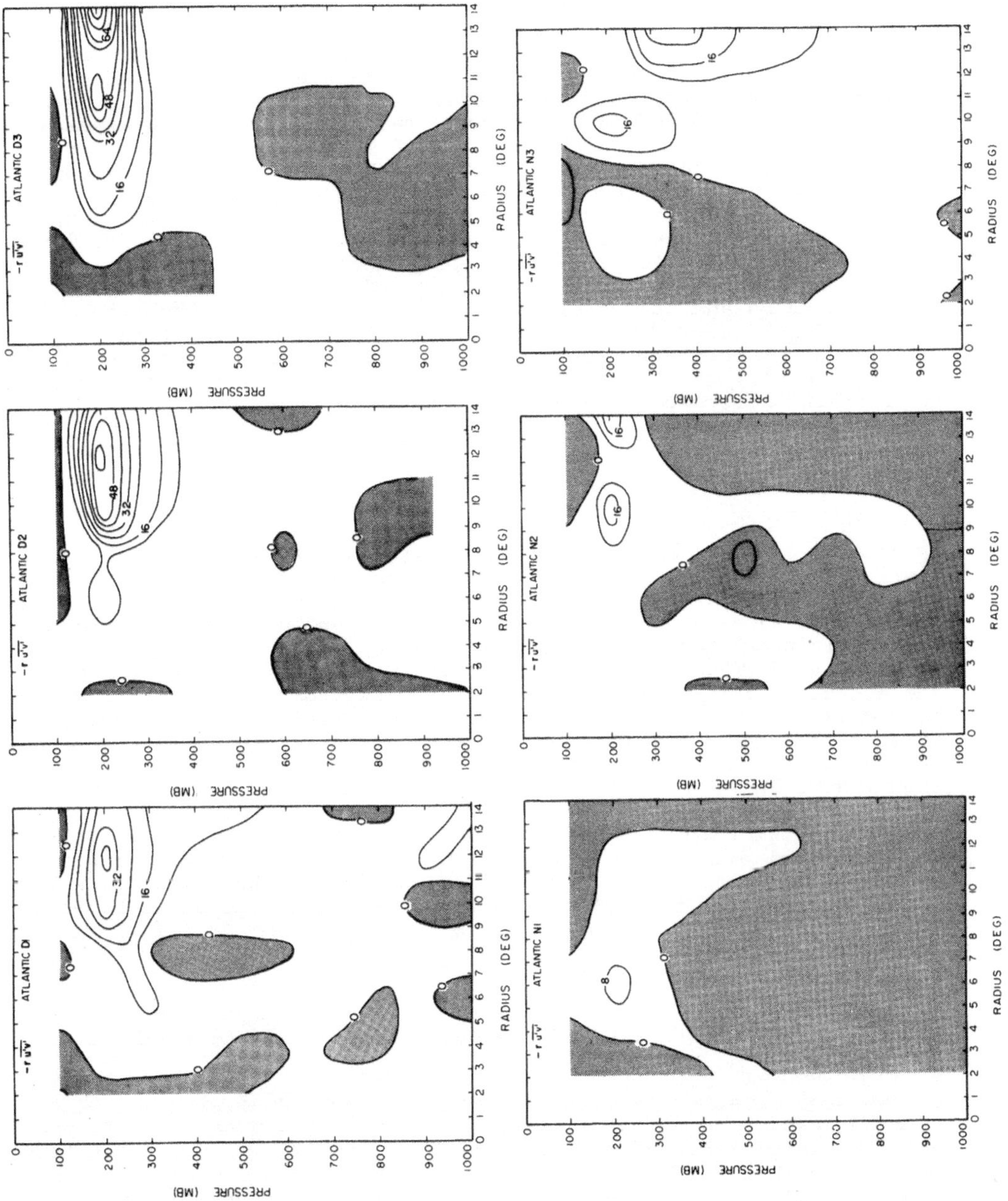

Fig. 2. Distributions of eddy flux of momentum $-r\overline{u'v'}$ in Atlantic developing and non-developing tropical disturbances using data extracted from MCBRIDE's tabulations at 100 mb intervals from 1000 to 100 mb and at 2 deg lat intervals from r = 2 to 14 deg. Shaded (negative) regions represent outward fluxes. Units are 10 deg m^2 s^{-2}

1000 to 100 mb (which our numerical model requires) and at radii r = 2, 4, 6, 8, 10, 12 and 14 deg lat extracted from MCBRIDE's tabulations for the D1, D2, D3, N1, N2 and N3 disturbances. Whereas the momentum fluxes are weak and poorly organized in the non-developing tropical disturbances, they are intense and well organized in the developing (prehurricane) disturbances. The intensity and the radial extent of the maximum inward eddy flux of momentum increases with each successive stage of development from D1 to D2 to D3. The eddy momentum flux distribution in these disturbances is such that there is a substantial vertical derivative of the flux convergence in the upper troposphere. According to equation (4), such a configuration represents a forcing function for the symmetric circulation (ψ) in the r-p plane which must draw moist air radially toward the center in the troposphere and pump drier air radially outward in the stratosphere, thereby providing fuel for the explosive growth of the hurricane. It is the efficacy of this mechanism at each stage in the evolution of Atlantic hurricanes that the numerical integrations to be presented here are designed to test.

4 Grid Network and Initial Conditions

The numerical model contains 10 equally spaced levels in the vertical direction (Δp = 100 mb) in the range $1000 \geq p \geq 100$ mb. The radial extent of the model is 1500 km, with a grid interval Δr = 25 km.

The initial tangential wind and specific humidity distributions for our integrations, taken from MCBRIDE's data are shown in Fig. 3. Here, we used MCBRIDE's data at 100 mb intervals from 1000 to 100 mb and at radius r = 0.5 deg lat as well as at 2 deg intervals from 2 to 14 deg lat. A notable feature of the tangential wind distributions is that, except in the case of the already intensifying cyclone (D3), the maximum cyclonic wind speed at 900 mb is smaller in magnitude than the value required by most numerical models to form a hurricane by virtue of Ekman layer inflow and cooperative instability.

The initial temperature distribution was taken to be that of the mean tropical atmosphere (Jordan, 1958), adjusted for thermal wind balance with the initial tangential wind field. The sea surface temperature in all of our integrations was held fixed at 300.5 K, which is the critical value required for SUNDQVIST's symmetrical model to form a hurricane.

5 Numerical Integrations for Atlantic Developing Disturbances

Two sets of numerical integrations were performed with the initial data corresponding to the Atlantic developing disturbances. In one set, the eddy fluxes of momentum in the region $2° \leq r \leq 14°$ latitude were prescribed as shown in Fig. 2. In the region $0 < r < 2°$, the required values of $\overline{u'v'}$ were obtained by linear interpolation using the observed values at r = 2° and setting $\overline{u'v'} = 0$ at r = 0. The prescribed values were held fixed for the entire length of the integration. In the other set, the eddy momentum fluxes were prescribed to be zero everywhere, in order to test whether the model would be capable of developing a hurricane from a purely symmetric initial vortex of observed strength (without eddy flux forcing), utilizing only the Ekman layer inflow and cooperative instability. In both sets of integrations, the initial tangential wind and specific humidity distributions in the region $0.5 \leq r \leq 14°$ were prescribed as

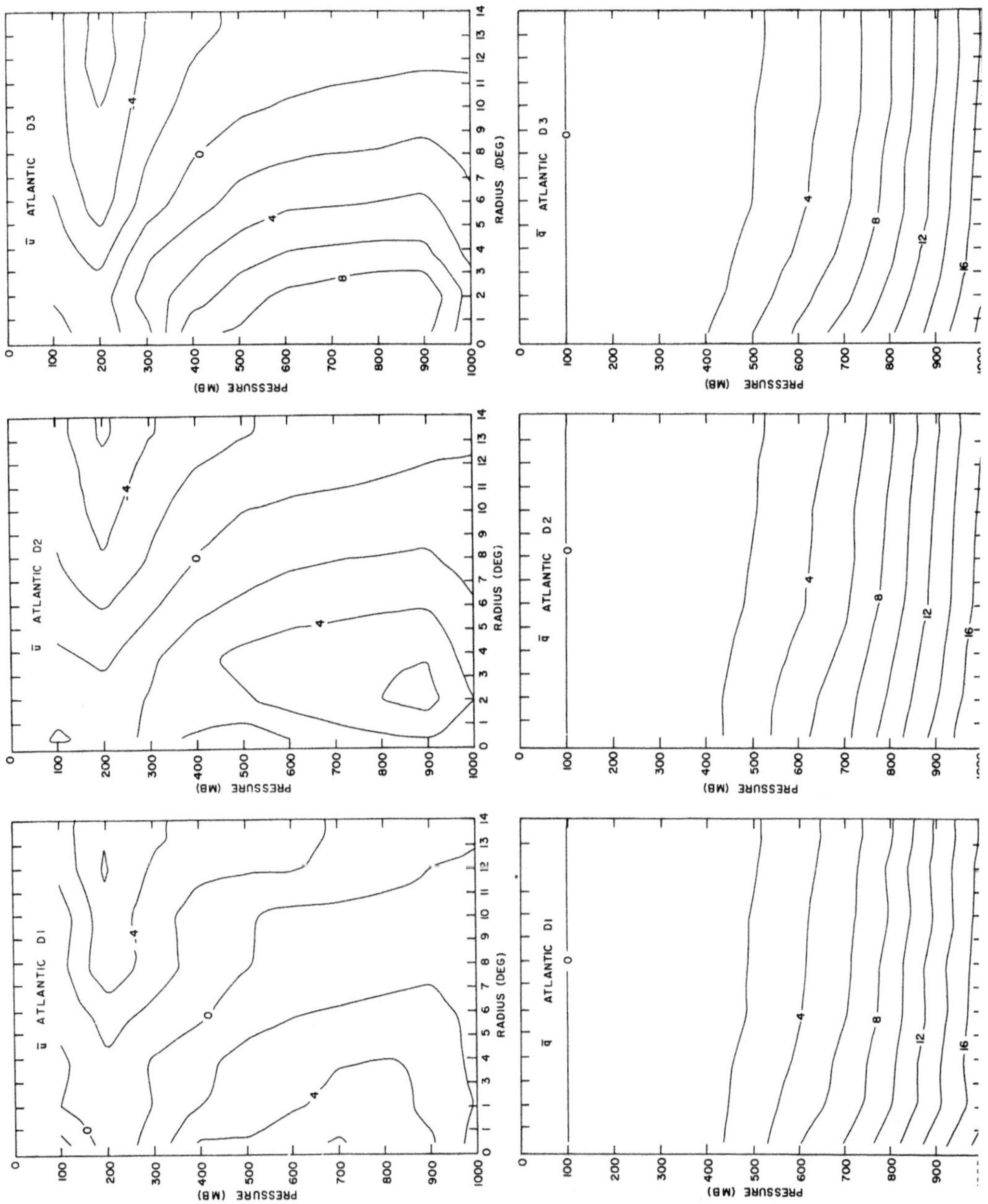

Fig. 3. Initial tangential wind ($\bar{u}$) and specific humidity ($\bar{q}$) distributions in Atlantic developing tropical disturbances, using data extracted from MCBRIDE's tabulations at 100 mb intervals from 1000 to 100 mb, at r = 0.5 deg lat and at 2 deg lat intervals from r = 2 to 14 deg. Units are m s^{-1} for $\bar{u}$ and gm kg^{-1} for $\bar{q}$

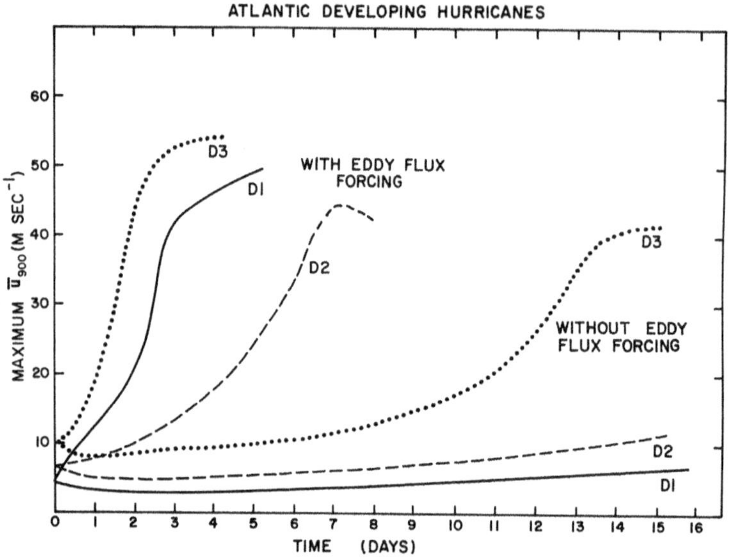

Fig. 4. Time variations of maximum tangential wind velocity at p = 900 mb in the numerical integrations corresponding to Atlantic developing hurricanes with and without eddy fluxes of momentum

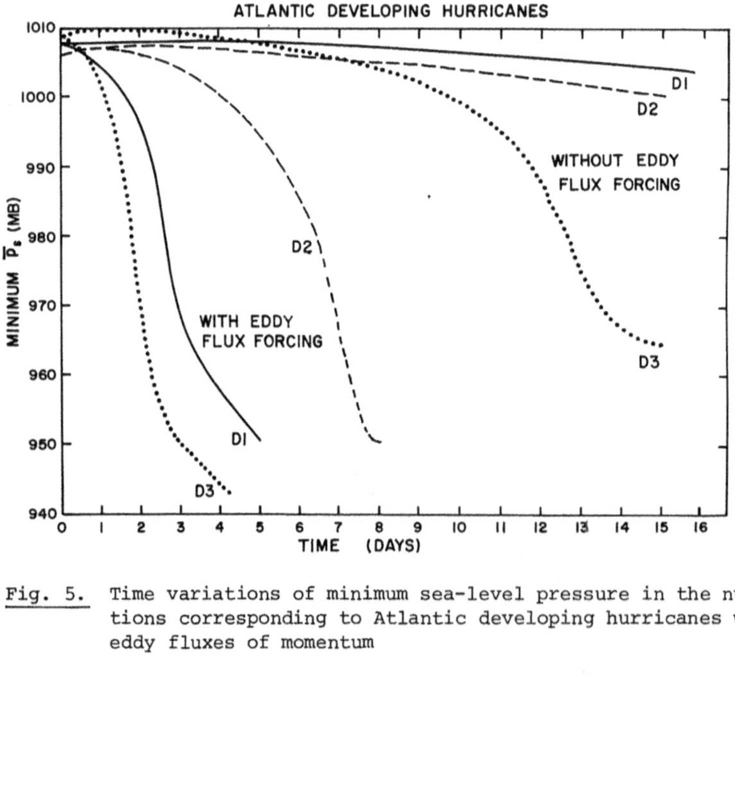

Fig. 5. Time variations of minimum sea-level pressure in the numerical integrations corresponding to Atlantic developing hurricanes with and without eddy fluxes of momentum

shown in Fig. 3. In the region $0 \leq r < 0.5°$ the specific humidity was taken to be equal to the value at $r = 0.5°$ and the tangential wind was obtained by linear interpolation using the observed values at $r = 0.5$ and setting $\bar{u} = 0$ at $r = 0$.

Figs. 4 and 5 show the time variations of the maximum tangential wind velocity at 900 mb and of the minimum sea level pressure in these integrations. Without eddy flux forcing, the prehurricane cloud cluster (D1) and the prehurricane depression (D2) do not develop into hurricanes, and the intensifying cyclone (D3) takes approximately 13 days to reach hurricane strength. With eddy flux forcing, all three initial disturbances develop rapidly into hurricanes, the D3 disturbance requiring less than 2 days to reach hurricane strength. The rate of development and final intensity of each disturbance is different, depending upon the details of the eddy momentum flux distribution. The marked difference of behavior with and without the eddy momentum flux forcing suggests that such forcing, which is associated with large-scale asymmetries and nonlinear interactions between the vortex and its surroundings, might be crucial to the development of hurricanes from tropical cloud clusters, depressions and cyclones. In nature, the result could be even more dramatic than implied by Figs. 4 and 5, since the intensity and radial extent of the inward eddy flux of momentum is observed to increase with each stage of development from the cloud cluster through the developing cyclone stage, as indicated in Fig. 2, whereas we have held the flux distribution and intensity fixed in our numerical integrations.

Fig. 6 shows the tangential wind field in the region $0 \leq r \leq 500$ km in the mature stages of the three integrations with observed eddy fluxes of momentum; Fig. 7 shows the radial and vertical velocities, and Fig. 8 shows the temperature departures from the mean tropical atmospheric values at the same stages. The final size, intensity and eye structure of each hurricane is seen to be different, depending upon the details of the eddy momentum flux distribution. The larger size of the D2 hurricane, for example, is apparently due to the weak outflux of momentum in the lower half of the troposphere at small radii (shaded region in Fig. 2) which is not present in the data for the other cases. Apparently, the rates of growth, final intensities, eye structures and sizes of the model hurricanes are sensitive even to such details in the eddy momentum flux distribution. Although we used MCBRIDE's data without modification and held the momentum flux fixed for the entire length of each integration, we believe that such details in the data are below the threshold of accuracy and are not representative of conditions in nature. Moreover, it is unlikely that such small magnitudes would persist throughout the development of a hurricane as would the larger, more organized, fluxes observed in the upper troposphere. It would be presumptuous, however, to set the fluxes in the lower troposphere to zero, or to alter the data in any subjective way. In an effort to increase the signal to noise ratio, therefore, the authors have formed a composite picture of the developing hurricane by linearly averaging the data for D1, D2 and D3 at every grid point and using the resulting data as a basis for further numerical integrations. A similar procedure was followed for the non-developing N1, N2 and N3 disturbances. The results, presented in the paper by PFEFFER and CHALLA (1981), are summarized in the introduction of the present paper.

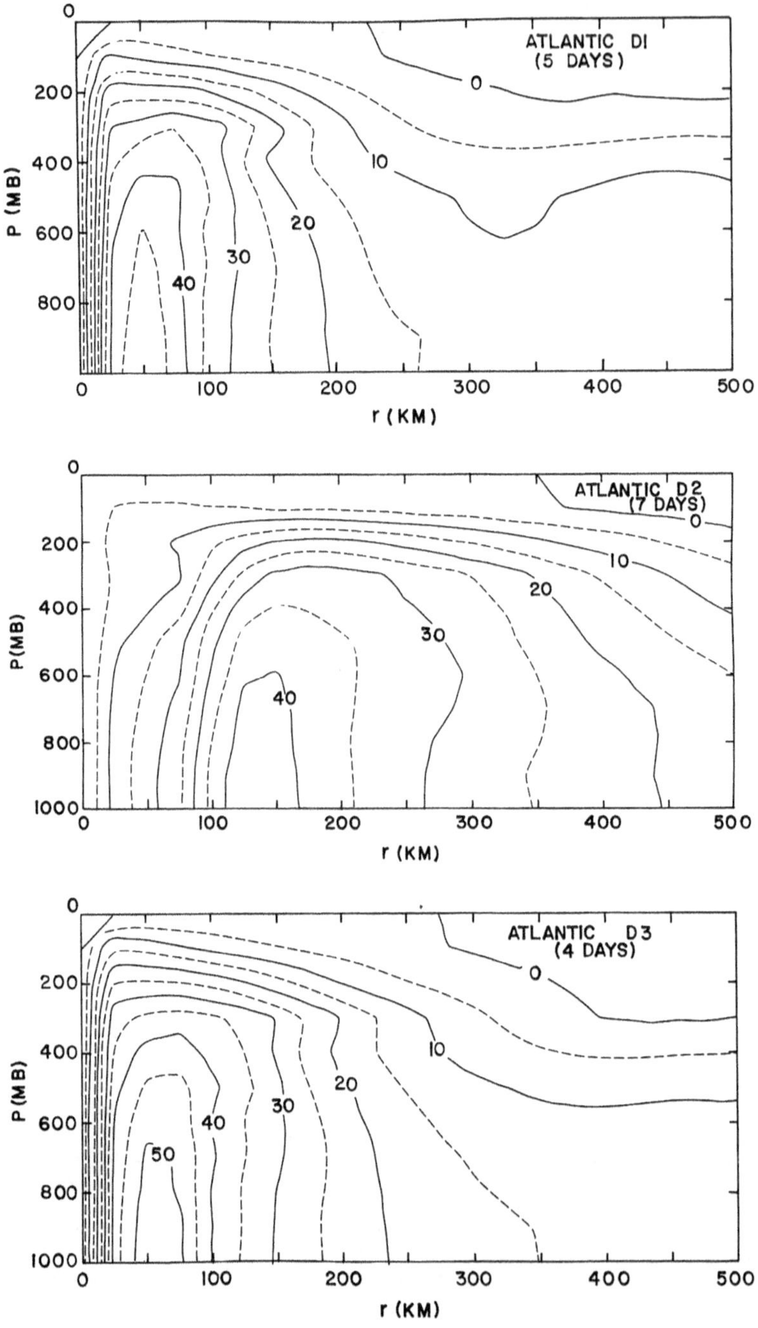

Fig. 6. Tangential wind field ($\bar{u}$) in the region $0 \leq r \leq 500$ km in the mature stages of the three model hurricanes formed by numerical integrations with observed eddy fluxes of momentum. Units are m s^{-1}

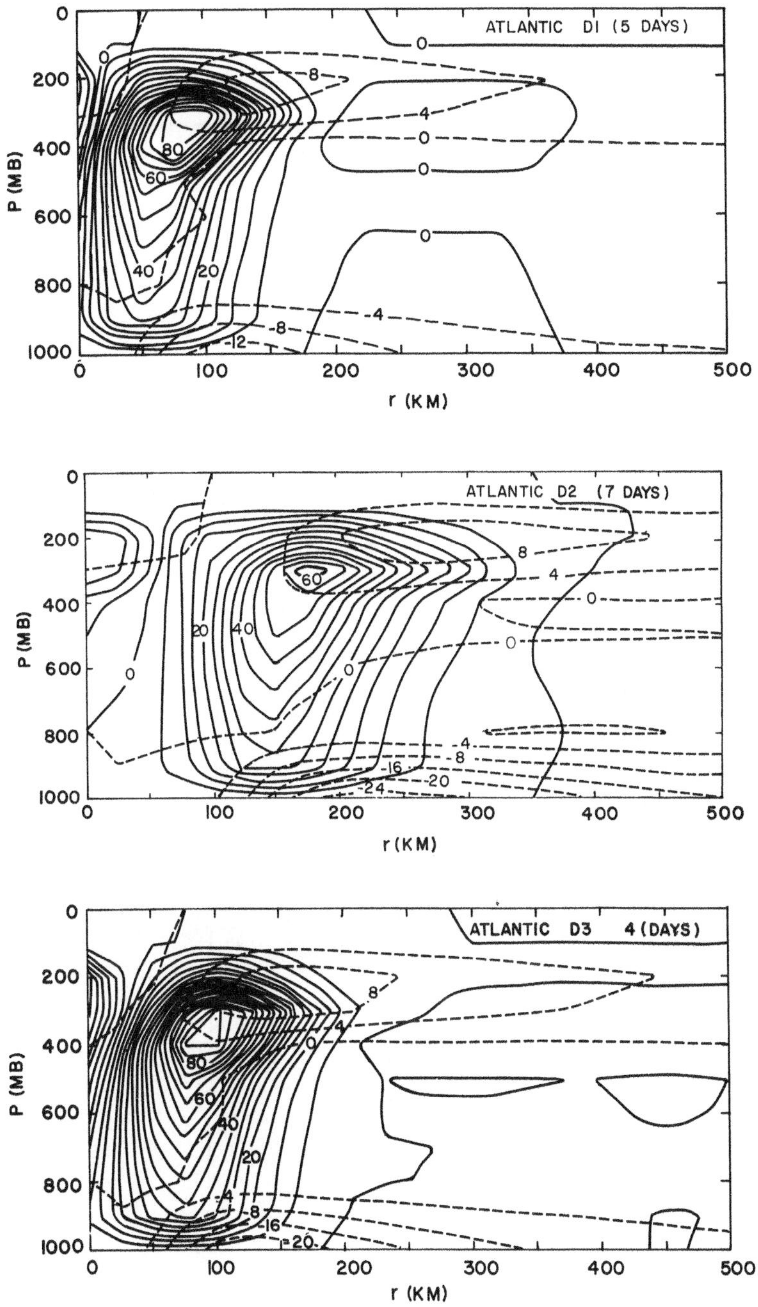

Fig. 7. Radial (dashed) and vertical (solid) velocity distributions in the mature stages of the three model hurricanes. Units of $\bar{v}$ and $\bar{w}$ ($\equiv -\bar{\omega}/g\bar{\rho}$) are m s^{-1} and cm s^{-1}, respectively

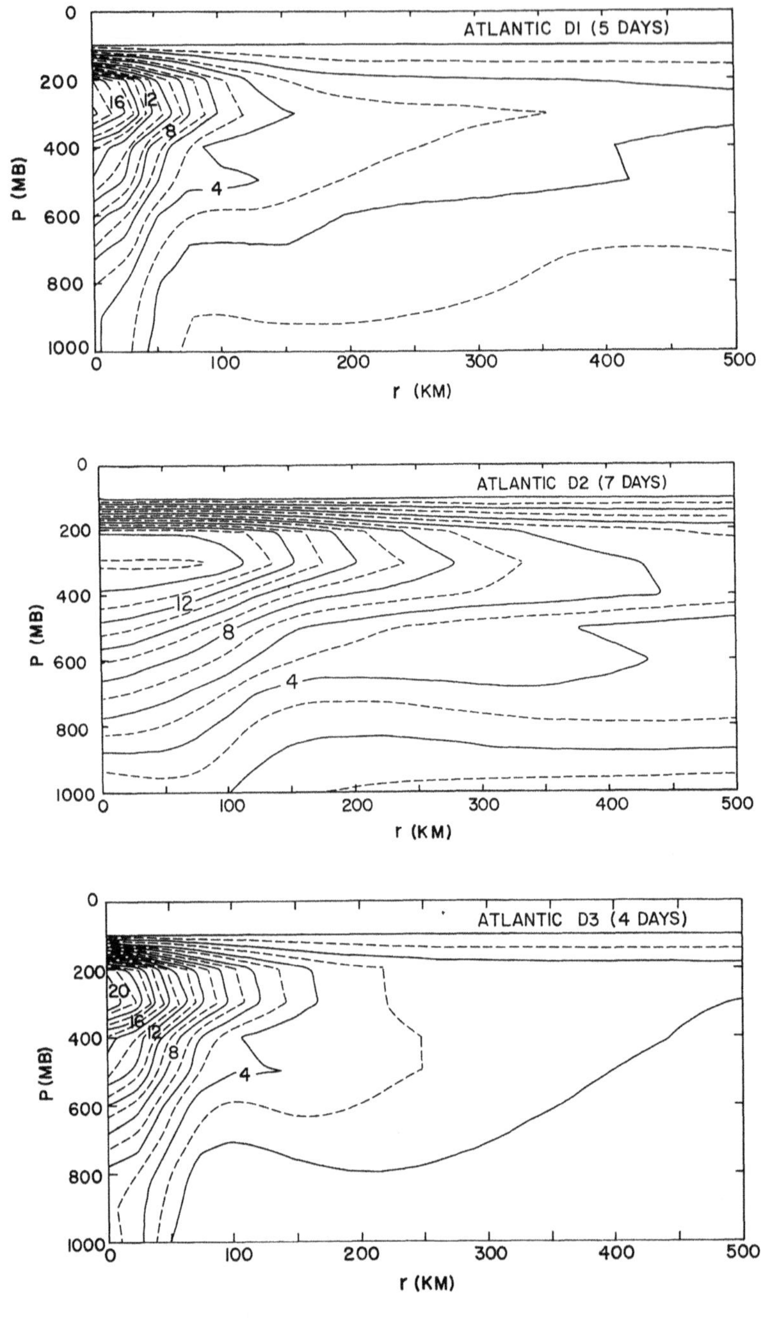

Fig. 8. Distributions of temperature anomalies (°C) in the mature stages of the three model hurricanes

6 Summary and Conclusions

The results presented in this paper give evidence of the importance of large-scale eddy fluxes of momentum in the development of Atlantic hurricanes from prehurricane cloud clusters, depressions and cyclones. SUNDQVIST's symmetric numerical model predicts that, in the absence of the eddy flux convergences of momentum which are observed in such disturbances, prehurricane cloud clusters and depressions are not capable of developing into hurricanes through the action of Ekman pumping and cooperative instability, and prehurricane cyclones require exceedingly long time periods to develop. With the observed fluxes specified in a parameterized way in the model, all such disturbances develop into mature hurricanes in realistic time periods. The rate of development of each disturbance, as well as the eye structure and the size and intensity of the mature hurricane, depend upon the details of the eddy momentum flux distribution. These results suggest that the observed inward eddy fluxes of momentum associated with synoptic scale tropical wave asymmetries may be crucial to the development of Atlantic hurricanes.

Acknowledgements

We are grateful to Professor Hilding Sundqvist for making available his symmetric hurricane model, and to Professor William Gray for making available McBride's composite data set which was prepared at Colorado State University under his supervision. We also appreciate the financial support we have received from the Office of Naval Research under ONR Contract N00014-77-0265 and from the computing Centers at Florida State University and at the National Center for Atmospheric Research which provided computer time for the calculations presented in this paper. The National Center for Atmospheric Research is supported by the National Science Foundation.

References

Black, P. G., and Anthes, R. A., 1971: On the asymmetric structure of the tropical cyclone outflow layer. J. Atmos. Sci., 28, 1348-1366

Challa, M., Pfeffer, R. L., 1980: Effects of eddy fluxes of angular momentum on model hurricane development. J. Atmos. Sci., 37, 1603-1618

Frank, W. M., 1977: The structure and energetics of the tropical cyclone II. Dynamics and energetics. Mon. Wea. Rev., 105, 1136-1150

Jordan, C. L., 1958: Mean soundings for the West Indies area. J. Meteor., 18, 91-97

McBride, John, 1981a: Observational analysis of tropical cyclone formation, I: Basic description of data sets. J. Atmos. Sci., 38, June (in press)

____________, 1981b: Observational analysis of tropical cyclone formation, III: Budget analysis. J. Atmos. Sci., 38, June (in press)

McBride, John and Raymond Zehr, 1981: Observational analysis of tropical cyclone formation, II: Comparison of non-developing vs. developing systems. J. Atmos. Sci., 38, June (in press)

Palmén, E. and H. Riehl, 1957: Budget of angular momentum and energy in tropical cyclones. J. Meteor., 14, 150-159

Pfeffer, R. L., 1955: A discussion of the balance of angular momentum in hurricanes. Bull. Amer. Meteor. Soc., 37, 234

______________, 1958: Concerning the mechanics of hurricanes. J. Meteor., 15, 113-120

______________, and M. Challa, 1981: A numerical study of the role of eddy fluxes of momentum in the development of Atlantic hurricanes. J. Atmos. Sci., 38 (to be published)

Sundqvist, H., 1970: Numerical simulation of the development of tropical cyclones with a ten level model. Part 1. Tellus, 22, 369-390

The Numerical Experiment on the Contribution of Landed Typhoon to the Rainfall in North China

X. An(A. Hsieh) and X. Wen-jun

1 Introduction

The weather of China is greatly affected by typhoon. Typhoons, which move over China or close Chinese coasts, usually cause gales and rainstorms in the adjacent sea areas. Furthermore, some of typhoons may generate heavy rainfall in North China when they are interacting with the cold air of westerlies. Such kind of rainfall process is considered as "indirect" typhoon rainstorm, which constituted the majority of the heavy rainfall (≥ 100 mm·day^{-1}) and flood. Since the "Aug.1975" heavy rain in Henan Province(1060 mm in 24-hour, 685mm in 6-hour and 189.5mm in one hour), Chinese meteorologists have paid great attention to these rainfall processes (Beijing Group, 1979; Xie An et al, 1979). The case selected in this study is a landed typhoon. Besides a short synoptic analysis of this case, some simple numerical experiments of real data forecast are made by using a five-level fine-mesh model. A better understanding to the physical mechanism of the rainfall may be expected.

2 Synoptic Study

Typhoon Billie is one of the strongest typhoons hiting China in 1976. It first landed on Taiwan in the evening of 9th Aug., then landed on the coast area of Fujian Province in the morning of 10th (Fig.2.1a). At the same time a remarkable cold front was just moving southeastwards from Great Bend of Yellow River and approached to Billie (Fig.2.1b). An extratropical cyclone formed at 00Z 12th in the inverse V trough of the typhoon in the middle of Huai River, then moved northeastwards to Shangdong (Fig. 2.1c).

Fig.2.2 illustrates the total precipitation from 9th Aug. (Billie started to affect on the weather of southeast coast of China) to 12th (the cyclone moved out of Shangdong). It is clear that the rain zone of North China is caused by the cyclone. It is also an indirect typhoon rainstorm, which will be discussed in the next section. The above-mentioned process is a typical case of typhoon-westerlies interaction.

A strong jet stream is situated between the typhoon and the subtropical anticyclone. Fig.2.3 shows the horizontal and vertical structures of the jet. The wind speed in the strong wind core which is located at the height of four km is larger than 30 m/s. The thickness of the jet is also larger than ordinary south-west jet usally located near 850 mb. From a large number of case studies, the jet stream accompanied by the typhoon played an important role in the heavy rainfall which occurred in North China.

Topics in Atmospheric and Oceanographic Sciences
Intense Atmospheric Vortices, ed. by Bengtsson/Lighthill

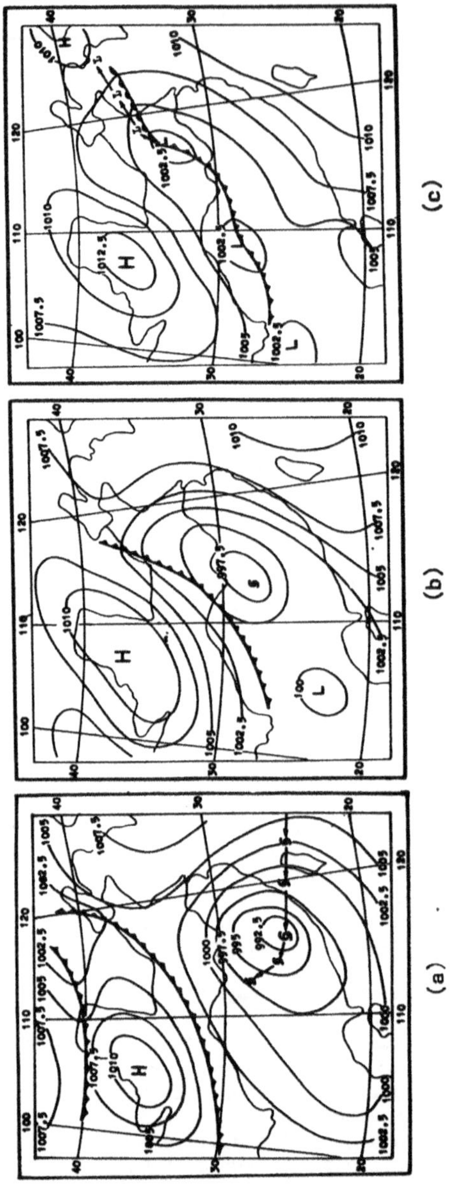

Fig.2.1. Maps of observed surface pressure

(a)--12 Z 10th Aug.
(b)--12 Z 11th Aug.
(c)--00 Z 12th Aug.

3 The Numerical Model and Control Run

The selected case has been predicted by using a fine-mesh primitive equation model (Zhang Yu-ling et al,1978). A brief descriptions of the model is given as follows. The model is 5 levels in σ-coordinate with the top at 100 mb. The grid distance is 150 km at standard latitudes 30°N and 60°N. The integrated domain is 19°N--45°N and 110°E--130°E(see Fig.3.1), with time step of 300 seconds. A fixed lateral boundary condition is used. The

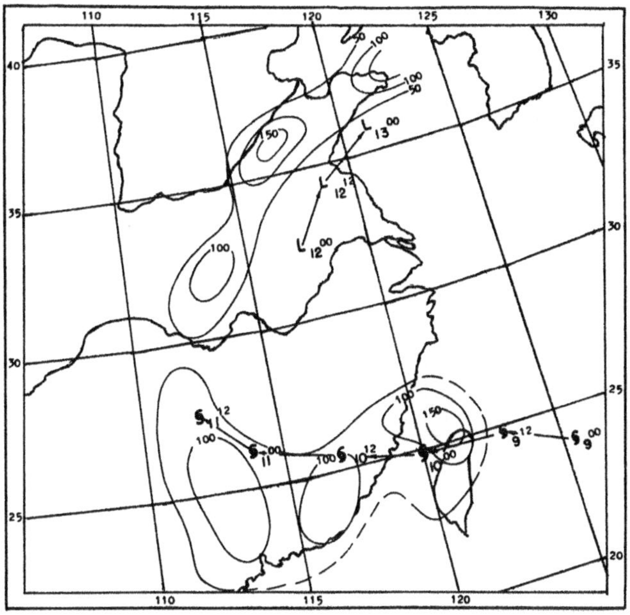

Fig.2.2. Total amount of precipitation (in mm) from 9th to 12th Aug

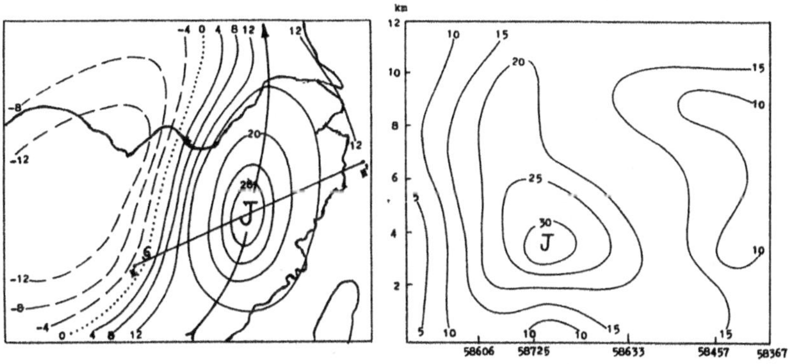

(a) (b)

Fig.2.3. The isotaches at 700 mb (a) and vertical cross section (b) of wind speed along NN' for 00Z 11th Aug. Units: m/sec

physical processes involve large-scale condensation and convective adjustment, surface friction, sensible- and latent-heat exchanges between the atmosphere and the sea. The geopotential height , observed wind which is adjusted by use of O'Brien's (1969) technique and dew-point depression ($T-T_d$) are taken as initial data.

The data at 12 GMT 11 Aug. are used as the initial step of integration. The 24-hour forecast surface chart and the accumulated amount of precipi-

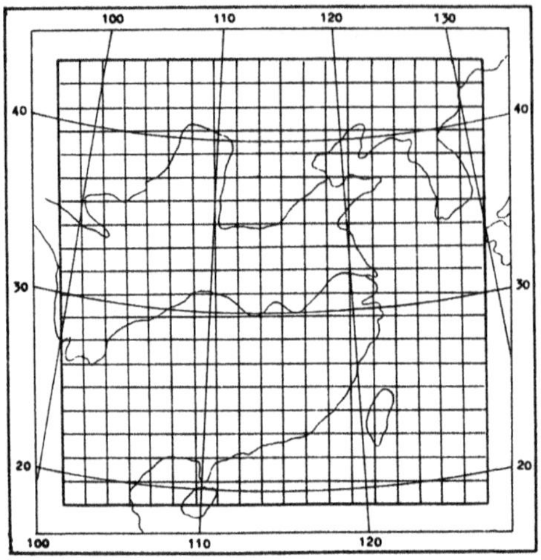

Fig.3.1. The integrated domain

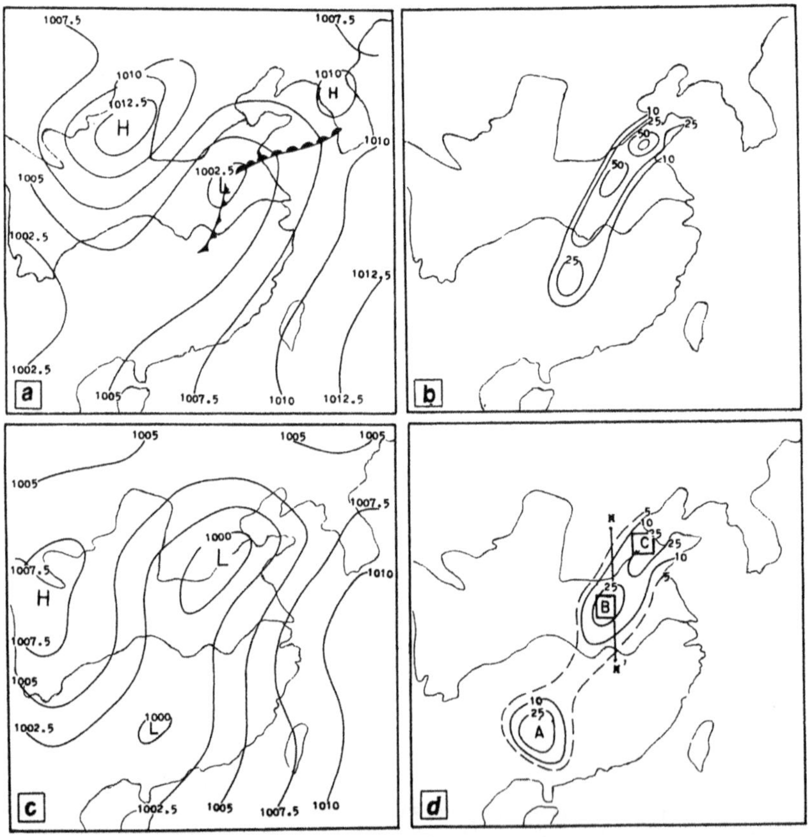

Fig.3.2. Surface maps and precipitations of 24-hour for 12 GMT 12th Aug.
a,b--observed, c,d--predicated

tation are given in Fig.3.2. In order to compare the prediction with observation, corresponding maps for 12 GMT 12 are also given in Fig.3.2. It can be seen that the predicted cyclone has formed in the middle of Huai River and then moved northeastwards to the downstream of Yellow River, only its location is in a distance slightly north to the observed. The predicted amount of precipitation is less than the observed. The distribution of the model precipitation is similar to the observed on the whole. Three centres of rainfall, indicated by A,B,C, can be found in Fig.3.2b and Fig.3.2d. The remarkable difference in the pressure system between the predicted and observed is the predicted high pressure which was in Shanxi on Fig.3.2C. Actually, this high pressure has moved to Great Bend of Yellow River. This difference, however, only has little effect on the process of the interaction between typhoon and westerlies. Concerning the rainfall in North China, the physical processes included in the model are similar to that of the actual processes. This prediction may be considered as a "control run" (CR) hereafter.

4 The Results of Numerical Experiments

In order to examine the influence of the lower tropospheric jet stream accompanied by the landed typhoon on the rainfall in North China, three runs of experiment are performed and the results are compared with CR.

EXP I --WLJ and CR:

The intensity of the lower tropospheric jet stream is weakened by using Shapiro's (1970) filter to the initial geopotential height and the longitudinal wind field to examine the function of the jet in the interaction of the typhoon and westerlies. The predicted result using above-mentioned initial data is called WLJ. In Fig.4.1b the wind speed within the interinal square has been smoothed eight times. The wind speed and its horizontal gradient is much decreased in Fig.4.1b. The 24-hour predicted maps for WLJ are shown in Fig.4.2.

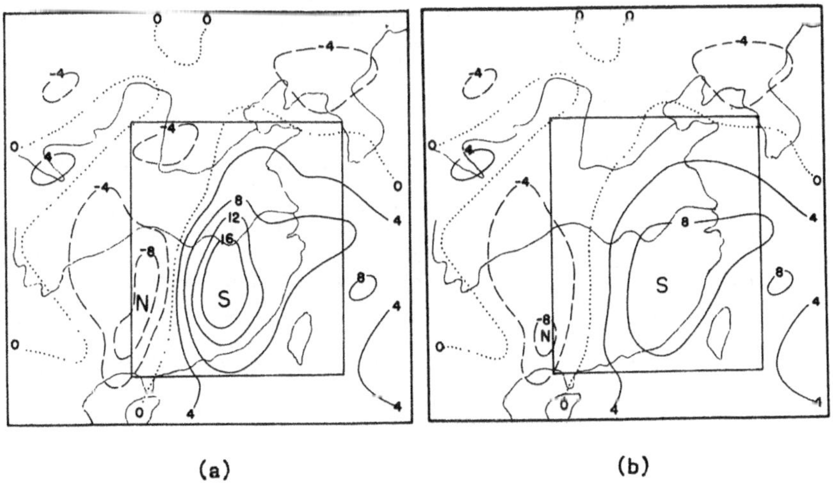

Fig.4.1. Initial data of v at 700 mb for CR(a) and WLJ(b). Units: m/sec

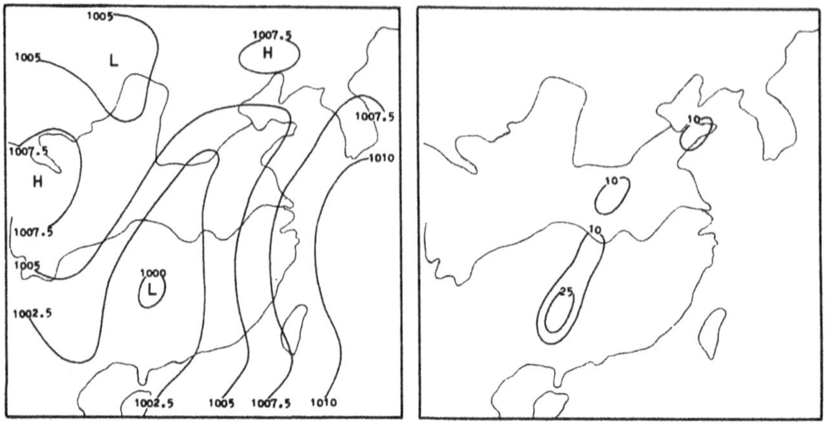

(a) (b)

Fig.4.2. Surface map(a) and 24-hour accumulated precipitation(b) for 12 Z 12th Aug. predicted by WLJ

Comparing with Fig.3.2, we can see that the differences between WLJ and CR are obvious in the pressure system and the amount of precipitation. For WLJ,the pressure pattern still maintains the shape of the inverse V trough, the extratropical cyclone has not formed in the discussed area. The rainfall of WLJ is less than that of CR by 50% approximately, especially in the rain areas B and C. Thus,it is clear that the rainfall in North China is closely related to the jet stream corresponding with typhoon coming from the south of China.

Fig.4.3 shows the latitudinal distributions of various elements which are 12-hour predicted for CR and WLJ through the rain area B. For CR, the jet stream and the moist tongue have reached the north part of China. The convergence of moisture flux is very strong and appears to the left of the jet. There is no doubt that the convergence of moisture is one of the most important conditions for the heavy rainfall. All the values of elements of WLJ are lower than that obtained from CR. Furthermore, the convective instabilites represented by difference of θ_e between $\sigma = 0.37$ and 0.73 levels ($\Delta\theta_e$ (σ: 0.37-0.73)) in Fig.4.3 are quite different for CR and WLJ cases. Near the rain area B the value of this element reaches -19°C for CR, but the value of $\Delta\theta_e$ for WLJ is only -5°C, at some individual grids the positive value even occurs. The difference between CR and WLJ in the convective instabilityis result from the difference of warming and moistening in the lower troposphere. It was caused by the jet stream (see $\Delta\theta_e$(WLJ-CR) in Fig.4.3). The major influence of this jet on the heavy rainfall process in North China can be seen clearly from comparing the curves in Fig.4.3.

The 12-hour forecasts of $\nabla\cdot\frac{1}{g}\vec{V}q$ for CR and WLJ are illustrated in Fig. 4.4. In each prediction there is a long convergence zone of moisture flux run from northeast to southwest. The main difference between CR and WLJ in North China lies in the convergence centre corresponding to the rain area B. The negative centre -7.2 for CR is caused by the superposition of the two effects which are produced by the cold front and the jet stream. It is why we emphasize the interaction between typhoon-westerlies.

EXP II--DCR and CR:

The cases of prediction experiment by use of dry model are called DCR hereafter. Fig.4.5 shows the ageostrophic wind speed for two cases of

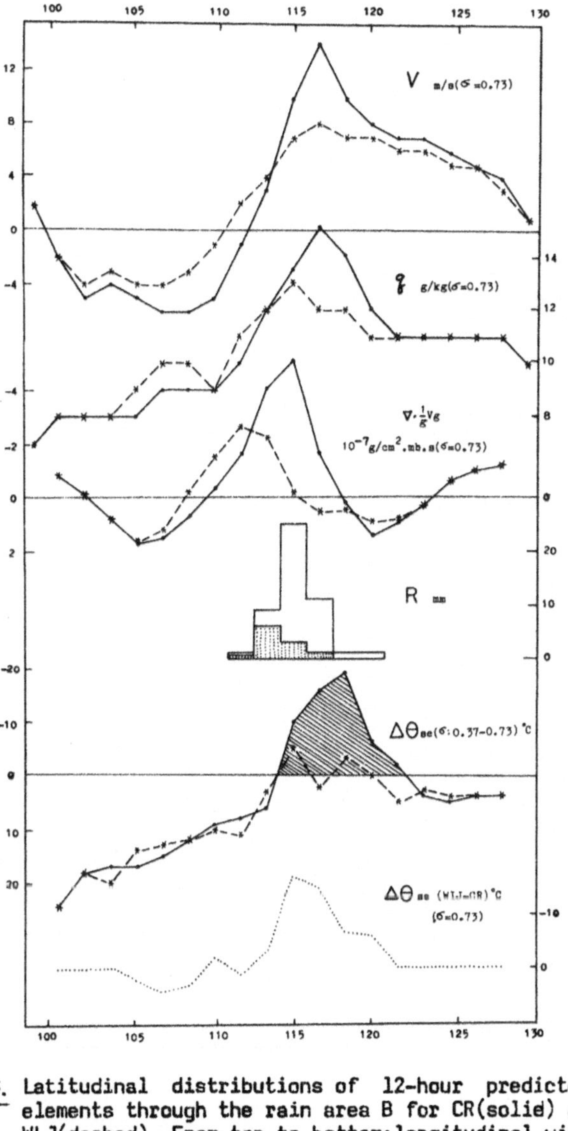

Fig. 4.3. Latitudinal distributions of 12-hour predicted elements through the rain area B for CR(solid) and WLJ(dashed). From top to bottom: longitudinal wind v, specific humidity q, divergence of moisture flux $\nabla\cdot\frac{1}{g}\vec{V}q$ (the above all at σ=0.73), amount of precipitation R, the differenve of θ_e between σ=0.37 and 0.73, and the difference of θ_e between WLJ and CR at σ=0.73

prediction. In the south of Yangtse River, the situation of predictions are very similar to each other. The difference of the ageostrophic wind between DCR and CR lies in the vicinity of the jet in North China. In this region for CR the jet moves obviously northward, the ageostroptic wind increases and the cyclone develope. These phenomena are directly related to the moist model. The more attention has been paid to the influence of latent heat on the development of system(Chen Jia-bin et al, 1978).

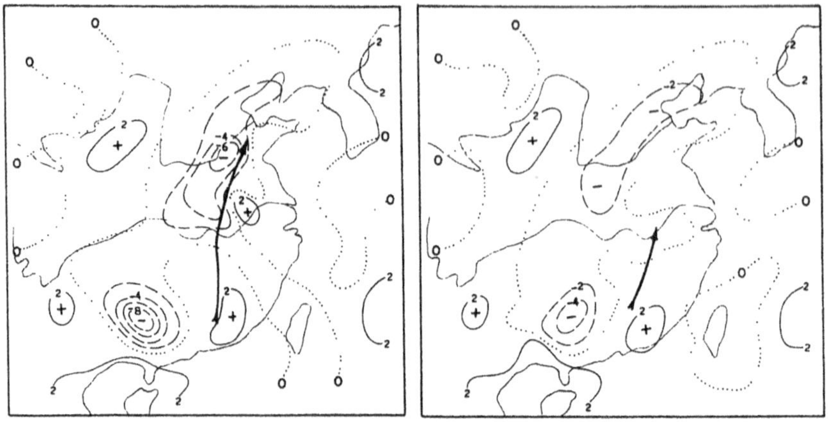

(a) (b)

Fig.4.4. 12-hour forecasts of $\nabla\cdot\frac{1}{g}\vec{V}q$ at σ =0.73 level for CR (a) and WLJ(b). Units: 10^{-7}g.cm^{-2}.mb^{-1}.s^{-1}

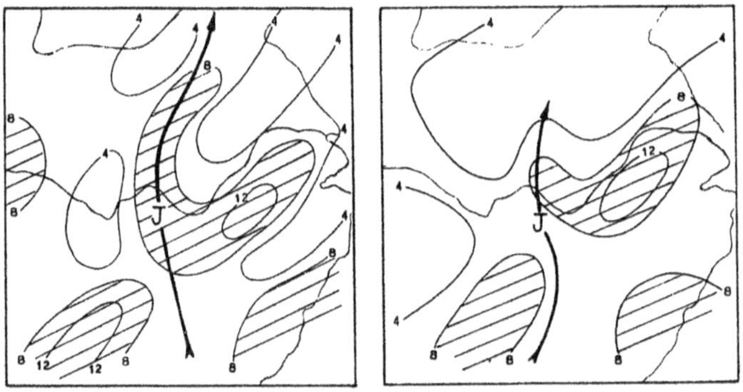

(a) (b)

Fig.4.5. 12-hour predicted ageostrophic deviation (m/s) and jet axis ($\geq$ 12 m/s) at σ=0.73 level for CR (a) and DCR(b)

EXP Ⅲ--DWLJ and DCR:

The experiment with the weak lower tropospheric jet stream as initial data in dry model is named DWLJ. It can be seen from comparing DCR with DWLJ that their results are very similar each other (see Fig.4.6). More-ever, the relation coefficient of the predicted $\dot{\sigma}$ at σ =0.64 level for DCR and DWLJ is 0.94. It is interesting to note that the function of the jet stream without the moisture is not important. In other words,the pure baroclinicity without moisture may not be very important in the development of above-mentioned synoptic process. The interaction between the typhoon and westerlies must be accompanied by moisture. Otherwise the interaction could not be so strong and rainfall would be not so heavy as it should be. These results are coincided with the theory of moist baroclinic instability suggested by Xie Yi-bing(Hsieh,1978).

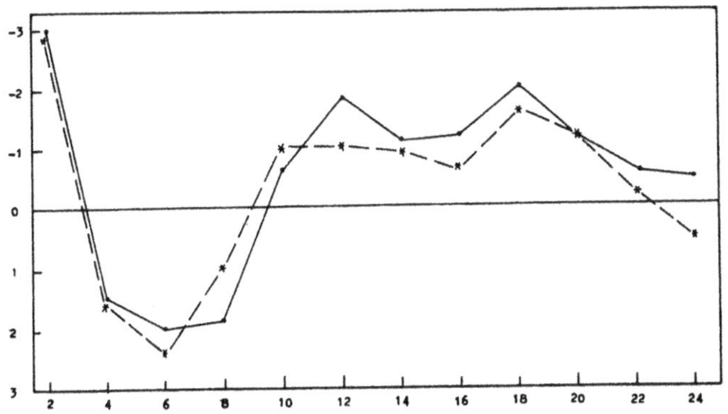

Fig.4.6. The variation with time for the predicted vertical motion $\dot{\sigma}$ (10^{-6} sec^{-1}, at σ=0.64 level), i.e. the average value on four grids in the rain area B for DCR (solid) and DWLJ (dashed). Abscissa is the model time (hour)

5 Conclusion

The real data forecast experiment made in this study has indicated the important effect of the jet stream accompanied by typhoon on the rainfall process in North China. It is clear that the moist baroclinicity can strongly effect on the development of the rainfall process. Our study may be helpful to understand the mechanism of the interaction between typhoon-westerlies and to improve the forecast of heavy rain.

In this paper, only a case study is presented. Though no final conclusion has been arrived on this problem, the results are interesting. A more sophisticated model and more case experiments are needed and now we are preparing for further investigations.

ACKNOWLEDGEMENTS

The authors would like to express their hearty thanks to Prof. Xie Yi-bing(Y. P.Hsieh) and Prof. Chen Shou-jun for helpful comments on the manuscript.

REFERENCES

Beijing Group, "75.8" Heavy Rain Cooperative Research, 1979: A diagnostic analysis about the "75.8" heavy rain in Henan. Acta Met. Sinica, 37,4,45-55

Chen Jia-bin et al., 1978: A numerical simulation of the genesis of an extratropical cyclone of rain. Selected Papers on Rainstorms. 94-102

Hsieh Yi-ping, 1978: Some aspects of the dynamics of the moist baroclinic atmosphere. Selected Papers on Rainstorms. 1-15

O'Brien,J.J., 1970: Alternative solutions to the classical vertical velocity problem. J. Appl. Met., 9,197-203

Shapiro, 1970: Smoothing, filtering and boundary effects. Rev. of Geo. and Space Sci. 8,359-387

Xie An et al., 1979: A comparative analysis of two cases of rain bearing synoptic process. Meteorological Monthly. NO:6,26-28

Zhang Yu-ling et al., 1978: A five-level primitive equation model for prediction of precipitation. Second Conference on Numerical Weather Prediction. Science Press, 1-12

Estimation of Radial Distribution of Eddy Viscosity in a Mature Hurricane

D. R. Davies and H. P. Evans

Summary

This paper describes an analytic solution relating the radial distribution of the radial component, K_1, of eddy viscosity to the observed radial distribution of tangential velocity in a mature hurricane. It is an extension of a previous paper by EVANS and DAVIES (1979), which was concerned with the vertical distribution of the eddy viscosity components. Over most of the region between the eye boundary and the position of maximum tangential velocity the radial distribution of K_1 is found to be very close to a linear function of the radial distribution of tangential velocity, but increasing sharply in the peak velocity region. The corresponding radial distribution of vertical velocity is then found to be very similar to typical observed profiles.

1 Introduction

In a previous paper, EVANS and DAVIES (1979) have shown that an analytic solution (in p co-ordinates) can be obtained of the dynamical equations describing a mature hurricane system, in an assumed quasi-stationary and axisymmetric phase. This was used to connect the vertical (p) distribution of K_1 and K_2, describing respectively horizontal and vertical eddy transfers of momentum, with a typical observed vertical distribution of tangential velocity. This showed that (i) K_1 and K_2 fell off quite rapidly with height above the surface, and (ii) their numerical values were similar to those obtained from available observational data, and of the order of 10 times larger than the values of K_1 and K_2 which are found empirically to give generally good results in numerical models of hurricanes. This result reflected one of the interesting differences between a spatial grid resolution and a continuous distribution.

It was also noted in the Conclusions of the paper that the analysis is capable of improvement in a relatively simple form by taking K_1 to be a function of radial distance, r, with consequent effects on the radial distribution of the vertical velocity field. For example, one of the deficiencies noted in the previous paper was that the maximum of the vertical velocity field, w, appeared at the position of the eye itself, whereas most observational analyses place the maximum of w between the radius of the eye, r_o, and the position of peak velocity, r_m.

Following conventional turbulence formulations the spatial distribution of K_1 is related to $\overline{u'v'}$, where u' and v' are eddy velocities in the radial and tangential directions respectively, and the averaging is taken spatially over the circumference of a circle radius r and also over a suitable time period. The component K_2

Topics in Atmospheric and Oceanographic Sciences
Intense Atmospheric Vortices, ed. by Bengtsson/Lighthill

is similarly related to $\overline{w'v'}$, where w' is eddy velocity in the vertical direction. As in all turbulence problems a closure theory is needed to express these correlations in terms of statistical averages of the basic dynamical characteristics of the individual cloud elements. There is a recent growth in the general theoretical development of this basic requirement (e.g. Moncrieff, 1981). In hurricane conditions, however, it is of course particularly difficult, and a knowledge of the approximate dependence of eddy viscosity on r by an inverse calculation (following EVANS and DAVIES, 1979) could be helpful. The K_1 , $\overline{(u'v')}$, component is likely to depend sharply on typical values of the radial, u' , eddy velocity component produced by the invididual cumulo-nimbus clouds of the hurricane system. This will of course depend on the large inward pressure gradient, with u' varying in magnitude sharply from a peak value near r_m to zero at r_o. On the other hand the K_2 , $\overline{(w'v')}$, component depends strongly on w', the vertical eddy velocity component. In the lower region of the hurricane this is largely controlled by the frictionally created turbulence, depending much more sharply on height above sea surface than on radial distance. Consequently as a first approximation K_2 is taken to be a function of p only, and K_1 to be radially dependent as well as a function of p .

A relatively simple mathematical analysis, using separability in p and r of the describing equations, can then be applied, leading to an interesting result.

2 The Model Equations and Analytic Solution Method

The tangential equation of mean motion in pressure, p , and radial, r , co-ordinates is

$$u\frac{\partial v}{\partial r} + \omega\frac{\partial v}{\partial p} + \frac{uv}{r} + fu = K_1(p)\frac{\partial}{\partial r}\left[K_1(r)\left(\frac{\partial v}{\partial r} + \frac{v}{r}\right)\right] + \frac{\partial}{\partial p}\left[K_2(p)\frac{\partial v}{\partial p}\right] \quad (1)$$

where the mean tangential and radial velocity components are denoted by v and u , dp/dt by ω , f is the coriolis parameter, $K_1(r,p)$ is compartmentalised into the form $K_1(r)\,K_1(p)$. The continuity equation is

$$r\frac{\partial u}{\partial r} + u + r\frac{\partial \omega}{\partial p} = 0 \; . \quad (2)$$

Following EVANS and DAVIES, 1979, ω is eliminated from (1) and (2), solutions are looked for in the form

$$u = z(r)\,u(p) \; , \quad (3)$$

$$v = z(r)\,v(p) \quad (4)$$

and the equation giving radial dependence becomes

$$\frac{d}{dr}\left[K_1(r)\left(\frac{dz}{dr} + \frac{z}{r}\right)\right] = z\left(\frac{dz}{dr} + \frac{z}{r}\right) \quad (5)$$

replacing equation (15) of the previous analysis. This equation can be simplified in form by writing for mathematical convenience

$$K_1(r) = \hat{K}_1(r)\,z(r) \; , \text{ and } \; y(r) = rz(r) \; , \quad (6)$$

giving

$$\hat{K}_1(r)\ (y\,y'' + y'\,y') + (\hat{K}_1' - \frac{2\hat{K}_1}{r} - 1)\,y\,y' = 0 . \tag{7}$$

Replacing $y\,y'$ by $\frac{d}{dr}(y^2/2)$ and integrating then gives

$$y^2 = A \int \frac{r^2}{\hat{K}_1(r)} \exp\left[\int dr/\hat{K}_1(r)\right] dr . \tag{8}$$

Writing for mathematical convenience

$$\hat{K}_1(r) = \frac{1}{f'(r)} \tag{9}$$

and integrating from r_o to r, using $z(r) = 0$ on $r = r_o$, leads to the integral form

$$z(r) = \frac{A}{r}\left\{r^2 e^{f(r)} - r_o^2 e^{f(r_o)} - 2\int_{r_o}^{r} e^{f(r)}\, r\,dr\right\}^{\frac{1}{2}} \tag{10}$$

where A is a numerical constant. The previous analysis by EVANS and DAVIES (1979, equation 25) gave

$$\frac{\omega(r,p) + C}{H(p)} = \frac{dz}{dr} + \frac{z}{r} ,$$

where the constant C was shown to be generally very much smaller than ω, and $H(p)$ is known in terms of $u(p)$, $v(p)$ and $K_1(p)$. This gives the r dependence of ω in the form

$$\frac{\omega + C}{H(p)} = e^{f(r)}\, f'(r)/z(r) . \tag{11}$$

3 Asymptotic Behaviour of $\hat{K}_1(r)$, $z(r)$ and $\omega(r)$ as $r \to r_o$

Writing $\xi = \frac{r}{r_o} - 1$ and $f(r) = a_o\,\xi^q$ in equations (9), (10), (11), for the region of the eye edge ($r \to r_o$, $\xi \to 0$) with the conditions that $z(r) \to 0$, $\omega(r) \to$ finite value, and $K_1(r) \to$ finite value, as $r \to r_o$, the value of q can be determined for this inner condition. The limiting value of $z(r)$, as $\xi \to 0$, is found to be $Aa_o\,\xi^{q/2}$, the $\omega(r)$ behaviour is given

(taking $C = 0$) by $H(p).\frac{1}{2}Aa_o^{\frac{1}{2}}q\ \xi^{q/2-1}$, and $\hat{K}_1(r) = \frac{1}{f'(r)} = \frac{1}{q\,a_o}\,\xi^{1-q}$.

The only solution is <u>$q = 2$</u>, giving z, ω, $\hat{K}_1(r)$ and $K(r)$ to be respectively proportional to ξ^1, a finite constant, ξ^{-1}, and a finite constant. This result is a useful guide in evaluating a more general relationship in Section 4.

4 Evaluation of $K_1(r)$ in terms of $v(r)$

Given an observed radial profile, $v(r)$, of the tangential velocity, the r dependence of the radial component of eddy viscosity, $K_1(r)$, can now be deduced. Equation (10) can be expressed in the form

$$\frac{z}{A} = z^* = \frac{1}{(1+\xi)}\left\{e^{f(\xi)}(1+\xi)^2 - 1 - 2\int_0^{\xi} e^{f(\xi)}(1+\xi)\,d\xi\right\}^{\frac{1}{2}} \tag{12}$$

and conveniently re-arranged by writing

$$q(\xi) = e^{f(\xi)} \quad :$$

this leads to

$$q(1+\xi)^2 - 1 - 2\int_0^{\xi} q(1+\xi)\,d\xi = z^{*2}(1+\xi)^2 = z^{**} \quad .$$

Differentiation with respect to ξ gives

$$q'(1+\xi)^2 = z^{**\prime} \quad ;$$

hence

$$q(\xi) = \int \frac{z^{**\prime}}{(1+\xi)^2}\,d\xi + B \ ,$$

and finally

$$q(\xi) = z^{*2} + 2\int \frac{z^{*2}\,d\xi}{(1+\xi)} + B \ , \tag{13}$$

where B is a numerical constant. Given z^*, $q(\xi)$ and hence $\hat{K}_1(r)$ can be evaluated by integration of equation (13).

Observed radial profiles of the tangential velocity component are typically of the general form

$$\frac{z}{A} = z^* = \xi\sqrt{(1+\xi)}\,e^{-b\xi} \ , \tag{14}$$

with b acting as a radial scaling factor. This gives an asymptotic variation for z of ξ^1 as $\xi \to 0$ in accordance with Section 3.

Figure 1 shows a typical radial profile of z^* with $b = 1/4$; for an r_o value of 10 km., r_m is at about 65 km. for this case. This can easily be extended for any r_o, r_m values by adjustment of the numerical b value. Substitution of the form (14), as a typical profile, into (13) leads on integration to the result

$$q(\xi) = e^{-2b\xi}\left[\xi^2(1+\xi) - \frac{1}{b}\xi^2 - \frac{1}{b^2}\xi - \frac{1}{2b^3}\right] + D = e^{f(\xi)}, \tag{15}$$

where $D - 1/(2b^3) = 1$ to satisfy the condition that $f \to 0$ as $\xi \to 0$. This leads to the form

$$\hat{K}_1(\xi) = \frac{\left\{e^{-2b\xi}\left[\xi^3 + (1-\frac{1}{b})\xi^2 - \frac{1}{b^2}\xi - \frac{1}{2b^3}\right] + (1+\frac{1}{2b^3})\right\}}{e^{-2b\xi}\left[-2b\xi^3 + (5-2b)\xi^2 + 2\xi\right]} \tag{16}$$

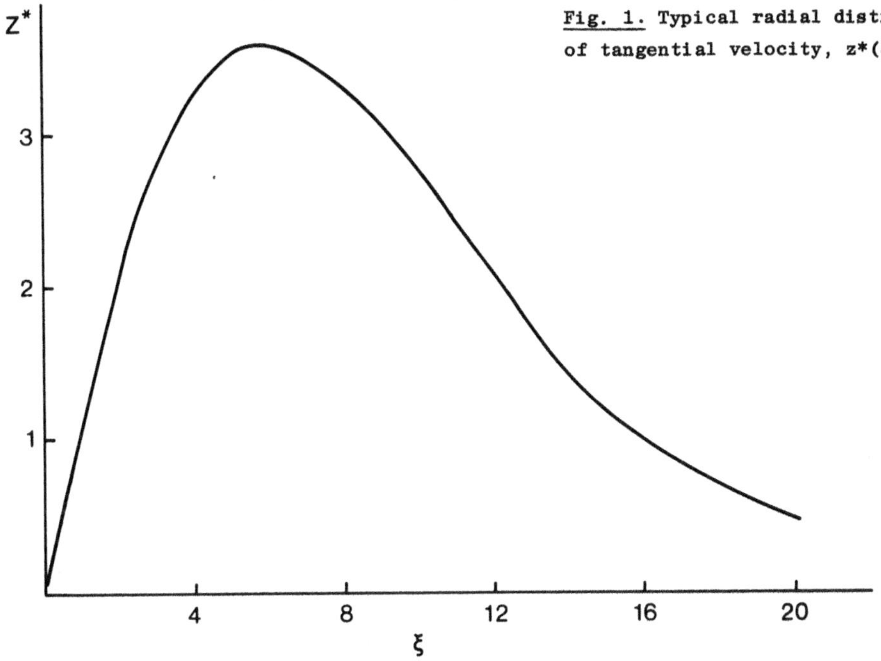

Fig. 1. Typical radial distribution of tangential velocity, $z^*(\xi)$

and, by using $b = 1/4$, this gives

$$\hat{K}_1(\xi) = \frac{\left\{ e^{-\frac{1}{2}\xi} \left[\xi^3 - 3\xi^2 - 16\xi - 32 \right] + 33 \right\}}{e^{-\frac{1}{2}\xi} \left[-\frac{1}{2}\xi^3 + \frac{9}{2}\xi^2 + 2\xi \right]} \tag{17}$$

the constant D having been chosen to give $\hat{K} \to \xi^{-1}$ as $\xi \to 0$ in accordance with Section 3 : $z(\xi)$ then behaves like ξ^1 and consequently $K_1(r) \to$ constant, as $\xi \to 0$.

Table 1 then shows computed values of $\hat{K}_1(\xi)$. These results show that over most of the range of r from r_o to r_m , $K_1(\xi)$ is remarkably independent of r (excluding the singularity as $\xi \to 0$) , but increasing rapidly in the region of peak tangential velocity. This means that the values of $K_1(r)$ are almost linearly dependent on the variation with r of the tangential velocity over most of the radial distance from r_o to r_m , but increasing rapidly as $r \to r_m$.

TABLE 1 Values of $\hat{K}_1(\xi)$ in arbitrary units (taking, e.g., r_o = 10 km r_m is at 65 km, corresponding to ξ = 5.5)

ξ	0	0.5	1	2	4	6
$\hat{K}_1$	∞	1.1	1.0	1.3	3.0	10.0

The associated radial distribution of the ω field is also interesting. Table 2 dhows the radial distribution of ω, corresponding to Table 1. The peak in ω lies between r_o and r_m : this is typical of many observed storms. The peak in ω calculated in the previous paper by EVANS and DAVIES (1979) appeared at the eye, $r = r_o$, and is not in accord with observations.

TABLE 2 Values of $\omega(\xi)$ in arbitrary units (corresponding to Table 1).

ξ	0	0.5	1	2	4	6
ω	1.0	1.7	3.3	3.2	1.9	1.1

5 Conclusions

A previous analysis by EVANS and DAVIES (1979) showed that $K_1(r,p)$ and $K_2(r,p)$ decreased rapidly with height above the sea surface, and their surface values were found to be similar in magnitude to available observational data analyses and were of the order of 10 times larger than those generally used in the grid resolutions of numerical models. This extension of the analysis (based on the limitations of the previous model) shows that over most of the inner region of the hurricane (from r_o to r_m) the radial distribution of the radial component of eddy viscosity is remarkably close to a linear dependence on the radial distribution of tangential velocity. The basic problem remains of expressing the eddy viscosities in terms of some kind of statistical averaging process over the dynamics of assemblies of the basic elements of the storm, i.e. the main cloud systems.

Acknowledgements

Thanks are due to Mr. G.A. Duller, of the Exeter Mathematics Department, for very helpful discussions.

References

Evans, H.P. and D.R. Davies 1979 On the use of an analytic solution in estimating eddy viscosity distribution and water vapour flux in a mature hurricane, Quart.J.Roy.Met.Soc., 105, 551-569

Moncrieff, M.W. 1981 A theory of organised steady convection and its transport properties, Quart.J.Roy.Met.Soc., 107, 29-50

Spontaneously Growing Hurricanelike Disturbances in a Simple Baroclinic Model with Latent Heat Release

A. E. Gill

1 Introduction

Tropical winds and their variations on time scales down to a week or two appear to have a baroclinic character in the sense that upper winds (say at 250 mb) are mostly in the opposite direction to lower-level winds (say at 750 mb). One can exploit this fact by exploring models where a fixed vertical structure with this character is assumed so that the mathematical problem reduces to one of finding variations in the horizontal and in time. This paper is about such models, the overiding aim being mathematical simplicity rather than realism so that for example, it should be easy for students to understand the basic processes involved. Despite this, the results can readily be compared with observations.

The equations that will be used are the shallow-water equations of Laplace, possibly modified by a forcing term representing heating effects and by simple representations of friction and processes tending to restore thermal equilibrium. These equations are assumed to apply to a baroclinic disturbance with the appropriate vertical structure. You are free to take whatever solution of the vertical structure equations you like, but a convenient one is that for an incompressible atmosphere of constant buoyancy frequency N with a rigid lid at height πH. Then, if z is height from the ground, horizontal velocities go like $\cos(z/H)$ while vertical velocity, temperature perturbation and the heating function go like $\sin(z/H)$. Calculations of diabatic heating in the tropics tend to give a vertical structure like this, with a maximum near 500 mb, and that is presumably why the winds tend to have the corresponding baroclinic structure with upper level winds opposed to lower level winds - just as HALLEY envisaged in his original tropical circulation model in 1686. All I'm going to do is to add to Halley's concept the dynamics inherent in LAPLACE's equations and an elementary representation of moisture effects.

It's not difficult to find solutions with the heating specified if one chooses distributions of heating with mathematically-convenient forms. Fig. 1 shows some steady solutions obtained in this way, and observed features to which they may be related. Steady solutions have been obtained (see GILL 1980 for details) by the simple device of replacing $\partial/\partial t$ in LAPLACE's equations by $(\partial/\partial t + r)$, where r^{-1} is a decay time, and then placing $\partial/\partial t$ equals zero. The solutions look very similar to some obtained by GEISLER (1981) when calculating the forced response in the presence of cumulus friction, so one can think of the constant r as representing a simple parameterisation of this process.

Topics in Atmospheric and Oceanographic Sciences
Intense Atmospheric Vortices, ed. by Bengtsson/Lighthill

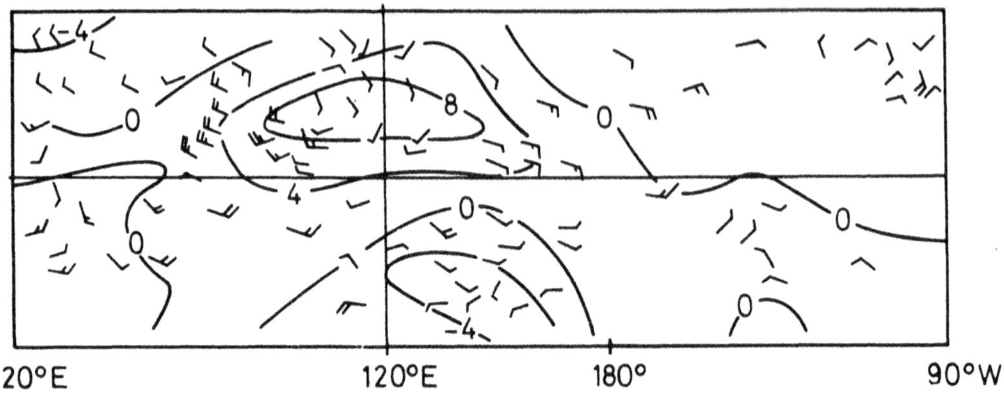

Fig. 1 (a). (i) The observed response to forcing over the Pacific and Indian Oceans in July. The contours are of vertical velocity at 500 mb, which is approximately proportional to diabatic heating rate. The arrows are 850 mb winds (from GILL 1980 fig. 4)

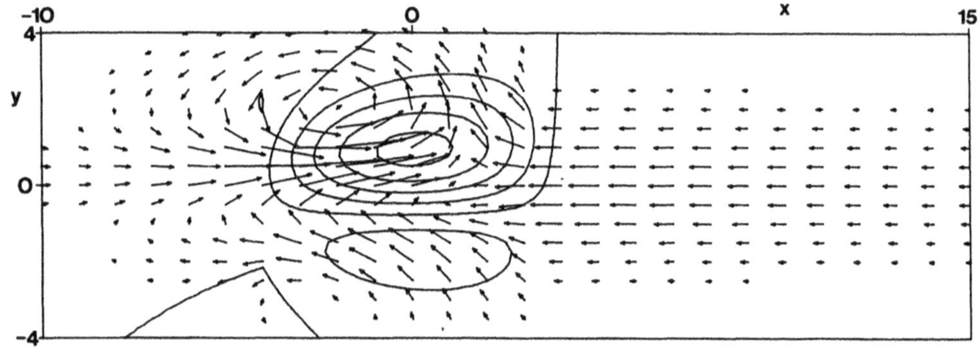

(ii) An analytic solution with vertical velocity shown by contours and the low-level wind response shown by arrows. The proportions of the diagram are the same as those for (i) if it is assumed that the length scale (the equatorial Rossby radius) is 10°. (This is the same as the solutions shown in GILL 1980 fig. 3c, but with heating in the region $|x| < 4$)

Although these solutions show remarkable similarity to observed patterns, the fact that heating needs to be specified seems somewhat unsatisfactory. Since the heating is due to latent heat release, it would seem desirable to add a moisture equation to the shallow-water set in order to give a simple representation of moist processes. The equations will be considered in the next section for the non-rotating case with motion in one horizontal direction only. In later sections, they will be generalised to (a) the f-plane and then (b) the equatorial beta-plane.

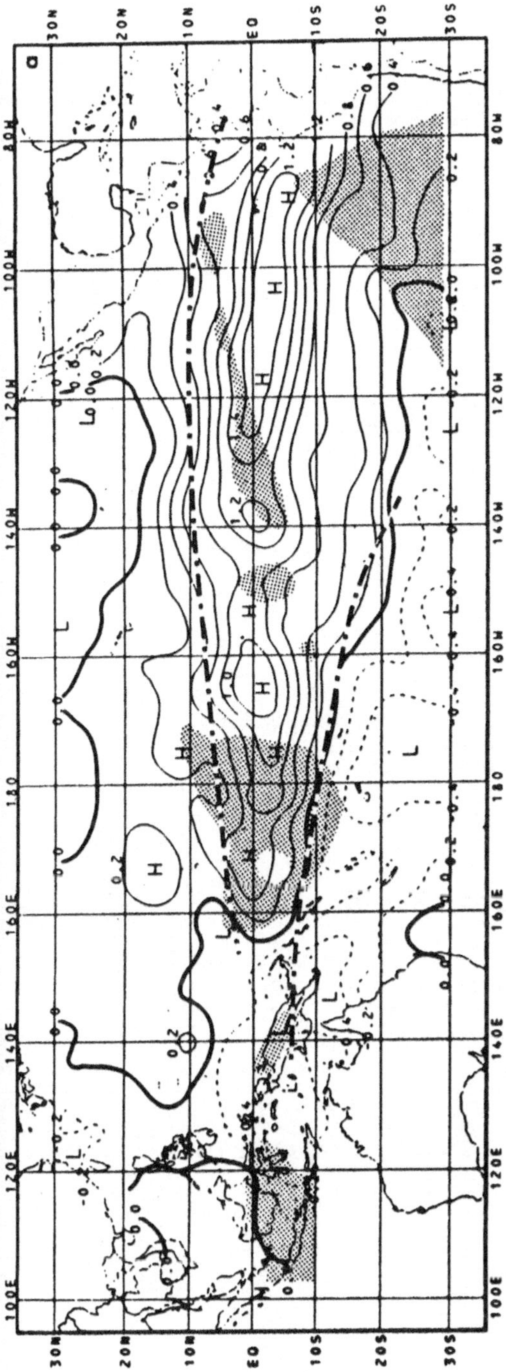

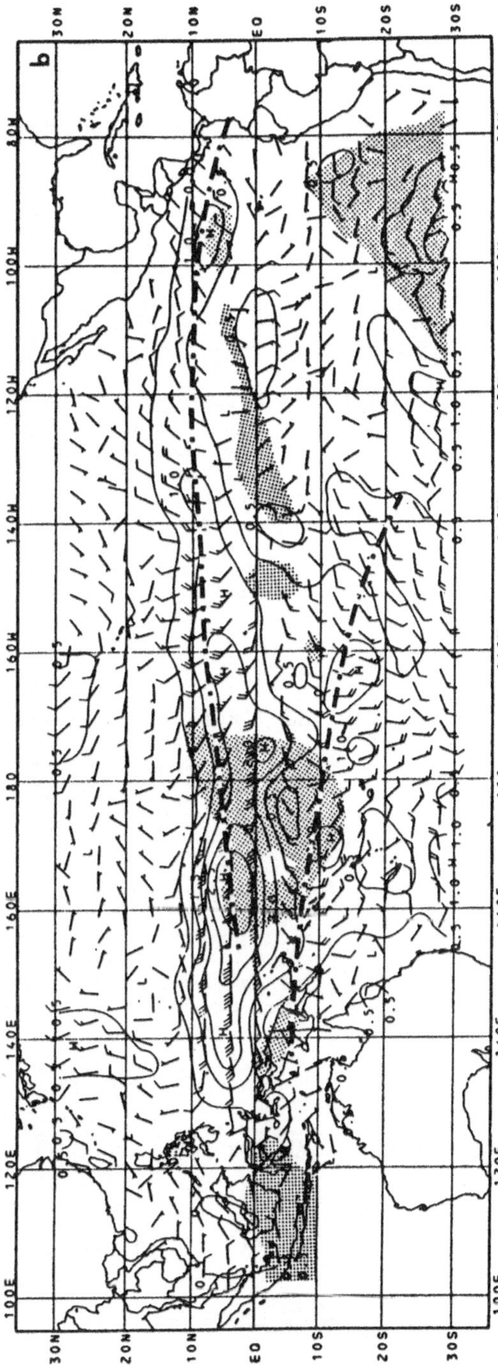

Fig. 1 (b).

(i) Observed surface temperature anomaly

(ii) Observed surface wind anomaly forme by compositing data for Aug-Oct of sever El Nino years (1951, 1953, 1957, 1965, 1969, 1972) (from RASSMUSSON and CARPENT 1981)

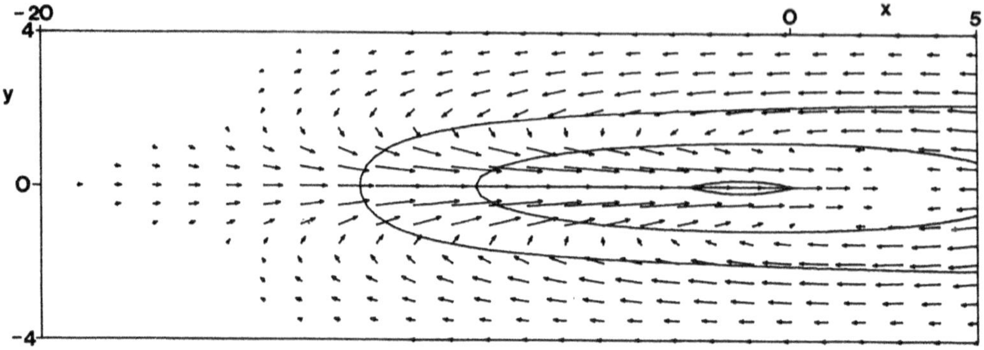

Fig. 1 (b). (iii) **An analytic solution for a long symmetric heating zone. The vertical velocity is shown by contours and the low level wind response by arrows. The scale is the same as in (a) (ii). The region to the right of the picture is dominated by easterlies whereas the presence of the American Continent may have a strong effect on this region in practice. (This is the same as the solution shown in GILL 1980 fig. 1 (a) but with heating in the region $|x| < 12$)**

2 The Equations

Consider the linear equations for small perturbations to an incompressible atmosphere of constant buoyancy frequency at rest. Neglecting decay processes for the moment, suppose the horizontal motion is in the x-direction with velocity u, that the potential temperature perturbation from the mean value θ_0 is θ', and that u, θ' have the form

$$u = \hat{u}(x,t)\cos(z/H) \qquad \theta' = \hat{\theta}(x,t)\sin(z/H) \tag{2.1}$$

It follows from the hydrostatic equation and incompressibility relation that the pressure perturbation divided by density ϕ (or equivalently the geopotential perturbation) and vertical velocity are given by

$$\phi = -g\theta_0^{-1}H\hat{\theta}(x,t)\cos(z/H) \qquad w = -H\hat{u}_x(x,t)\sin(z/H) \tag{2.2}$$

The heating function Q (in degrees per unit time) is assumed to have the form

$$Q = \hat{Q}(x,t)\sin(z/H) \tag{2.3}$$

Now the horizontal momentum equation gives

$$\hat{u}_t = g\theta_0^{-1}H\hat{\theta}_x \tag{2.4}$$

while the potential temperature equation gives

$$\hat{\theta}_t - H(d\theta_0/dz)\,\hat{u}_x = \hat{Q} \tag{2.5}$$

Equations (2.4) and (2.5) are the version of the shallow water equations with heating that will be used in this paper.

Now consider the equation for the total precipitable moisture q per unit area. If the moisture scale height is small compared with H, the velocity which advects the moisture is equal to $\hat{u}(x,t)$, the velocity at the ground, and this velocity may be regarded as independent of height for the levels where the bulk of the moisture is found. Thus the moisture equation is simply

$$q_t + (\hat{u}q)_x = -P \tag{2.6}$$

where P is the loss due to precipitation. Evaporation is not included in the simplest version of the model.

The next stage is to linearise this equation, since the shallow water equations are already in a linearised form. What is the appropriate equilibrium state to linearise about? Presumably a resting atmosphere would be saturated, so this will be taken as the unperturbed state for which $q = \bar{q}$. If q' is the perturbation from this state, and $\bar{q}$ is constant, then (2.6) becomes

$$q'_t + \bar{q}\,\hat{u}_x = -P \tag{2.7}$$

The precipitation will be assumed to be zero unless the air is already saturated and converging at the lower levels ($w > 0$, $u_x < 0$). In that case, the air is assumed to remain saturated, i.e. $q' = 0$ and so (2.7) gives

$$\bar{q}\,\hat{u}_x = -P \qquad \text{in wet regions} \tag{2.8a}$$

Elsewhere, there is no precipitation so (2.7) gives

$$q'_t + \bar{q}\,\hat{u}_x = 0 \qquad \text{in dry regions} \tag{2.8b}$$

Finally, the heat released by condensation is equated to the heating function, i.e.

$$L_v P = \rho c_p H \hat{Q} \tag{2.9}$$

Combining (2.8a) and (2.9) and substituting in (2.5), one obtains in wet regions

$$\hat{\theta}_t - \left\{H\,d\theta_0/dz - L_v\bar{q}/\rho c_p H\right\}\hat{u}_x = 0 \qquad \text{in wet regions} \tag{2.10a}$$

whereas in dry regions, the equation is simply (2.5) with zero forcing, i.e.

$$\hat{\theta}_t - H(d\theta_o/dz)\,\hat{u}_x = 0 \tag{2.10b}$$

So the governing equations are simply (2.4) and (2.10). In the dry regions, they reduce to the wave equation

$$\hat{\theta}_{tt} = c^2\,\hat{\theta}_{xx} \qquad \text{in dry regions} \tag{2.11b}$$

where

$$c^2 = g\,\theta_o^{-1} H^2\, d\theta_o/dz \tag{2.12}$$

The value of c for the tropics is about 60 ms^{-1}. In the wet regions, (2.4) and (2.10) reduce to

$$\hat{\theta}_{tt} = c^2(1-\bar{q}_*)\,\hat{\theta}_{xx} \qquad \text{in wet regions} \tag{2.11a}$$

where

$$\bar{q}_* = L_v\,\bar{q}/(\rho c_p H^2\, d\theta_o/dz) \tag{2.14}$$

Thus the governing equation is (2.11) which has the two forms (2.11a) and (2.11b) depending on whether or not precipitation is occurring. The whole domain may be divided up into wet and dry zones in each of which the governing equation is linear. However the system as a whole is non-linear because the boundaries between the zones can move about. It is a very interesting non-linear system to explore with classical features like shocks (at the leading edge of a rainy zone) and expansion fans (which can occur at the trailing edge of a rainy zone). More details can be found in GILL (1981), where the "stable" case

$$\bar{q}_* < 1$$

is examined. In this regime, heating effects reduce buoyancy restoring forces in the moist zones, but only sufficiently to slow down the propagation of gravity waves (see 2.11a) - not to make them unstable.

3 The Unstable Case

In this paper, I want to look at some simple solutions in the "unstable" case

$$\bar{q}_* > 1$$

where the coefficient in the "wave" equation (2.11a) has the opposite sign from normal. This is because the buoyancy due to latent heat release on rising is greater than the ordinary buoyancy restoring forces. An alternative interpretation of (2.10a) which brings out this fact is

$$\hat{\theta}_t - H(d\theta_e/dz)\,\hat{u}_x = 0 \qquad \text{in wet regions} \tag{3.1}$$

where

$$d\theta_e/dz = d\theta_c/dz - L_v\bar{q}/\rho c_p H^2 \tag{3.2}$$

represents a suitably-averaged gradient of equivalent potential temperature. The unstable regime thus corresponds to conditions which are moist unstable in the sense that the gradient given by (3.2) is negative.

Before going further, it is convenient to put the equations in non-dimensional form. There is no natural length scale, so this can be chosen to suit the problem, but there is a natural wave speed c given by (2.12), so the time scale will be c^{-1} times whatever length scale is chosen. The non-dimensional equations are

$$\hat{u}_t = \hat{\theta}_x \tag{3.3}$$

$$\left.\begin{aligned} \hat{\theta}_t &= \hat{u}_x && \text{in dry regions} \\ \hat{\theta}_t &= -\gamma^2\hat{u}_x && \text{in wet regions} \end{aligned}\right\} \tag{3.4}$$

which come from (2.4) and (2.10) respectively, where γ is the non-dimensional stability parameter given by

$$\gamma^2 = \bar{q}_* - 1 \tag{3.5}$$

In addition, the moisture variable can be scaled to give

$$q'_t + \hat{u}_x = 0 \qquad \text{in dry regions} \tag{3.6}$$

These equations have solutions which grow exponentially in time. They have an inner precipitating core where the spatial structure is sinusoidal and an outer dry region where the spatial structure is one of exponential decay with distance from the core. One can think of this solution as the one for a "hot tower". Choosing the length scale as the one appropriate to the dry region, x > a, the solution is

$$\theta = -u = -q' = \exp(a-x+t) \qquad \text{for } x > a. \tag{3.7}$$

$$\left.\begin{aligned} \theta &= \cos(x/\gamma)\sec(a/\gamma)\exp(t) \\ u &= -\sin(x/\gamma)\operatorname{cosec}(a/\gamma)\exp(t) \\ q' &= 0 \end{aligned}\right\} \qquad \text{for } |x| < a. \tag{3.8}$$

where

$$\tan(a/\gamma) = \gamma \tag{3.9}$$

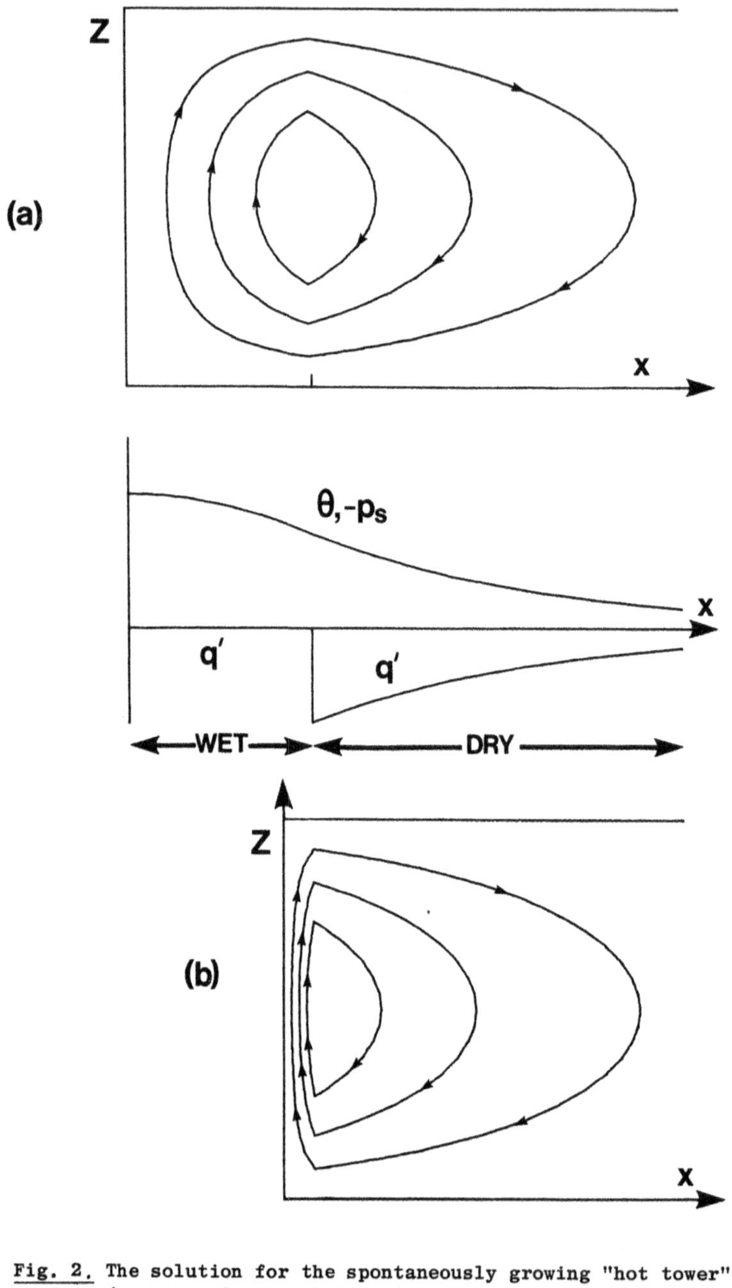

Fig. 2. The solution for the spontaneously growing "hot tower" (non-rotating case) for (a) $\gamma = 1$ and (b) γ small. The upper panel shows the circulation with inflow into the "wet region" or precipitating core and downflow in dry outer regions. The lower panel of (a) shows the profiles of temperature θ at mid-level, which is the negative of the profile for surface pressure p_s. The moisture profile q' is the perturbation from saturated conditions. The profiles in the dry region for case (b) are the same as in (a)

Fig. 2 shows the solution for two cases namely (a) $\gamma = 1$ and (b) γ small. In the latter case, where the atmosphere is only weakly unstable to moist processes, the inner moist core is relatively narrow. The interpretation is very simple. In the inner precipitating core, air rises and acquires buoyancy through latent heat release. This draws in moist air from outside the core and this keeps the cell going - in fact the system is self-exciting and everything grows at an exponential rate. The air is continually warming up and the surface pressure decreasing. However, the air outside the core is descending and drying all the time, so the process cannot continue indefinitely when it reaches a finite amplitude.

In dimensional terms, if L is the size of the region affected by the convection (i.e. the scale of the dry outer solution rather than that of the precipitating core) the growth rate is c/L and it gets larger as the scale of the disturbance gets smaller. This is a familiar phenomenon in moist convection. Another way of looking at the same result is to say that the time to reach maturity is of order L/c suggesting that the lifetime of a convection cell is proportional to its size. Thus small cells certainly grow quickly, but they also decay quickly and their main effect on larger scale motions is apparently the cumulus friction mentioned earlier, this representing the net effect of the many small cells which grow and decay in the lifetime of a much larger cell. For the large cells, the rotation of the earth is important, and this will be examined in the next section.

4 The f-Plane Solution

When rotation effects are included, there is a natural time scale, namely f^{-1} and a natural length scale, namely the Rossby radius c/f. Choosing these scales, the non-dimensional equations now take the form

$$\hat{u}_t - \hat{v} = \hat{\theta}_x \tag{4.1}$$

$$\hat{v}_t + \hat{u} = 0 \tag{4.2}$$

$$\left.\begin{aligned} \hat{\theta}_t &= \hat{u}_x && \text{in dry regions} \\ \hat{\theta}_t &= -\gamma^2 \hat{u}_x && \text{in wet regions} \end{aligned}\right\} \tag{4.3}$$

These may be combined to give the Klein-Gordon equation in the dry regions, namely

$$\hat{\theta}_{tt} + \hat{\theta} - \hat{\theta}_{xx} = 0 \qquad \text{in dry regions} \tag{4.4b}$$

and a modified form in the precipitating zones, namely

$$\hat{\theta}_{tt} + \hat{\theta} + \gamma^2 \hat{\theta}_{xx} = 0 \qquad \text{in wet regions} \tag{4.4a}$$

Now consider a spatially exponentially growing solution in the moist zone, i.e. one of the form

$$\hat{\theta} = \sin(kx)\exp(\sigma t) \tag{4.5}$$

Then (4.4a) gives

$$\sigma^2 = \gamma^2 k^2 - 1 \tag{4.6}$$

which reveals a very important fact, namely that there is a limit to the size of self-exciting systems, this size being $k^{-1} = \gamma$. In dimensional terms, limiting size is

$$(-g\,\theta_0^{-1}\,d\theta_e/dz)^{1/2}\,H/f \tag{4.7}$$

which I like to call a <u>moist Rossby radius</u> as it has the same expression as for the ordinary Rossby radius, but with $-d\theta_e/dz$ replacing $d\theta/dz$.

Equation (4.6) applies only in the moist region. To find the structure of a growing disturbance, the inner solution has to be patched on to a spatially decaying solution for the dry outer region. The result is

$$\theta = -q' = -\mu u/\sigma = \mu v = \exp[\mu(a-x)+\sigma t] \quad \text{for } x>a \tag{4.8}$$

$$\left.\begin{aligned} \theta &= \cos(kx)\sec(ka)\exp(\sigma t) \\ \sigma v &= -u = (\sigma/\mu\gamma)\sin(kx)\sec(ka)\exp(\sigma t) \end{aligned}\right\} \text{for } x<a \tag{4.9}$$

where

$$\mu^2 = k^2\gamma^2 = \sigma^2 + 1 \tag{4.10}$$

and

$$\tan(ka) = \tan(\mu a/\gamma) = \gamma \tag{4.11}$$

There are two limiting cases which can be considered, namely small quickly-growing cells and large slowly-growing cells. The former limit corresponds to $\sigma \to \infty$ in which case $\mu \sim k\gamma \sim \sigma$. This is the non-rotating "hot tower" limit which was studied in the last section and there is no need to examine it further. The other limiting case $\sigma \to 0$ will now be studied, this being one where rotation effects are very important. It is, in fact, a quasi-geostrophic limit.

5 The Quasi-Geostrophic Limit

The first property to note from (4.10) is that the <u>scales are fixed</u>

$$\mu = k\gamma = 1 \tag{5.1}$$

i.e. the outer scale is the dry Rossby radius and the inner scale is the moist Rossby radius. The size of the precipitating core is given by (4.11). It is $\pi/4$ for $\gamma = 1$ and γ^2 for small γ.

The next property to note is the relation between the dependent variables. (4.8) and (4.9) show that as $\sigma \to 0$, θ, q' and v remain fixed but u is relatively of order σ. It follows that the first term in (4.1) is of order σ^2 compared with the rest. This is the basis of the quasi-geostrophic approximation where (4.1) is replaced by

$$-v = \hat{\theta}_x \tag{5.2}$$

(4.2) and (4.3) are unchanged, but it is instructive to combine (4.2) with (5.2) to give

$$\hat{u} = \hat{\theta}_{xt} \tag{5.3}$$

which shows that the inflow is on isallobaric wind (the relative flow could be called an isallothermal wind).

The structure of the solution is found by using (5.1) in (4.8) and (4.9). It is formally the same as for the non-rotating case, but with t replaced by σt and u replaced by u/σ (which now equals -v). Fig. 2 gives the meridional circulation, the temperature, pressure and moisture fields. The new feature is the swirl velocity V which is proportional to -u, i.e. where there is inflow, the swirl is cyclonic and where there is outflow the swirl is anticyclonic. This is an inevitable consequence of angular momentum conservation in a linear system where particles have no memory. Thus, when the swirl is added, fig. 2 becomes a model of a developing hurricane rather than a hot tower. Contours of swirl velocity are shown in fig. 3. For comparison

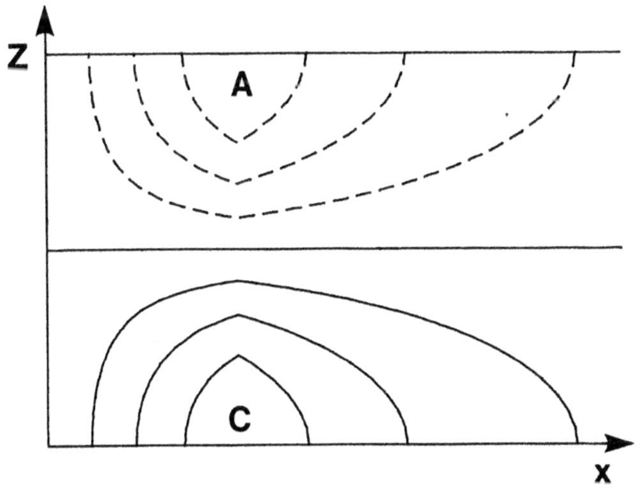

Fig. 3. The solution for a spontaneously-growing "hurricane" (rotating case) for $\gamma = 1$. The circulation in the vertical plane and variations of θ, p_s and q' are the same as in 2(a), except that the horizontal scale is much larger. The swirl velocity distribution is as shown in this figure with cyclonic rotation in the lower half and anticyclonic rotation in the upper half. The maximum swirl velocity is at the outer edge of the precipitating core

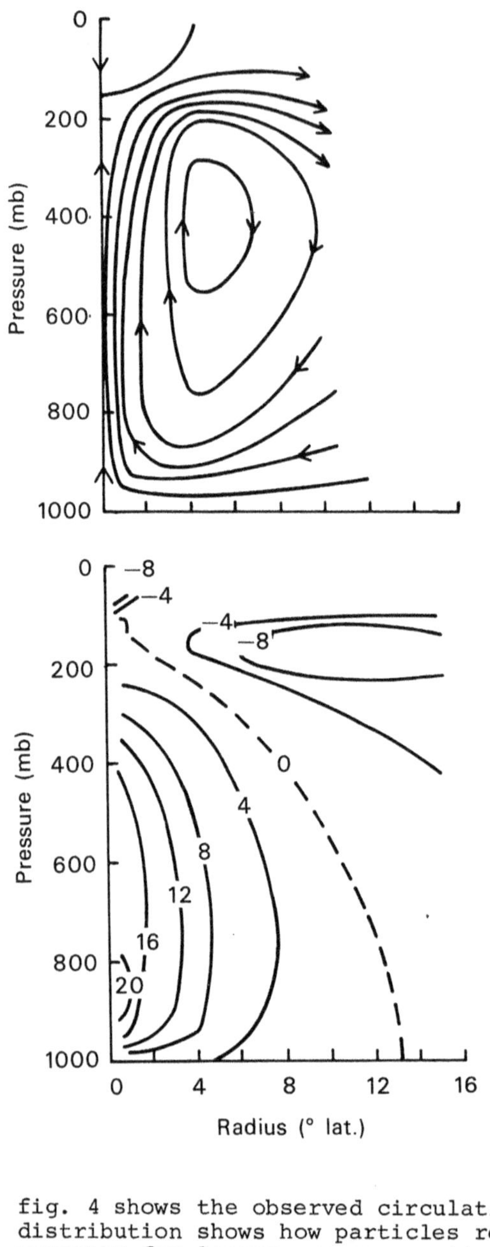

Fig. 4. The observed vertical circulation (upper panel) and swirl velocity (lower panel) in a hurricane (adopted from GRAY 1979 figs. 48 and 50). The contour interval in the upper panel is 1/3 megaton per second, and in the lower panel is 4 ms^{-1}

fig. 4 shows the observed circulation and swirl. The observed swirl distribution shows how particles retain a memory of their past angular momentum for longer, so can continue to have cyclonic rotation for a short distance beyond the level where the flow starts to diverge.

Another feature to note is that, to the quasigeostrophic order of approximation, there is no distinction between the marginally stable case, the weakly unstable case and the weakly stable case. In fact, σ may be set arbitrarily (provided it is small) and the difference between the three cases just mentioned is that u is proportional to σ so that there is low-level inflow in the unstable case and low-level outflow in the stable case. In order to regain the distinction between cases, (4.10) and (4.11) show the effects of order σ^2 need to be included. Then the scale of the weakly unstable case is slightly smaller than that for the weakly stable case. In fact the growth rate goes up very quickly as the scale is reduced. This is an important fact to remember when interpreting quasi-geostrophic solutions.

6 Relation to Previous Solutions with Ekman Convergence

In "Planetary Fluid Dynamics", CHARNEY (1973) gives a basic CISK solution (based on concepts developed by CHARNEY and ELIASSEN 1964 and OOYAMA 1969) for which the basic fields have the same form as in the above quasi-geostrophic solution, i.e. sinusoidal variations in an inner precipitating core region and exponential spatial decay in the dry region outside. So what is the difference between this solution and the above one? Perhaps the most important physical difference is that the CISK solution only allows latent heat release when there is Ekman convergence so if bottom friction is removed, so is the source of instability. In the above solution, the latent heat release occurs instead when there is isallobaric convergence, which is more like the situation envisaged in "wave-CISK" studies.

The effects of Ekman convergence can be added to the model considered previously, but the calculations are more involved because of the more complicated vertical structure. The easiest method is to follow Charney and use a layered structure i.e. an upper layer of depth H where quantities are labelled by a suffix 2, a lower layer of the same depth where suffix 1 is used, and a very much thinner Ekman layer on the surface. The basic calculation that first needs to be done is to find out how the mass and moisture fluxes are partitioned between these layers.

First consider the mass flux. The Ekman transport U_E is given by

$$U_E = - H_E v_1 \tag{6.1}$$

where H_E is a height which may be estimated from $H_E \approx C_D |v_1|/f$ where C_D is the drag coefficient (U_E is stress over f). For $C_D = .0015$, $v = 20\ ms^{-1}$ $f = .5 \times 10^{-4} s^{-1}$, $H_E = 600$ m. Since the x-component of the momentum equation gives

$$\sigma u_1 + v_1 = 0 \tag{6.2}$$

where σ is the non-dimensional growth rate used before (i.e. given in units of f),

$$U_E = H_E u_1/\sigma \tag{6.3}$$

This may now be compared with the fluxes Hu_i due to the isallobaric winds in the other layers. The total flux is zero so

$$Hu_1 + Hu_2 + H_E u_1/\sigma = 0 \tag{6.4}$$

This equation gives the partition of mass flux between the layers.

Next consider the moisture flux, which is assumed to be entirely in the lower layer and the Ekman layer. The flux due to the isallobaric wind u_1 has already been estimated in section 2 to be given by

$$\bar{q} u_1 \tag{6.5}$$

where $\bar{q}$ is the total precipitable moisture per unit area. Now, if the Ekman layer is assumed thin, compared with the moisture scale height, H_m, the moisture level in this layer will be equal to the value at the ground. If the moisture content per unit volume falls off exponentially with height on the scale H_m, this surface level is $\bar{q}/H_m$ and the Ekman moisture flux is

$$U_E \bar{q}/H_m = (H_E/H_m \sigma) \bar{q} u_1 \tag{6.6}$$

This information on the partition of fluxes can now be put into the equations, which will be written in non-dimensional form as before. The form of the equations is no different, except for factors depending on the ratios of fluxes in the various layers, and these ratios have been calculated above. The two-layer analogue of the quantity $\hat{u}$ used previously is

$$\hat{u} = \tfrac{1}{2}(u_1 - u_2) \tag{6.7}$$

the factor ½ being included so that when the velocities in the two layers are equal and opposite (as previously assumed), $u = u_1 = -u_2$. It follows from (6.3) and (6.4) that, in terms of $\hat{u}$, the velocities are given by

$$u_1 = \frac{2\sigma}{2\sigma+\epsilon}\hat{u}, \quad u_2 = -\frac{2(\sigma+\epsilon)}{2\sigma+\epsilon}\hat{u}, \quad \frac{U_E}{H} = \frac{2\epsilon}{2\sigma+\epsilon}\hat{u} \tag{6.8}$$

where

$$\epsilon = H_E/H \tag{6.9}$$

is a small parameter (about 0.1).

The equations are the thermal wind equation (5.2) which is unaltered and the momentum equation (4.2) which is unaltered. The temperature equation is, however, changed because the vertical motion between the layers is related to the divergence of u_2 rather than of u. Thus the appropriate factor from (6.8) is introduced, so giving in place of (4.3)

$$\hat{\theta}_t = \left\{\frac{2(\sigma+\epsilon)}{2\sigma+\epsilon}\right\}\hat{u}_x \qquad \text{in dry regions} \tag{6.10b}$$

The equations (5.2), (4.2) and (6.9) for the dry regions are the same as Charney's.

The moisture equation is altered by the factor arising when the two moisture fluxes (6.5) and (6.6) are summed. The result is

$$q'_t + \frac{2(\sigma+A\epsilon)}{2\sigma+\epsilon}\bar{q}_* \hat{u}_x = -P \tag{6.11}$$

where

$$A = H/H_m \tag{6.12}$$

is greater than unity (about 2). Now using (6.11) to put in the effect of latent heating in the wet regions, there results

$$\hat{\theta}_t + \left\{ \frac{2\sigma(\bar{q}_* - 1) + 2\epsilon(A\bar{q}_* - 1)}{2\sigma + \epsilon} \right\} \hat{u}_x = 0 \quad \text{in wet regions} \tag{6.10a}$$

Solutions can be constructed as before. The e-folding scale μ for the dry region is now given by

$$\mu^2 = \frac{2(\sigma + \epsilon)}{2\sigma + \epsilon} \tag{6.13}$$

while the wavenumber k for the wet region is given by

$$k^2 = \frac{2\sigma(\bar{q}_* - 1) + 2\epsilon(A\bar{q}_* - 1)}{2\sigma + \epsilon} \tag{6.14}$$

The size a of the precipitating core region is given by

$$k \tan(ka) = \mu \tag{6.15}$$

The only constraint on being able to find a solution is that μ^2 and k^2 be positive. Also σ has to be small for the quasi-geostrophic approximation to be valid.

There are two limiting cases. For $\sigma \gg \epsilon$, the isallobaric fluxes dominate over the Ekman fluxes, and the solution is the one found in the previous section. In the other limit $\sigma \ll \epsilon$, corresponding to very small growth rates, Ekman fluxes dominate over isallobaric fluxes. The scales are fixed as before, being given in this limit by

$$\mu^{-2} = 2, \quad k^{-2} = 2(A\bar{q}_* - 1) \tag{6.16}$$

Because A is greater than unity, Ekman driven instability can occur at lower saturation moisture levels. The instability is, however, weak, because $\sigma \ll \epsilon$. The reason for Ekman driving giving enhanced instability is that it gives convergence where moisture levels are highest, i.e. at the surface.

7 Summary and Further Work

The device of adding a moisture equation to the usual shallow water equations gives a simple mathematical system for demonstrating the nature of moist convection. In particular, it gives rise to a model of a developing hurricane with the low-level convergence being due to the isallobaric wind. Ekman effects can be added, but at the cost of increased complexity. These can give rise to a weak instability at lower saturation moisture levels.

Solutions can also be found on an equatorial beta-plane. They are not too complicated if there is no dependence on longitude, the sines and exponentials being replaced by special functions which are readily available in tables (parabolic cylinder functions of order ½ - see GILL 1980). The result is a model for the intertropical convergence zone, and fig. 5 shows an example. The precipitating region does not need to be placed at any particular latitude.

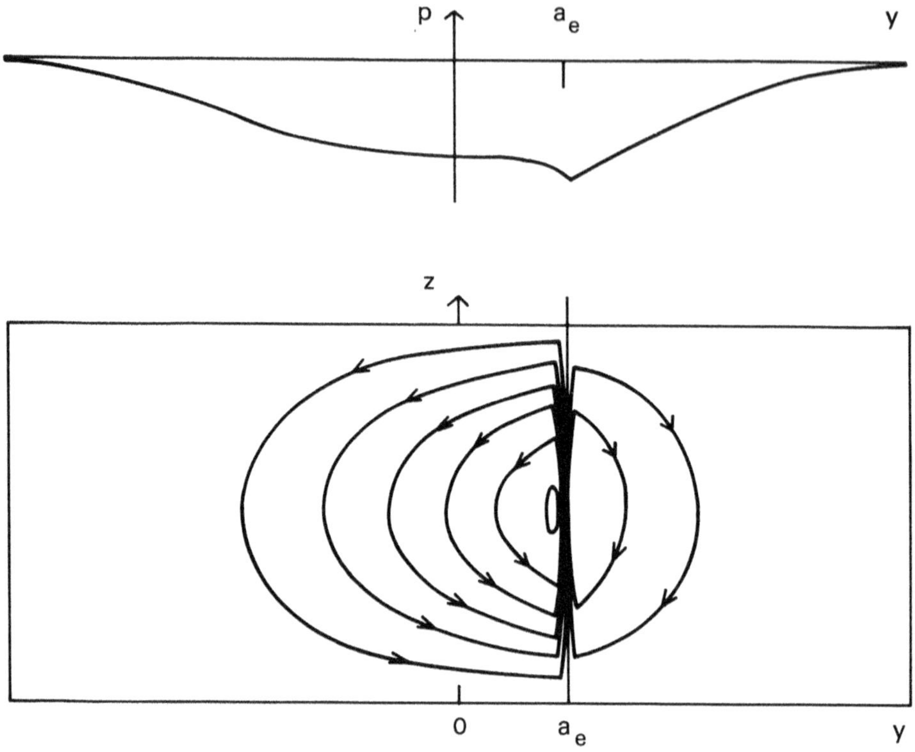

Fig. 5. **Solution for γ small for the case where the precipitating core is one equatorial Rossby radius a_e from the equator. The upper panel shows the surface pressure distribution (which is the negative of the mid-level temperature distribution) and the lower panel the circulation in the vertical plane. The form of solution is given in GILL 1980, equation (6.2)**

One would like to do numerical work with a simple model like this, but in the unstable regime, one is confronted with the problem that small scale disturbances grow the fastest, and there are no non-linear effects in the simplest version of the model to stop them. Thus the only way to study hurricane-scale disturbances is to suppress the small scale ones. A simple way of doing this is discussed in the appendix by J.M. Smith, to whom I am also indebted for the computer-drawn diagrams.

References

Charney, J.G. 1973 Planetary fluid dynamics in "Dynamical Meteorology" pp. 97-351. P. Morel (ed.) Reidel, Dordrecht

Charney, J.G., and A. Eliassen 1964 On the growth of the hurricane depression. J.Atmos.Sci., 21, 68-75

Gill, A.E. 1980 Some simple solutions for heat induced tropical circulation. Quart.J.Roy.Met.Soc., 106, 447-462

Gill, A.E. 1981 Studies of moisture effects in simple atmospheric models: the stable core. Geophys. Astrophys. Fluid Mech. (in press)

Gray, W.M. 1979 Hurricanes: their formation, structure and likely role in the tropical ocean. In "Meteorology over the tropical oceans" D.B. Shaw (ed.) Roy.Met.Soc. London pp.151-218

Ooyama, K. 1969 Numerical simulation of the life cycle of tropical cylones. J.Atmos.Sci., 26, 3-40

Rasmusson, E.M. and T.H. Carpenter 1981 Variations in tropical sea surface temperature and surface wind fields associated with the southern oscillation/El Niño. Month.Weath.Rev. (in press)

Appendix: Model with Lateral Viscosity

by J.M. Smith

A simple method of suppressing the small scales is by introducing a lateral viscosity, or rather, replacing $\partial/\partial t$ by $\partial/\partial t - \nu\nabla^2$ in equations (4.1), (4.2) and (4.3). Then for solutions of the form given in the paper, the decay rate μ in the dry region is given by

$$\nu^2\mu^4 - (2\sigma\nu+1)\mu^2 + (1+\sigma^2) = 0 \tag{A.1}$$

so there are 2 positive values for each value of σ given by

$$\mu_\pm = (M_1 \pm M_2)/2\nu \tag{A.2}$$

where

$$M_1^2 = 1 + 2\nu((1+\sigma^2)^{1/2} + \sigma) \tag{A.3}$$

$$M_2^2 = 1 - 2\nu((1+\sigma^2)^{1/2} - \sigma) \tag{A.4}$$

Similarly, the wavenumber k in the wet region satisfies

$$\nu^2 k^4 + (2\sigma\nu - \gamma^2)k^2 + (1+\sigma^2) = 0 \tag{A.5}$$

and has two positive values given by

$$k_\pm = (L_2 \pm L_1)/2\nu \tag{A.6}$$

where

$$L_1^2 = \gamma^2 - 2\nu((1+\sigma^2)^{1/2} + \sigma) \tag{A.7}$$

$$L_2^2 = \gamma^2 + 2\nu((1+\sigma^2)^{1/2} - \sigma) \tag{A.8}$$

There is also a part of the solution with decay rate μ_0 in both regions determined by

$$\mu_0^2 = \sigma/\nu \tag{A.9}$$

which corresponds to the solution $\hat{u} = 0$, $\hat{v} = -\hat{\theta}_x$. The solutions are necessary in order to satisfy the extra patching conditions at x = a, namely that $\hat{v}$, $\hat{u}_x$, $\hat{v}_x$ and $\hat{\theta}_x$ (as well as $\hat{u}$ and $\hat{\theta}$) be continuous. Since the flow is convergent in the wet region and divergent in the dry zone, and since u_x is continuous, u_x must vanish at x = a.

The solution for $\hat{v}$ has the form

$$\hat{v} = \begin{cases} A_+ \sin(k_+ x) + A_- \sin(k_- x) + A_0 \sinh(\mu_0 x) & x \leq a \\ B_+ \exp(-\mu_+ x) + B_- \exp(-\mu_- x) + B_0 \exp(-\mu_0 x) & x \geq a \end{cases} \tag{A.10}$$

and similar for $\hat{\theta}$. The solution for $\hat{u}$ is the same except that $A_0 = B_0 = 0$. The coefficients can be found by using the continuity conditions. The equation for σ that results is

$$\tan(\pi L_1 / 2 L_2) = M_1 / L_1 \tag{A.11}$$

while the half width a of the core is given by

$$a = \pi \nu / L_2 \tag{A.12}$$

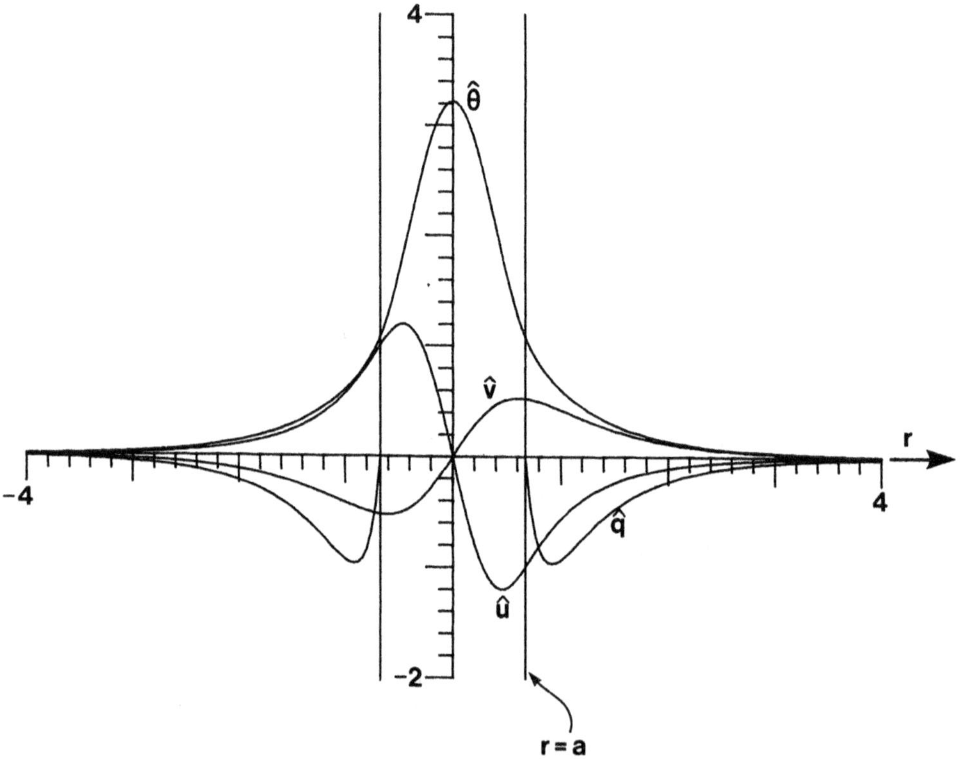

Fig. A1. **Axisymmetric solution for $\nu = 0.15$, $\gamma = 1$. The value of σ is 1.001 and the radius of the core is 0.674 units**

The same procedure can be applied to the axisymmetric case where the equivalent of (A.10) involves Bessel functions and modified Bessel functions of order one. The equations determining growth rate σ and radius a are more complicated than for the two dimensional case, and they contain Bessel functions and modified Bessel functions. If these functions are replaced by their asymptotic forms for large arguments, the approximate equation for σ that results is

$$\tan(3\pi L_1/4L_2) \approx M_1/L_1 \tag{A.13}$$

while a is approximated by

$$a \approx 3\pi\nu/2L_2 \tag{A.14}$$

Equation (A.13) yields a larger value of σ than does (A.11) for the two dimensional case. For example, for ν = 0.15, γ= 1, the value of σ is 0.494 for the two dimensional case and 1.001 for the axisymmetric case. (The approximate value given by (A.13) is 0.983). The axisymmetric solution is shown in fig. A1.

Time-dependent numerical calculations were also done for both these cases confirming the exponential growth. Such calculations were also done starting with moisture levels below saturation. A 'hurricane' still resulted provided there was enough convergence or heat in the initial field to produce saturation and establish the growing solution. Also calculations were made with a non-linear moisture equation, resulting in a limit to the amplitude of the motion.

The Factors Determining Radial Flow and Eye Formation in an Intensifying Tropical Cyclone

R. P. Pearce

ABSTRACT

Moist static energy and momentum budgets for an intensifying tropical cyclone indicate that the radial motion determines the angular momentum source and at the same time the moist static energy sink. On the other hand, surface transfers determine the angular momentum sink and the moist static energy source. It was pointed out by Pearce [1981] that whether or not intensification can occur depends crucially on the vertical distributions of the moist static energy and angular momentum of the lower and middle tropospheric inflow air.

A simple axisymmetric model of intensification, in which the air flowing through the system in a contracting convective annulus is treated separately from the air in the interior, is briefly described. The interior may be regarded as responding, through gravity wave propagation, to dynamical forcing by the air flowing through the clouds.

The maintenance of the radial flow (or secondary circulation) is discussed, with the buoyancy of the inflow air highlighted as the main controlling factor. The dynamics of cumulus convection in axisymmetric flow and the factors determining contraction (intensification) are described.

It is shown that the mass adjustment to the primary (tangential) flow, accomplished by gravity waves, is rapid but that gradient balance cannot be achieved in an axisymmetric system. Instead the central pressure fall is less than that required for balance, i.e. the tangential winds are super-gradient. This is the result of the existence of the secondary flow with its large non-linearities.

These concepts are briefly contrasted with ideas based on CISK.

1 INTRODUCTION

Hurricanes and typhoons have long been recognised as systems in which low-level radial inflow and upper tropospheric outflow (sometimes referred to as the secondary circulation in contrast to the 'primary' tangential or swirl flow) play

Topics in Atmospheric and Oceanographic Sciences
Intense Atmospheric Vortices, ed. by Bengtsson/Lighthill

crucial roles in their angular momentum balance. The role of the lowest-level inflow in supplying energy to these systems through the import of moisture is also well-recognised. What is not properly understood is the precise nature of the factors which control radial motion and how these determine development. Many authors, for example Ooyama [1969], Ogura [1964], have developed the so-called CISK theory proposed by Charney and Eliassen [1964] in which boundary-layer friction is regarded as the main factor determining the radial inflow and associated moisture supply to a developing system. Recent studies by Gray [1978] and McBride [1979] have shown that the inflow layer extends through the middle troposphere (this feature is also present in models based on CISK theory) and also that tropical systems which intensify into hurricanes are characterised by a larger tropospheric relative vorticity than those which do not (see pp. 3-22). It is easily seen from angular momentum considerations that the secondary circulation acts so as to intensify the primary circulation in this case; more absolute angular momentum is imported by the low-level inflow than is exported in the outflow. The boundary-layer dissipation acts as a momentum sink. Indeed, as is shown by Pearce [1981], the secondary circulation, while tending on the one hand to increase the total angular momentum of the system, tends on the other to deplete, i.e. export, its moist static energy; this is a consequence of the import (entrainment) of middle layer air of low moist static energy.

It was in order to explore the implications of the observed lower tropospheric vorticity maximum that Pearce [1981] constructed a simple axisymmetric model of an intensifying system in which the air involved was considered in three categories (see Fig. 1). The first is that which rises through the clouds, the

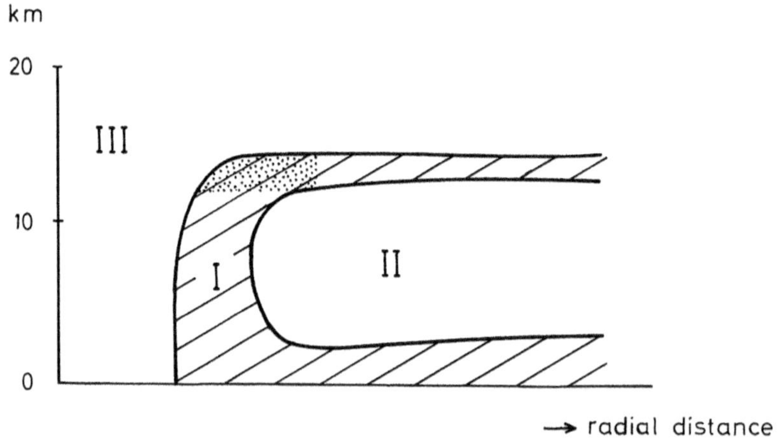

Fig. 1. Schematic vertical section showing the three main air categories involved in tropical cyclones. The air channelled through the clouds is shaded. The outflow injection region is stippled

second is the air between the clouds in the inflow region, some of which is entrained, and the third category is the remaining air occupying the central region and the lower stratosphere. These categories are very similar to those of Carrier et al. [1971], but the resemblance of their analysis to that presented here goes no further than this.

The model treats explicitly angular momentum transport (down the gradient) of the air flowing through the clouds and, by incorporating a parameter (β) which defines the radial outflow component of the system in terms of this transport, simulates intensification or no intensification, depending on the value of β and the initial conditions. The intensifying model system consists of two quite separate regions, (a) a contracting annulus of convection located by the maximum surface winds and (b) a subsiding central region, which eventually becomes the eye, from which mass is continually removed by gravity waves. The flow in the central region is computed as the mass-adjustment response to the momentum transports, both vertical and radial, occurring in the convection annulus.

This chapter develops this approach further and suggests that many essential structural features of hurricanes, including the eye and super-gradient winds surrounding it, are a consequence of the inter-relation between gravity-wave activity which dominates the central region and the transports of momentum and energy in the cumulus convection surrounding it. Section 2 presents a summary of the main results from the model described by Pearce [1981] and states its limitations. Section 3 then discusses in some detail the dynamics of cumulus convection in an annulus, including the crucial role of buoyancy, and the factors which determine contraction and intensification. This is followed in Section 4 by a brief analysis of gravity wave propagation in intensifying axisymmetric convective systems and emphasises their crucial role in determining the mass adjustment to the flow and, because of the secondary circulation, to the occurrence of super-gradient winds. Finally, in Section 5, the results of the model integration presented in Section 2 are interpreted in terms of these processes and related to real, asymmetric, intensifying systems.

2 A SIMPLE HURRICANE MODEL

The main features of the model described by Pearce [1981] are first briefly presented. The radial wind is assumed to be of the idealised form shown in Fig. 2(a), defined by a single parameter $\delta u(r,t)$, and the initial angular momentum to be of the form shown in Fig. 2(b), defined by two parameters $m_s(r,t)$ and $\delta m(r,t)$. Each of these profiles is based on the distributions observed by McBride [1979]. The air rising through the clouds is assumed to entrain middle level air, and to emerge at the outflow level with angular momentum m_i (specified by an entrainment

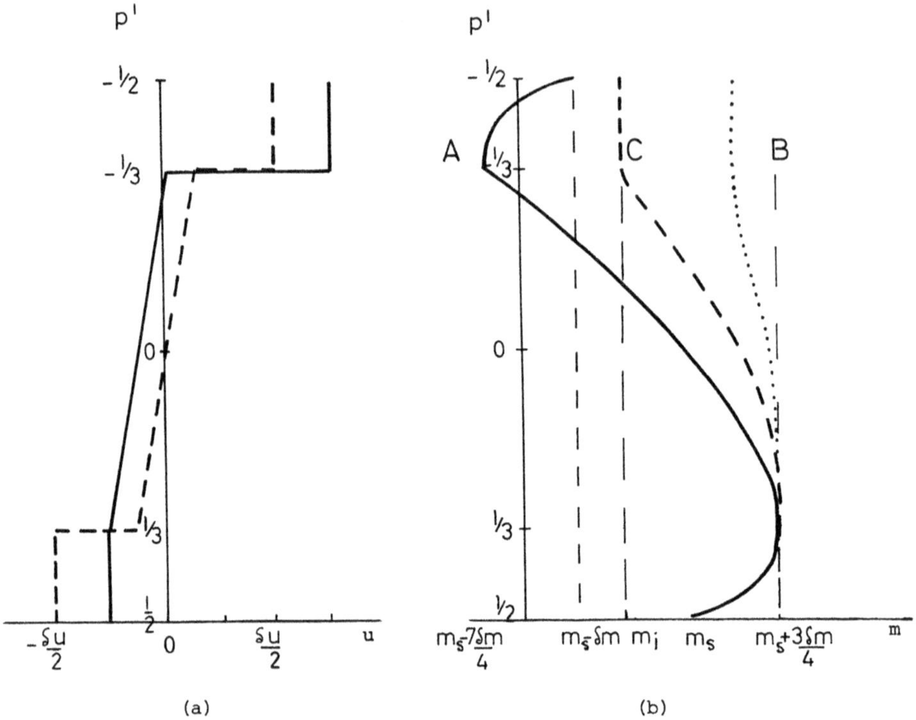

Fig. 2. (a) Model profiles of radial wind (u), The full line is the profile used over the main part of the convective region. The dashed line is the profile used near the radius of maximum wind.

(b) Model profiles of absolute angular momentum. The full curve refers to the inflowing air at middle levels and that originally present at the outflow levels; the dashed curve refers to air rising in the clouds which both entrains and detrains; the dotted curve refers to air rising through the clouds simply entraining middle layer air. (P' is a non dimensionalised pressure coordinate)

parameter) larger than that of the air it replaces. The angular momentum equation is found to be essentially of the form

$$\frac{\partial m_s}{\partial t} = \frac{\delta u}{4} \frac{\partial m_s}{\partial r} - S \tag{1}$$

where S denotes the surface frictional sink.

The radial outflow is assumed to satisfy an equation of the form

$$\Delta\left\{\frac{\partial u}{\partial t} + \frac{1}{r}\frac{\partial}{\partial r}(ru^2)\right\} = \Delta(v^2/r + fv) - \frac{\partial}{\partial r}(\Delta\phi) \tag{2}$$

where u(r,t) is the mean outflow speed $\left(= \frac{3\delta u}{4}\right)$, ϕ is the geopotential at the mean outflow level and Δ denotes the change (over time interval Δt) resulting from the injection into the outflow layer of air flowing through the clouds; v is the tangential velocity at the outflow level and f the Coriolis parameter $\left\{\text{i.e. } m = rv + \frac{fr^2}{2}\right\}$.

The device used to close the system was to partition the mass and flow adjustments at the outflow level empirically by introducing a parameter $\beta(r,t)$ $(0 < \beta < 1)$ and putting

$$\frac{\partial}{\partial r}(\Delta\phi) = (1-\beta)\Delta\left(\frac{v^2}{r} + fv\right)$$

and (3)

$$\Delta\left\{\frac{\partial u}{\partial t} + \frac{1}{r}\frac{\partial}{\partial r}(ru^2)\right\} = \beta\Delta\left(\frac{v^2}{r} + fv\right) .$$

The form adopted for $\beta(r)$ was such that its maximum value (β_m) was located in the vicinity of the surface wind maximum, tailing rapidly to zero on either side. No time variation was included.

Several integrations of the system were performed on a computational grid of about 27 km. Intensification and contraction from an initial surface wind maximum of 10 m s^{-1} at radius 250 km occurred with $\beta_m > 0.3$; otherwise there was no intensification, the secondary flow not being sufficiently strong to concentrate the angular momentum more rapidly than it was destroyed by surface friction. Also it was necessary to modify the radial flow profile in the vicinity of the surface wind maximum to the form shown by the dashed line in Fig. 2(a) - allowing a somewhat larger low-level inflow - to obtain intensification.

The model primary and secondary circulations after a simulated $2\frac{1}{4}$ days, with $\beta_m = 0.3$, are shown in Figs. 3(a),(b). These represent a rapidly rotating convective annulus with an inner (eye) radius (where $u = 0$) of about 30 km. The associated geopotential and mean temperature fields are shown in Fig. 3(c). The maximum temperature is found to be in the eye region which has not been penetrated by the convection. The eye warming can only take place as a result of subsidence associated with mass adjustment to the convective region flow, a gravity wave phenomenon. Neglecting the gravity wave transients, the flow associated with this subsidence is simply calculated from the thermodynamic equation, assuming an idealised (parabolic) vertical motion profile. The flow component calculated in this way is shown in Fig. 3(d). The total secondary flow is thus a superposition of the flows in Figs. 3(b) and 3(d).

The model clearly has considerable limitations, but these are to a large extent intentional, having been imposed in the interest of simplicity in order to highlight essential processes. It demonstrates that by making certain assumptions about the mass-flow adjustment process [embodied in the form of $\beta(r,t)$ and the

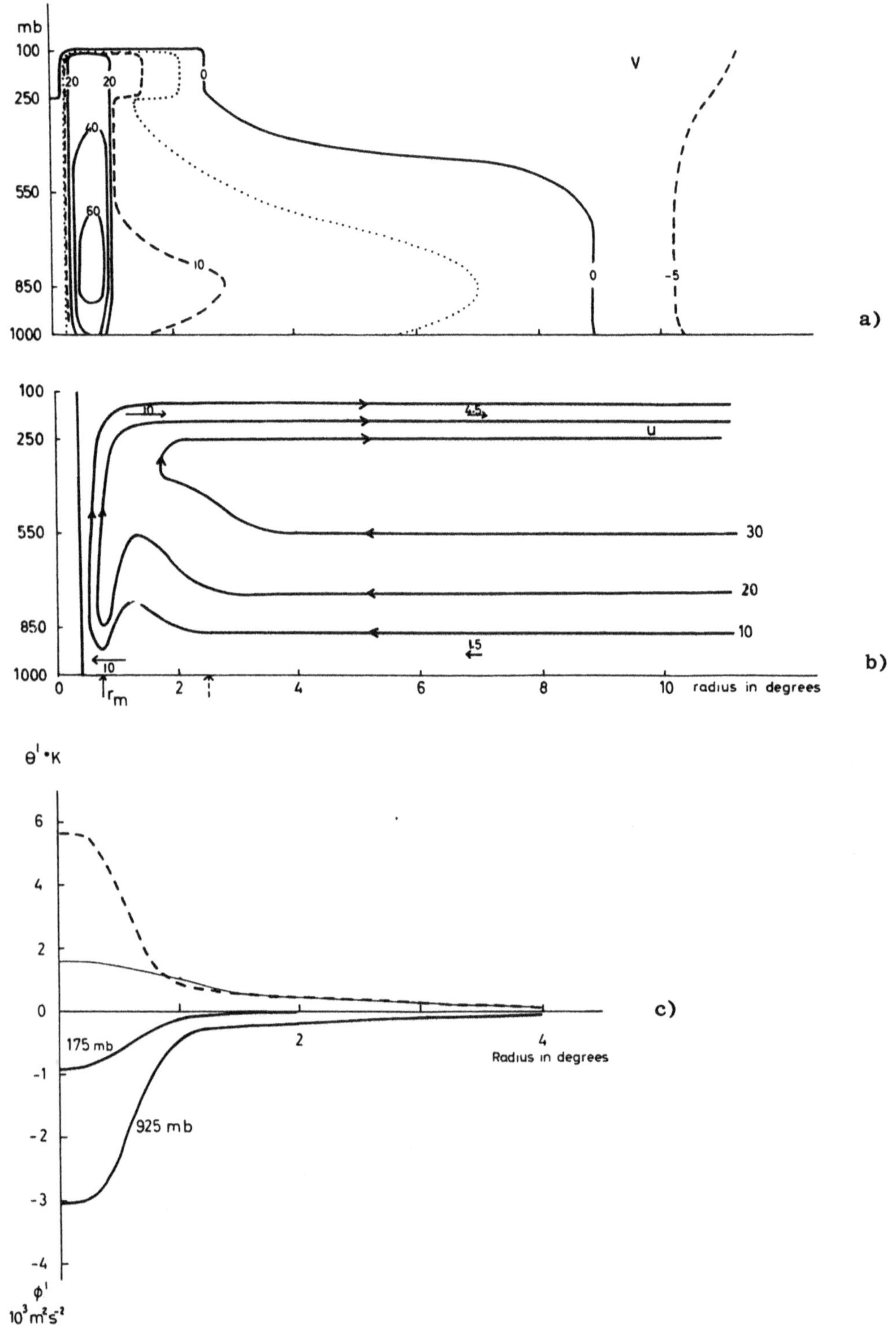

mb
100
250
550
850
1000
V
20
20
40
60
10
0
0
-5
a)
u
10
4.5
30
20
10
1.5
10
b)
0
2
4
6
8
10
radius in degrees
r_m
θ' °K
6
4
2
0
-1
-2
-3
-4
φ'
$10^3 m^2 s^{-2}$
175 mb
925 mb
2
4
Radius in degrees
c)

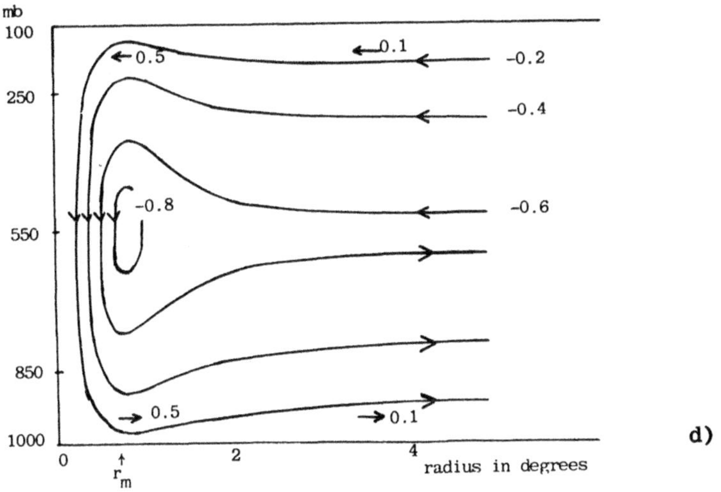

Fig. 3. Distributions of (a) tangential wind (m s^{-1}) and (b) radial wind and vertical motion represented as isopleths of a streamfunction $\psi(r,p')$ in m s^{-1} (100 km) (100 mb) with $ru = -\partial\psi/\partial p$ and $r\omega = \partial\psi/\partial r$. The ordinate p' is a non-dimensionalised pressure. The numbered arrows in (b) indicate values of u in m s^{-1}. The maximum value of $-\omega$ is 140 mb hr^{-1} (averaged over a grid length).

Also (c) deviations from the area mean of geopotential ϕ' (full curve) in m^2 s^{-2} and of the mean tropospheric temperature θ' (dashed) in K as functions of radial distance after $2\frac{1}{4}$ days.

and (d) streamlines of the gravity-wave flow in the (r,p) plane at time t = $2\frac{1}{6}$ days for the model system in (a), (b) and (c). The numbered arrows indicate the values of u_g in m s^{-1}. Streamfunction values are indicated in the same units (m s^{-1} (100 km) (100 mb) as in (b) above. The maximum descent rate is 13 mb hr^{-1} (averaged over a grid length)

cumulus entrainment assumptions], contraction and intensification of a convective annulus can be simulated purely from angular momentum considerations. Furthermore this process leads to the formation of a central eye region in which the subsidence associated with the mass adjustment can readily be calculated. Experiments with the model indicate that intensification will only occur (a) if there is, initially, a larger relative vorticity in the lower troposphere than in the upper troposphere (ensuring $\Delta\left(\frac{v^2}{r} + fv\right) > 0$ in Eqs. (3)) and (b) $\beta > \beta_c$ where β_c is a lower bound for β with a value of about 0.25. It is also necessary to assume that convection takes place one or two grid points inwards from the location of the surface wind maximum.

It is clear from Eq. (3) that the larger the value assumed for β (< 1) the larger is the resulting secondary flow and the smaller the implied change in the geopotential at the outflow level. The larger too is the intensification rate. As already pointed out in the introduction, the physical constraint which limits the strength of the secondary flow is the need to maintain the thermodynamic energy balance - a too intense circulation would entrain the low energy middle tropospheric air more rapidly than it can be replenished by the surface heat and moisture source. The model, however, deals only with the angular momentum balance and cannot therefore be used in its present form to determine the upper limit for β_m. Obviously it is also desirable to relate the empirical form assumed for $\beta(r)$ to actual physical processes determining the development, i.e. to those determining the relative rates of mass and flow adjustment. This requires consideration of (a) the maintenance of the cumulus convection and (b) gravity wave activity. These are the topics of the next two sections.

3 BUOYANCY AND THE MAINTENANCE OF THE SECONDARY CIRCULATION

3.1 The Mature Hurricane

It is useful first to consider the flow in the convective region in the steady mature hurricane. Fig. 4 shows schematically the observed form of the streamlines of the secondary flow. [A streamfunction $\psi(r,p)$ is defined by writing $ru = -\frac{\partial\psi}{\partial p}$ and $r\omega = \frac{\partial\psi}{\partial r}$.] The dotted streamline in the inner region is included

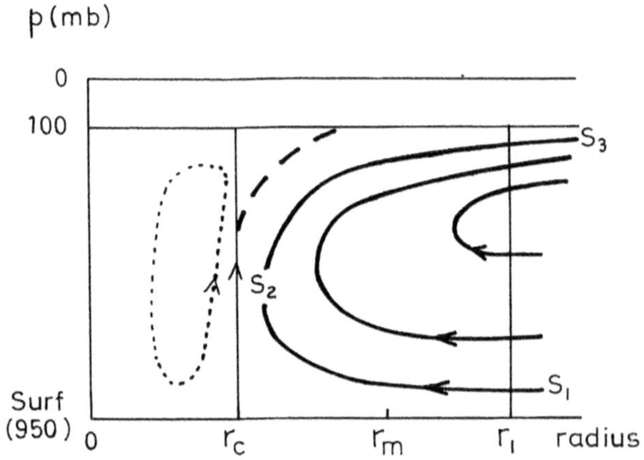

Fig. 4. Schematic secondary circulation in a mature hurricane. The full curves are streamlines, and the dotted curve is a weak circulation in the eye region associated with eye-wall entrainment. The dashed curve denotes the observed sloping eye-wall at upper levels. The radius r_c denotes the inner limit of convective penetration ($u = 0$), r_m the radius of maximum wind and r_1 the radius of maximum inflow rate [$\partial(ru)/\partial r = 0$]

to represent the circulation there resulting from entrainment at the eyewall cloud, although this is not an essential feature. The radius $r = r_1$ is chosen to be a little beyond the radius r_m of maximum wind and where the secondary flow can be assumed horizontal with ascent between $r = r_c$ = constant (where $u = 0$) and $r = r_1$ and relatively slow descent for $r > r_1$. $r_1 - r_c$ is typically the width of a spiral cloud band near the eyewall (10 - 20 km).

Define the equilibrium (gradient balance) geopotential ϕ_e by

$$\phi_e(r,p) = -\int_r^R \left(\frac{v^2}{r} + fv\right)dr \ . \qquad (4)$$

Then the radial momentum equation may be written (neglecting friction) as

$$\frac{du}{dt} = -\frac{\partial}{\partial r}\left(\phi - \phi_e\right) \ . \qquad (5)$$

The geopotential ϕ may be expressed as the sum of three components,

$$\phi(r,p,t) = \phi_0(p) + \phi_e(r,p,t) + \phi'(r,p,t)$$

where $\phi_0(p)$ denotes the undisturbed initial state (a fourth, gravity wave transient, component is considered in the next section but is neglected here). Thus (5) may be expressed as

$$\frac{du}{dt} = -\frac{\partial \phi'}{\partial r} = -\frac{1}{\rho_0}\frac{\partial p'}{\partial r} \qquad (5')$$

where the last expression uses height coordinates. Since $\frac{du}{dt} > 0$ for all the air parcels flowing through the clouds, $\frac{\partial \phi'}{\partial r} < 0$ and the flow is everywhere <u>supergradient</u>. The non-linear terms are of particular importance in contributing to this characteristic of the flow. The vertical momentum equation may be written (in height coordinates) as

$$\frac{dw}{dt} = -\frac{1}{\rho_0}\frac{\partial p'}{\partial z} - \frac{1}{\rho_0}\frac{\partial p_e}{\partial z} - \frac{g\rho'}{\rho_0} \qquad (6)$$

where p_e is the gradient-balance pressure deviation. Integrating eqns. (5') and (6) along a streamline gives the Bernoulli equation

$$\frac{u^2 + w^2}{2} + \int \frac{dp'}{\rho_0} + \int \frac{g\rho'}{\rho_0}dz + \int \frac{1}{\rho_0}\frac{\partial p_e}{\partial z}dz = \text{constant}. \qquad (7)$$

Applying this equation to the streamline $S_1S_2S_3$ in Fig. 4 gives

$$\delta\left[\frac{u^2 + w^2}{2} + \frac{p'}{\rho_0} + \frac{p_e}{\rho_0}\right] = -\frac{g\overline{\rho'}}{\rho_0}\delta z$$

where δ denotes the difference between S_1 and S_3 and the bar an average value. Since $\frac{\delta p_e}{\rho_0} = \delta\phi_e$ and v generally decreases with height in the troposphere (Fig. 2(b)), this term is positive; so also is $\delta(u^2 + w^2)$ when the outflow is concentrated at upper levels. Both of these terms are of order 100 $m^2 s^{-2}$. On the other hand, the flow may be assumed to depart only slightly from gradient balance at $r = r_1$, i.e. $\delta p' \cong 0$. Thus the flow pattern can only be maintained where there is a positive mean buoyancy ($\overline{\rho'} < 0$) in the air flowing through the clouds. Taking $\delta z = 10$ km and the expression on the left as 100 $m^2\ s^{-2}$ gives $(-\ \overline{\rho'}/\rho_0) = 10^{-3}$. This requires $(\overline{T'}/T_0) = 10^{-3}$ in isobaric surfaces, i.e. a temperature excess of about 0.3 K. The inflow air must thus have a sufficiently large moisture (and sensible heat) content to rise along a saturated adiabat which is about 0.3K warmer than the environment.

3.2 An Intensifying Hurricane

The schematic circulation shown in Fig. 4 is still applicable except that r_c is larger and decreases with t instead of remaining constant. However air on the axis side of $r = r_c$ (dotted streamline) is now an important part of the convective circulation and, with r_c decreasing, some of the trajectories extend into the general outflow. The air forming the main inflow passes through the region of maximum wind and, as in the mature storm, must possess sufficient buoyancy if the flow is to be maintained (the adverse pressure gradient is weaker in this case).

It is now possible to appreciate why contraction occurs and how it is eventually inhibited. Consider the streamlines converging from the exterior and interior at $r = r_c$ (Fig. 4). Those from the outer region (e.g. S_1S_2) represent flow down the gradient of absolute angular momentum (m), when m decreases with height as well as with r, and hence, if the inflow is sufficient to dominate the frictional sink, this region is one in which the maximum (tangential) wind, surface moisture transfer and buoyancy all increase. The radius of maximum wind r_m decreases. [Neglecting friction, m is conserved along the inflow trajectories and so the tendency is for m to increase everywhere in this region.] The main updraught region ($r_c < r < r_m$) cannot shrink to a width which is much less than the inflow depth (unless a large increase in buoyancy occurs to give the large increase in w - Eq. (7)) and so the only way in which the flow can respond to a general increase in m and decrease in r_m is for r_c to decrease along with r_m, i.e. for the system to contract.

The streamlines from the interior represent flow up the gradient of m and hence this is a region in which the tangential wind (v) decreases, along with the surface moisture transfer and buoyancy. The contraction thus arises as a result of contrasting tangential momentum advection in the presence of deep buoyant convection when this occurs in a flow about a vertical axis of symmetry.

As the contraction proceeds the contrast between the two flow regimes increases. The surface friction in the main inflow increases (as r_m increases), m ceases to increase and r_m ceases to decrease. The air in the interior is no longer buoyant and plays no significant role in the convective circulation which, in the presence of an adequate moisture source, can maintain itself in a steady state with supergradient winds.

An essential requirement for intensification is thus an adequate source of buoyancy (i.e. moisture and sensible heat) in the low-level inflow.

4 ADJUSTMENT OF THE MASS FIELD TO THE FLOW FIELD BY GRAVITY WAVES

The analysis in the last section essentially relates to the development of a region of convection in two dimensions, p' defining the (non-hydrostatic) pressure distribution and u and w the velocity field. The existence of the tangential velocity (v) field and the associated balanced pressure field (p_e) was involved in only two ways, through $\partial p_e/\partial z$ and p' ($\cong \rho_0\phi'$) in Eq. (7). The role of $\partial p_e/\partial z$ was shown to imply that ϕ was larger than ϕ_e in the interior region, ϕ' thus being greater than zero there, implying supergradient flow. This statement will now be justified by considering how the geopotential (pressure) in the system, in particular that in the interior region, is determined. Classical theory shows that the surface pressure tendency is equal to the vertical integral of the horizontal mass convergence so that the surface pressure fall associated with intensification must imply mean mass divergence in vertical columns (i.e. external gravity wave propagation). This is not accounted for by the u field so far considered since its vertical integral is zero - the secondary circulation is an isolated internal inertial-gravity wave of large amplitude. Letting an asterisk denote a gravity wave component, an overbar (vertical average) the external wave and suffix i an internal wave, the full linearised radial momentum equation gives, subtracting Eq. (5),

$$\frac{\partial u^*}{\partial t} = - \frac{\partial \phi^*}{\partial r} . \tag{8}$$

First consider the external gravity wave $(\overline{u^*}, \overline{\phi^*})$. The analysis is straightforward if, for this purpose, the atmosphere is taken to be isothermal and the expansion isothermal rather than adiabatic. Then both $\overline{u^*}$ and $\overline{\phi^*}$ are independent of p with

$$\frac{\partial \overline{u^*}}{\partial t} = - \frac{\partial \overline{\phi^*}}{\partial r} . \tag{8'}$$

Integration w.r.t. p of the continuity equation

$$\frac{\partial \omega}{\partial p} + \frac{1}{r}\frac{\partial}{\partial r}(ru) = 0 \tag{9}$$

and linearising gives the pressure tendency equation

$$\frac{\omega_s}{p_s} \simeq \frac{1}{p_s}\frac{\partial p_s}{\partial t} = - \frac{1}{r}\frac{\partial}{\partial r}(\overline{ru^*})$$

where suffix s denotes a surface value. But $\frac{\partial p_s}{\partial t} = \rho_s \frac{\partial \phi_s}{\partial t}$ so that $\phi_s = \overline{\phi^*}$ and

$$\frac{\partial \overline{\phi^*}}{\partial t} = -\frac{RT}{r}\frac{\partial}{\partial r}(r\overline{u^*}) - \left(\frac{\partial \phi_s'}{\partial t} + \frac{\partial \phi_{es}}{\partial t}\right) \quad . \tag{10}$$

Eqs. (8´) and (10) describe external wave propagation with phase speed $(RT)^{\frac{1}{2}}$ ($\cong$ 250 m s^{-1}) forced by $\partial\phi_s'/\partial t + \partial\phi_{es}/\partial t$, i.e. by the primary and secondary flows. Since this phase speed is so large compared with the system contraction rate, ($\cong$ 2 m s^{-1}), the external gravity wave transients may be neglected and the ϕ_s field regarded as instantaneously set up. It may thus be calculated using Eqs. (4) and (5´) as in the model integration. (This general conclusion is not affected if adiabatic expansion in a typical tropical atmosphere is considered - the phase speed involved is then that of sound waves.)

In Fig. 5 are shown some distributions of $\phi_e(p)$ obtained from the model described in section 2 (those of $\phi_e(p)$ + $\phi'(p)$ are similar but of small magnitude). A general decrease through the troposphere is evident. The effect of this is to require, in addition to the external gravity wave (which lowers all the isobaric surfaces in the interior), internal gravity wave propagation to warm the interior (and lift the isobaric surfaces there, leaving the stratosphere virtually undisturbed and the surface pressure unchanged).

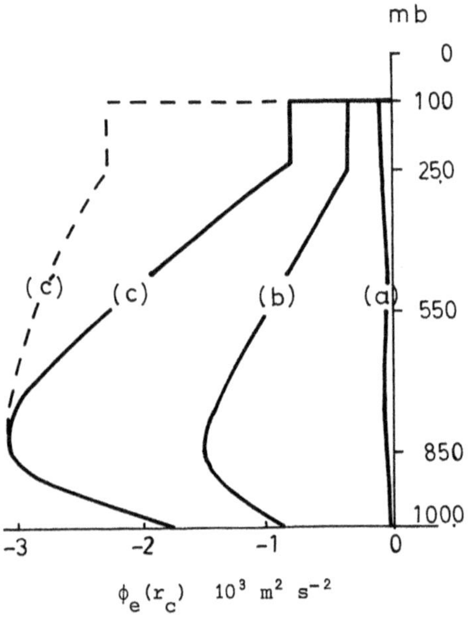

Fig. 5. Approximate vertical distributions of $\phi_e(r_c)$ from the model integration described in Section 2 at (a) t = 0 when the convection commences, (b) t = $2\frac{1}{12}$ days and (c) t = $2\frac{1}{4}$ days. Curve (c´) indicates the distribution at $2\frac{1}{4}$ days in the case of no middle level detrainment i.e. no recycling of air at middle and lower levels

The equations describing the forcing and propagation of the internal gravity waves are as follows:

Taking the derivative w.r.t. p of Eq. (8) and using the hydrostatic equation gives

$$\frac{\partial}{\partial t}\left(\frac{\partial u_i^*}{\partial p}\right) = A(p)\,\frac{\partial \theta^*}{\partial r} \tag{11}$$

where $A(p) = \frac{R}{p}\left(\frac{p}{p_0}\right)^{\kappa}$.

The continuity equation is

$$\frac{\partial \omega_i^*}{\partial p} + \frac{1}{r}\frac{\partial}{\partial r}(r u_i^*) = 0 \tag{12}$$

where, by definition, $\overline{u_i^*} = 0$, and, assuming adiabatic flow in the interior, and neglecting horizontal advection, the thermodynamic equation is

$$\frac{\partial \theta_i^*}{\partial t} + \omega_i^*\,\frac{\partial \theta}{\partial p} = 0 \quad , \quad r \leqslant r_c \quad . \tag{13}$$

On the other hand, in the convective region the flow is dominated by the heat sources and secondary circulation advection, although gravity wave effects may be identified at upper levels in a lowering of the upper limit of convection (sloping eye-wall - Fig. 4). The usual thermodynamic equation is used here, i.e.

$$\frac{\partial \theta}{\partial t} = -\,u\,\frac{\partial \theta}{\partial r} - \omega\,\frac{\partial \theta}{\partial p} + Q \quad , \quad r \geqslant r_c \quad . \tag{13$'$}$$

where Q denotes the diabatic contribution to $\partial\theta/\partial t$ and θ is inferred from the dynamics (Eqs. (4) and (5$'$)).

Continuity of ϕ at $r = r_c$ implies continuity of θ there. Expressing $\partial\theta/\partial t$ at $r = r_c$ in the form $-\,\omega_c\,\partial\theta/\partial p$, the lateral boundary condition at $r = r_c$ may be expressed as

$$\omega_i^* = \omega_c(p,t) \quad \text{at} \quad r = r_c \quad . \tag{14}$$

By definition $\omega_i^* = 0$ at $p = p_s$. Also $\omega_c = \omega_i^* = 0$ at $p = 0$. Thus ω_i^* can be expressed as a sum of Fourier components $\sum \beta_k \sin(k\pi p/p_s)$. Combining Eqs. (11), (12) and (13) the kth component $\omega_i^*(k)$ (denoted by ω_k) is seen to satisfy the wave equation

$$\frac{\partial^2 \omega_k}{\partial t^2} = \frac{c_k^{\,2}}{r}\frac{\partial}{\partial r}\left(r\,\frac{\partial \omega_k}{\partial r}\right) \quad , \quad r \leqslant r_c \quad , \tag{15}$$

where $$c_k^{\,2} = -\,\frac{R q^{\kappa-1}}{\pi^2 k^2}\,\frac{\partial \theta}{\partial q} \;\left(\text{noting } \frac{\partial \theta}{\partial q} < 0\right) \text{ with } q = \frac{p}{p_s} \quad .$$

The internal gravity waves may thus be calculated for $r \leqslant r_c$ using b.c. (14) and Eq. (15). The phase speed c_k is typically about 60 m s^{-1} for the first mode so that for the higher modes the transients may not be negligible. Subsidence is forced in the interior, with largest amplitude near the tropopause where $\omega_i^* \left(\simeq \rho \frac{\partial \phi_i^*}{\partial t} \simeq - \rho \frac{\partial \bar{\phi}^*}{\partial t} \text{ since } \phi^* \simeq \phi_e \simeq 0 \text{ in the stratosphere}\right)$ is a maximum.

5 DISCUSSION

The arguments presented in the last two sections have strongly contrasted the processes which determine the structure of the central region of an intensifying cyclone and those of the convective region. There is a semantic point which has already been referred to. This is that the secondary and primary flow constitute themselves a large amplitude inertial-gravity wave. However, it has characteristics quite different from those which determine the mass redistribution in the interior, the former resulting from the dominance of buoyancy effects associated with latent heat release. The semantic distinction drawn here is justified on these grounds.

To return to the model described in Section 2 it is seen that specifying the parameter β essentially defines the secondary flow. The quoted model results simply indicate that provided such a secondary flow can be maintained energy-wise, intensification will result, i.e. this flow is sufficient to enable enough angular momentum to be imported to overcome the frictional sink. It emphasises the important role of the cumulus convection in carrying the low-level momentum into the upper troposphere where the mass field cannot readjust to a gradient-balanced value, and so the strongly super-gradient flow in this region accelerates outwards. This process is defined (empirically) by Eqns. (3). The model convective system contracts because the numerical scheme allows the secondary flow to be initiated at the grid-point two grid points in from $r = r_m$ at each time step until the grid point next to the origin is reached. The radius of maximum wind r_m contracts in the model and the difference $r_m - r_c$ remains at 2 grid lengths until convection reaches the grid point next to the origin when it reduces to one grid length. This is qualitatively consistent with the principle stated in Section 3, that the width of the main convective updraught does not vary much, this implying that, with r_m decreasing, and sufficient inflow buoyancy to maintain the convection, r_c must also decrease. This is strictly only a hypothesis, however, and it is necessary to carry out numerical experiments with a convection model to demonstrate its validity and to provide a quantitative measure of the exact circumstances under which contraction must occur. Such a model should also be used to study the balances of buoyancy, ϕ_e, ϕ' and friction in the mature state. Although it has been assumed here that friction exerts the primary limiting influence (through the tangential momentum balance) at this stage, the buoyancy must also be involved since it determines the strength of the secondary circulation and hence also the momentum source.

The approach described here has deliberately avoided considering the role of the boundary (Ekman) layer. This is not because it plays no role at all in intensification, but because the intention is to show that intensification could occur without it, provided the buoyancy required to maintain the secondary circulation was provided through the surface moisture and sensible heat flux, and the m(p) and h(p) distributions were of the appropriate form. It is implied that the role of the Ekman layer is not crucial but simply a modifying one. The intensifying tropical cyclone is a system in which the 'boundary' layer should be regarded as extending through the whole depth of the troposphere!

The whole of the above argument has been related to an axisymmetric model of intensification. Real intensifying systems show strong axisymmetric characteristics but there are substantial asymmetries as well, notably spiral banding. Fig. 6 shows the 850 mb wind field associated with an intensifying tropical cyclone.

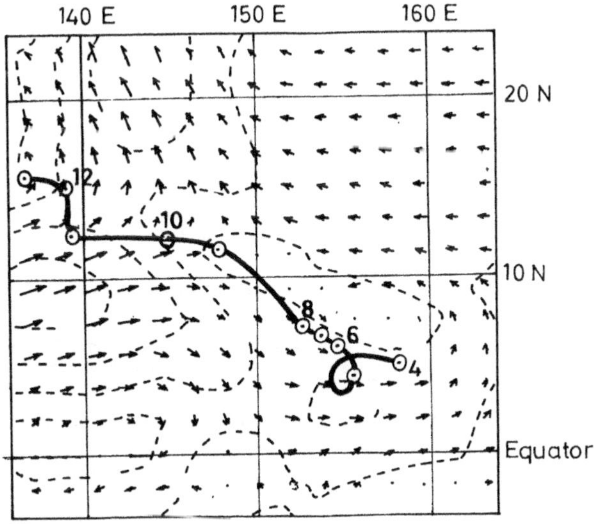

Fig. 6. 850mb vector winds and isotachs for 1200Z 3 October 1979. The full line shows the track of the developing Typhoon Tip which eventually became one of the most severe ever recorded. The circles indicate daily positions and the figures the dates. Note the isotach maximum of 10 m s^{-1} in the region to the SW of the cyclone position for 4 October (from ECMWF FGGE data analysis)

The isotachs show considerable asymmetry and the intensification process is one in which the region of maximum winds retains this structure as the band spirals inwards. The processes described above in an axisymmetric context (and the model results) should be related to a section normal to the isotach maximum, i.e. this is the region of super-gradient wind and maximum secondary flow. Thus the pressure at the centre is higher than the gradient value for this region, but could well be lower than the gradient value corresponding to the isotach minimum in the opposite (inactive) segment. This is the situation observed in the aircraft winds shown on p.42.

As already mentioned, further studies are needed if the roles of the individual physical processes involved in hurricane formation are to be properly quantified. Two- and three-dimensional numerical models with resolutions high enough to model the details of the convection, such as that described by Rosenthal (1978), are likely to provide extremely valuable tools for these investigations.

6 ACKNOWLEDGEMENTS

The author is much indebted to Herbert Riehl who first aroused his interest in hurricanes and to Professor William Gray and his colleagues at Colorado State University whose analyses of atmospheric data over a number of years have provided, through many valuable discussions, the main stimulus to explore the fascinating interplay of physical processes they represent. He is also most grateful to Mrs. Nan Spicer who typed the manuscript, Mrs. Valerie Daykin who drew the diagrams and Mr. Peter Hood who ran the computer model described in Section 2.

REFERENCES

Carrier, G.F., A.L. Hammond and O.D. George, 1971: A model of the mature hurricane. J. Fluid Mech., 47, 145-170

Charney, J.G. and A. Eliassen, 1964: On the growth of the hurricane depression. J. Atmos. Sci., 21, 68-75

Gray, W.M., 1978: Hurricanes: their formation, structure and likely role in the tropical circulation. Meteorology over the Tropical Oceans. Royal Meteor. Soc., 155-218

McBride, J.,1979: Observational analysis of tropical cyclone formations. Atmos. Sci. Paper No. 308, Colorado State Univ

Ogura, V., 1964: Frictionally controlled, thermally driven circulation in a circular vortex with application to tropical cyclones. J. Atmos. Sci., 21, 610-621

Ooyama, K., 1969: Numerical simulation of the life cycle of tropical cyclones. J. Atmos. Sci., 26, 3-40

Pearce, R.P., 1981: A dynamical analysis of hurricane intensification. Contributions to Atmos. Phys., 54, 19-42

Rosenthal, S.L., 1978: Numerical simulation of tropical cyclone development with latent heat release by the resolvable scales. I: Model description and preliminary results. J. Atmos. Sci., 35, 258-271

Tornadoes

The Development and Maintenance of Rotation in Convective Storms

D. K. Lilly

1 INTRODUCTION

The recent availability of doppler radar for observations of convective storms has produced convincing documentation of the importance of rotating storms and their close association with severe weather. Although not all severe thunderstorms rotate, an operational evaluation study conducted at NOAA's National Severe Storms Laboratory showed that most of the storms containing mesocyclones identifiable by doppler radar produced either tornadoes or large hail within an hour.

In this paper I will first review the scientific background and problems posed by the existence of rotating storms, the earlier attempts at resolving the problems and the new insights provided by numerical simulation. Then I will consider the two separate but interacting parts of the rotating storm problem, first what produces and maintains the rotation, and second what determines the anomalous propagation of the storms. The first part is relatively straightforward, though some uncertainty exists as to the relative importance of certain processes. The second part turns out to be considerably more difficult and some fundamental questions remain unresolved.

The fact that some convective storms move at distinctly different speeds and in different directions than the mean flow in which they are embedded was noted by Newton and Katz (1958), and the association of this lateral motion with rotation was established by Browning (1964) through development of an important and still fairly accurate conceptual model of "supercell" storms. Fujita and Grandoso (1966) noted the apparent association of rotational development with storm splitting and proposed a mechanism to account for both.

It has long been recognized that these severe, long-lived, rotating and anomalously propagating storms usually develop in a strongly sheared environment, even though, as will be discussed later, shear is known to be a deterrent to convective dynamics. Thunderstorm cells developing in an unsheared environment are known to have a limited life cycle, due apparently to the effect of downdrafts in cutting off the moist low level inflow. With the lack of any other obvious reason for maintenance of strong convection in shear, it was (and still is) generally believed that the principal favoring effect of shear is to allow a geometrical separation of the updraft and downdraft. Two-dimensional numerical simulations of clouds with a rain process were developed to test this hypothesis. In general the results tended to confirm and extend the results of linear theory. Takeda (1971) found that strong shear suppressed his simulated thunderstorm more than did a raining downdraft. Orville and Kopp (1977) were able to maintain a two-dimensional convective

Topics in Atmospheric and Oceanographic Sciences
Intense Atmospheric Vortices, ed. by Bengtsson/Lighthill

cell with topographic forcing only by reducing the wind speed to 1/5 that seen in an observational comparison. Hane (1973) was able to maintain a convective storm with shear but at a reduced amplitude.

For finite amplitude two-dimensional elements, the suppressing effect of shear can probably best be explained in terms of the theory of two-dimensional turbulence. It is known that in two dimensions the turbulent energy cascade works in reverse, with small scale energy tending to propagate to large scales and eventually to the mean shear. Thus a convective cell operating in a mean shear tends to lose its energy to the mean flow through upward momemtum flux.

Three-dimensional convective cloud and storm simulations were developed to avoid the limitations of two-dimensional models. The early numerical results were reviewed at some length by Lilly (1979) and the discussion will not be repeated fully here. In brief, Schlesinger (1978) and Klemp and Wilhelmson (1978a,b; Wilhelmson and Klemp, 1978) found that an initially hemispherical or ellipsoidal warm air element, embedded in a conditionally unstable environment with significant vertical shear, first grows into a cylindrical cumulus cloud, but the cloud rapidly develops a double vortex structure. As the rain process and a water-loaded downdraft become established the updraft splits in two, with the two halves apparently following the vorticity centers. Later the downdraft also splits and two independent storms emerge, both moving laterally to the mean wind hodograph. For the case of a rectilinear shear, the storm moving to the right of the mean shear vector develops cyclonic vorticity in the updraft and anticyclonic vorticity in the rain induced downdraft. These circulations are reversed in the left-moving storm.

Klemp and Wilhelmson found that the presence of a Coriolis acceleration enhances the right-moving cyclonically rotating storm slightly, but a curving mean hodograph produces much stronger asymmetry. If the shear vector rotates clockwise with height, the right-moving storm is enhanced strongly and the left-mover suppressed or eliminated, with reverse effects for a counterclockwise rotating shear vector. The hodograph rotation in the lower levels seems to exert the strongest effects. Thus we believe that the predominance of right-moving storms in nature is associated with Ekman turning in the boundary layer, since there seems to be no strong predominance of one sign of mean hodograph curvature above that layer in severe storm proximity soundings.

Further analyses and interpretations of his earlier simulations have been presented by Schlesinger (1980), and additional simulations with more varied environmental conditions have been carried out by Klemp and Wilhelmson and their associates. Detailed wind field analyses made from multiple doppler radar observations of a storm near Oklahoma City were closely compared with the results of a simulation carried out using wind and thermodynamic profiles from the storm's environment (Klemp et al., 1981). Many areas of agreement were found, although the simulated storm resolved less detail than was available in the observational analysis. Weisman and Klemp (1981) have conducted a series of simulations using idealized thermodynamic profiles and rectilinear shears, but with the amplitudes of the instability and shear varied over a substantial range. A large range of storm behavior was found in the results. Those storms which most closely resembled Browning's "supercell" tended to fall into a limited range of a characteristic Richardson number.

2 GENERATION AND CONCENTRATION OF VERTICAL VORTICITY

If the inviscid horizontal equations of motion are written in Boussinesq or anelastic form and then cross-differentiated, the thermodynamic terms drop out and an equation for rate of change of vertical vorticity results, in the vector form

$$\frac{\partial \zeta}{\partial t} + \underset{\sim}{V} \cdot \nabla\zeta = (f\,\underset{\sim}{k} + \nabla\times\underset{\sim}{V}) \cdot \nabla w \tag{1}$$

where the vertical Coriolis term is included and regarded as constant. The additional Coriolis term in the y-direction is neglected here because it is invariably smaller than any significant vertical shear. A version of the vertical vorticity equation which is convenient for some purposes is obtained by expressing (1) in horizontal natural coordinates, which follow the streamlines in the horizontal direction. In these coordinates the vorticity is given by $\nabla \times \underset{\sim}{V} =$ $(-V_H \frac{\partial\psi}{\partial z} + \frac{\partial w}{\partial n})\underset{\sim}{i}_s + (\frac{\partial V_H}{\partial z} - \frac{\partial w}{\partial s})\underset{\sim}{i}_n + (V_n \frac{\partial\psi}{\partial s} - \frac{\partial V_H}{\partial n})\underset{\sim}{k}$ where V_H is the horizontal velocity amplitude, ψ is the wind direction increasing counterclockwise, and s and n are coordinates along and orthogonal (to the left) to the local wind vector, respectively. The vertical vorticity equation is written in these coordinates as

$$\frac{\partial \zeta}{\partial t} + V_H \frac{\partial \zeta}{\partial s} + w \frac{\partial \zeta}{\partial z} = \frac{\partial V_H}{\partial z}\frac{\partial w}{\partial n} - V_H \frac{\partial \psi}{\partial z}\frac{\partial w}{\partial s} + (f + \zeta)\frac{\partial w}{\partial z} \tag{2}$$

The first two terms on the right are conventionally denoted as the tilting terms and the last as the vortex stretching term.

Linearized versions of equations (1) and (2) are obtained by assuming the mean flow to be horizontal but varying arbitrarily in the vertical. With the mean flow overbarred and the deviations from it primed, the linearized vector form of the vorticity equation is

$$\frac{\partial \zeta}{\partial t} + \underset{\sim}{\bar{V}} \cdot \nabla\zeta' = -(\underset{\sim}{k} \times \nabla w') \cdot \frac{d\underset{\sim}{\bar{V}}}{dz} \tag{3}$$

while in horizontal natural coordinates

$$\frac{\partial \zeta'}{\partial t} + \bar{V}_H \frac{\partial \zeta'}{\partial s} = \frac{\partial w'}{\partial n}\frac{d\bar{V}_H}{dz} - \frac{\partial w'}{\partial s}\bar{V}_H \frac{d\bar{\psi}}{dz} + f\frac{\partial w'}{\partial z} \tag{4}$$

We now look at solutions of simplified versions of these equations. In all cases the wind is defined relative to the motion of a defineable storm or disturbance, so that the remaining time derivative refers only to changes of storm structure or intensity. First, if the mean shear vanishes or is negligible, then the only generating term is the last on the right, which measures the stretching of the rotating atmosphere's vortex lines by a mean convergence. This situation is schematically illustrated in Figure 1. No steady state exists if convergence continues, and the solution to either (3) or (4) is

$$\zeta' = f\int \frac{\partial w'}{\partial z}\,dt \tag{5}$$

Note that if vertical velocity has a positive maximum in some midlevel of the troposphere, then vorticity maxima occur in low and high levels, with that in the low level having the same sign as the Coriolis acceleration, particularly hurricanes.

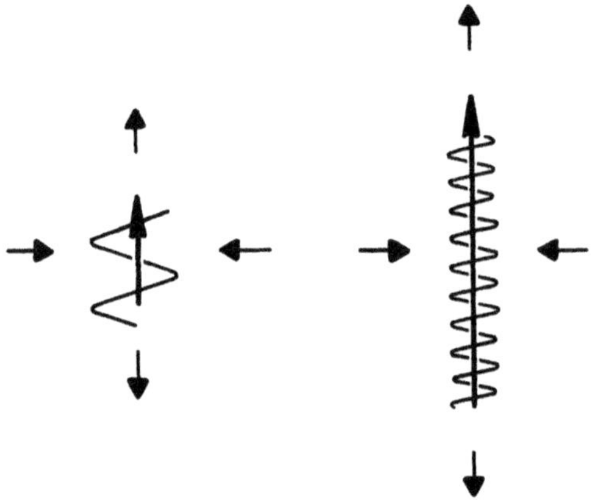

Fig. 1. Schematic illustration of stretching of planetary vorticity, by a vertical divergence field, as in Equation (5)

The second simplified case is that of a constant rectilinear vertical shear, e.g., $d\overline{V}_H/dz$ = constant, $d\overline{\psi}/dz = 0$. The Coriolis force is neglected, so that only the first term on the right of (4) is retained. If w is specified as a function of space and time, complete solutions can be found in some cases. A simpler and more revealing procedure is, however, to look at the solutions in the limits where either the first or second terms on the left can be neglected. If $\overline{V}_H t$ is much less than the length scale of the updraft region, i.e., for short times after initiation and/or for regions close to a critical layer, then the advection term can be neglected and the solution is, approximately,

$$\zeta' = \int \frac{\partial w'}{\partial n} \frac{d\overline{V}_H}{dz} dt \qquad (6)$$

For a confined updraft region this solution produces a double vortex, with positive vorticity to the right of the updraft center looking downshear. Figure 2 shows, schematically, the tilting of initially horizontal vortex tubes to produce both signs of vertical velocity.

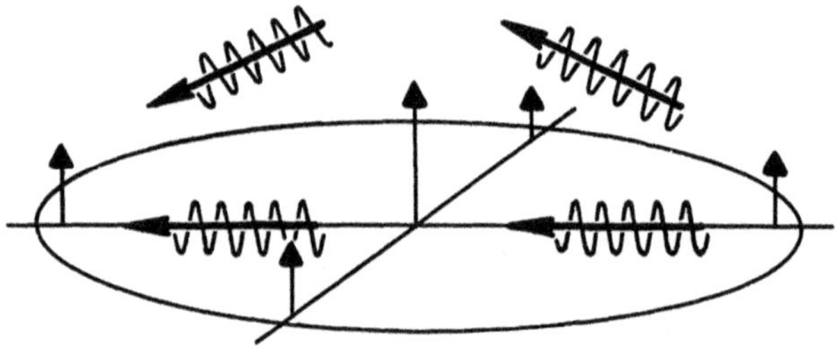

Fig. 2. Tilting of an initially horizontal vortex (from a rectilinear vertical shear) in an updraft, as in Equation (6)

The above flow solution is obviously applicable to Figure 3, which shows the horizontal and vertical velocity fields at an early stage of storm development at a level close to that where the storm motion is the same as the wind velocity (the critical level). The first frame of the sequence shows one member of the growing vortex pair on the edges of the central updraft, essentially as described by Equation (6). The later frames illustrate the early evolution of a

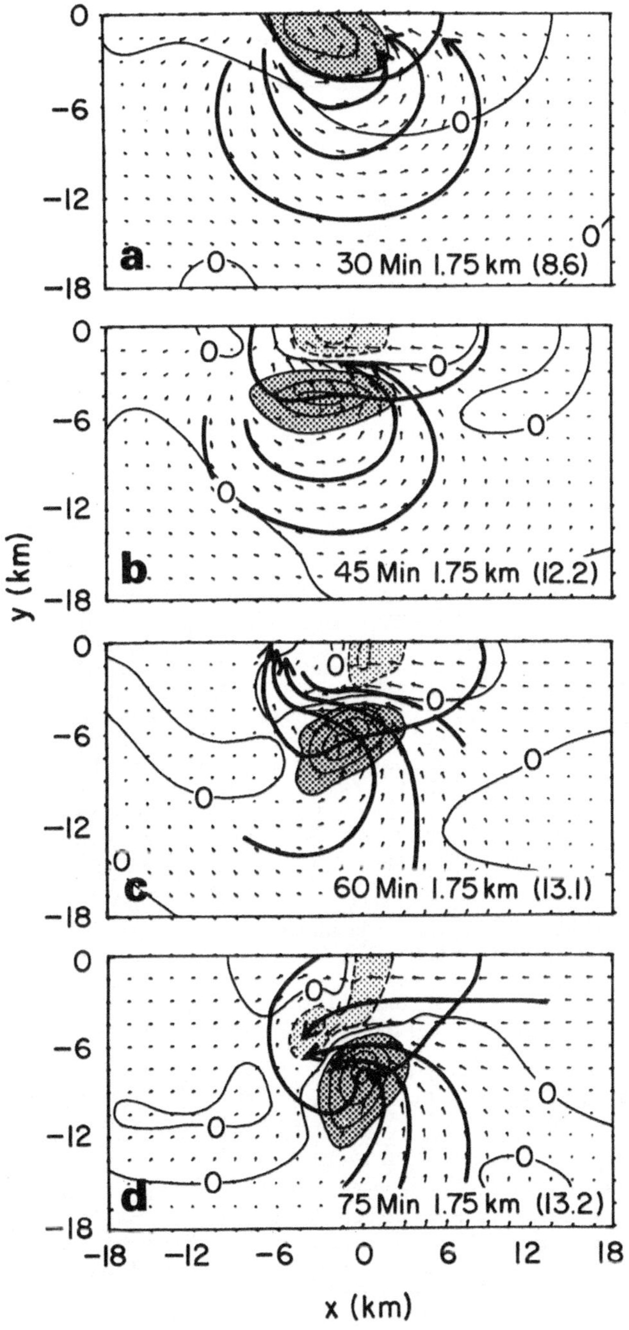

Fig. 3. Evolution of the flow field at the 1.75 km altitude in a Wilhelmson-Klemp (1978) simulation with rectilinear shear. The shaded areas are updraft and downdraft centers

splitting storm, with the vertical velocity maxima following the vorticity maxima. Rotunno (1981) has carried out a more detailed analysis of this early evolution, including the nonlinear effects of vortex stretching.

The other important case is that of a rotating mean hodograph, with $\bar{V}_H$ and $\partial\bar{\psi}/\partial z$ constant. Neglecting the Coriolis forces, Eq. (4) holds with only the second term on the right remaining. Again a transient solution may exist for $\bar{V}_H t$ much less than the horizontal scale of the updraft. Because no critical layer exists, the advective terms will everywhere become dominant in time, so that the steady state solution attains primary importance. Upon canceling $\bar{V}_H$ from both sides and integrating over s, (4) becomes simply

$$\zeta' = - w' \frac{d\bar{\psi}}{dz} \qquad (7)$$

Note that vorticity is still being generated upwind and downwind of the vertical velocity maxima. Because these maxima are propagating through the flow, however, vorticity centers coincide with the updraft and downdraft centers, with a clockwise rotating hodograph producing cyclonic vorticity in the updraft. Vorticity maxima are located in midtropospheric levels, in contrast to the predictions of Equation (5) of maxima at low and high levels, when these are produced from concentration of earth's vorticity. Figure 4 shows a vortex line first tilted upward and then back downward as it moves through the updraft center.

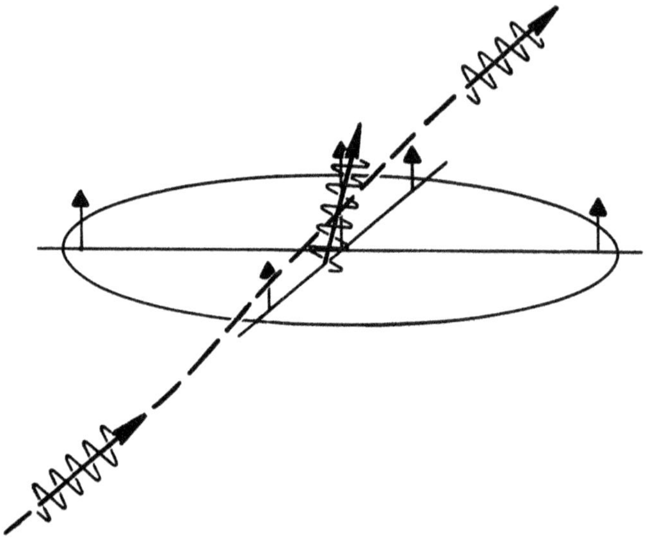

Fig. 4. Tilting of vortex tubes in a constant directional shear flow by an updraft, as in Equation (7)

The importance of this case arises from its qualitative resemblance and dynamic similarity to the vorticity patterns and evolution of mesocyclones, and also to the fact that it is not nearly so constricted by the linearization assumptions as the other cases. In particular the vertical advection and vortex stretching terms in Equation (2) cancel each other out (neglecting Coriolis effects). Real environmental hodographs do not usually closely resemble that assumed for this case, but for any storm which is propagating laterally to the hodograph there will be a level at which the wind shear is purely directional, and that level usually lies close to that of the maximum updraft. Thus if a storm propagates laterally to a strongly sheared hodograph it tends to generate vorticity in a way similar to that described for this case.

The above result can also be extended into an exact solution of the nonlinear equations of motion in the form of a Beltrami flow, with vorticity everywhere proportional to velocity. The mean rotating hodograph assumed above is itself a Beltrami flow. One can then superimpose a vertically bounded twisting solenoid flow with a structure given by

$$w' = W(x,y)\sin\mu z, \quad \nabla_H{}^2 W = \left[\mu^2 - \left(\frac{d\bar{\psi}}{dz}\right)^2\right] W \tag{8}$$

$$\zeta' = \frac{d\psi}{dz}\, w \tag{9}$$

This flow also satisfies the Beltrami condition without limitations on amplitude. The scale limitations and periodicity requirements are presumably associated with the fact that, although the Beltrami flow is a complete solution of the nonlinear equations of motion and continuity, it contains neither forcing nor dissipation terms and is unlikely to maintain a plausible buoyancy field. It seems, however, sufficiently revealing of the vorticity dynamics to warrant further investigation by means of numerical simulation. Therefore a simulation experiment was carried out with a circular hodograph and a thermodynamic sounding typical of severe storm environments. The simulation was successful, in that an intense quasi-steady storm developed with a propagation velocity close to the center of the rotating hodograph. Relative to the mean flow at each level this storm propagates at about 20 m s^{-1}, always in a direction approximately lateral to the local mean shear, and significantly faster than the lateral propagation of any of the rectilinear shear cases. Correspondingly the vorticity is greater, agreeing with Equation (9) to within about 20%.

To summarize the results of the above analysis of vorticity generation, the most important aspect seems to be the strong control exerted by the storm motion relative to the mean hodograph. The most important case is that for which vorticity maxima coincide with updraft and downdraft centers, and this will apparently always occur to some extent if the storm finds a way to propagate laterally to the mean hodograph. I will next consider the circumstances which favor such propagation.

3 THE DETERMINANTS OF STORM MOTION

A variety of theoretical approaches have been applied to predict the motion of convective storms. It now appears that both linear and nonlinear aspects are important but not yet fully understood. With a linearized approach we can apparently explain, if not precisely predict, the lateral movement of storms in a rotating hodograph. Storm motion lateral to a rectilinear shear remains somewhat unclear and apparently dependent on nonlinear effects.

I will first outline the results and conclusions of linear theory of buoyant instability in a shearing environment, including some rather recent results. Then I will review the recent analyses of storm motion by Rotunno and Klemp, based partly on consideration of simulation model results.

The growth rates and favored modes of instability of a buoyantly unstable flow with mean shear were investigated at some length by Asai (1970, 1973). He considered both linear and curved velocity profiles, unidirectional (1970) and with a rotating Ekman layer hodograph (Asai and Nakasuji, 1973). The model structure was always two-dimensional, but the horizontal wave vector could be directed at arbitrary angles from the shear vector. The results led to a generally straightforward interpretation. If the negative Richardson number was larger than order 10^{-2} the buoyant energy source dominated. The modes with maximum growth rates were of the form of roll convection, with axes along or close to the shear vector and minimal lateral movement. Mean shear across the convective rolls apparently always deterred growth. This general result was largely independent of the shape of the wind profile, with points of inflection nearly irrelevant. One small exception related to the current subject is noted. In a certain region of the parameter space, with $-Ri \sim 1$ and the horizontal wave number, $k \sim 3/H$, with H the convective layer depth, a secondary maximum of growth rate occurred for modes oriented and propagating diagonally to the shear vector. By combining two such modes, one moving forward faster than the vertically averaged mean flow and the other slower, a three-dimensionally cellular mode can be generated which moves laterally and contains rotation produced by the tilting mechanism discussed in the previous section. No evidence exists to my knowledge that this mode can become the primary instability for the kind of wind profiles considered by Asai.

More recently Marroquin and Raymond (1981) carried out similar linear theory calculations oriented more specifically to convective storms, with stability and shear profiles taken from particular storm environments. Their model seems rather successful in predicting storm motion in cases where strong rotation does not develop. Their storm motions always lie fairly close to a point on the hodograph, usually in the lower troposphere. The strong lateral motion characteristic of rotating storms is not predicted. They attributed their results to a qualitative extension of the Miles-Howard theorem, which requires that the phase velocity of an amplifying mode should coincide with the mean flow (component along the wave number vector) at a level near that of maximum vertical displacement.

The WAVE-CISK mechanism was applied by Raymond (1975, 1976) to predict storm motion and evolution, and was apparently successful in some cases of rotating storms. This mechanism was discussed further by Lilly (1979). Because a storm is modeled by WAVE-CISK as a self-induced gravity wave, it is predicted to propagate off the mean hodograph, with the control of that propagation rate heavily influenced by buoyancy forces.

Most recently, Rotunno and Klemp (1981) have carried out an analysis of some results of numerical simulation to determine the effect of mean shear on propagation. Their analysis is based on pressure gradient determination. The inviscid vertical equation of motion for a Boussinseq or anelastic system may be written

$$\frac{dw}{dt} + \frac{\partial(p/\bar{\rho})}{\partial z} = b \qquad (10)$$

where b is the buoyancy deviation from a mean hydrostatic state. The pressure may be solved from a Poisson equation derived by taking the divergence of the momentum equations, i.e.,

$$\nabla \cdot [\bar{\rho}\nabla(p/\bar{\rho})] = \frac{\partial}{\partial z}(\bar{\rho}b) - \nabla \cdot (\bar{\rho}\underset{\sim}{V} \cdot \nabla\underset{\sim}{V}) + f\bar{\rho}\zeta \qquad (11)$$

Buoyancy evolves on the basis of the equation

$$\frac{db}{dt} + N^2 w = 0 \qquad (12)$$

with N^2 negative over part of the volume. It is reasoned that without a mean shear all the fields will tend toward radial symmetry, or at least will remain radially symmetric if a disturbance is initially so. A mean shear will destroy this symmetry through the pressure field and through differential advection of the updraft and buoyancy.

The simplest analysis is based on linearizing the advection term on the right side of (11) and assuming that the sign of the pressure deviation is opposite to that of (11). This must be so on the average, since

$$\langle (p/\bar{\rho})\nabla \cdot [\bar{\rho}\nabla(p/\bar{\rho})]\rangle = -\langle \rho[\nabla(p/\bar{\rho})]^2\rangle \qquad (13)$$

where the angled brackets denote a horizontal average and we assume vanishing pressure deviation at horizontal infinity. The linearized elliptic equation is

$$\nabla \cdot [\bar{\rho}\nabla(p/\bar{\rho})] = \frac{\partial}{\partial z}(\bar{\rho}b) + \bar{\rho}f\zeta - 2\bar{\rho}\nabla w \cdot \partial\underset{\sim}{\bar{V}}/\partial z \qquad (14)$$

Assuming that an initially symmetric buoyancy disturbance is imposed, the first term on the right will not immediately affect the pressure symmetry but it reduces the updraft, essentially through exchange between horizontal and vertical kinetic energies. The Coriolis term produces a pressure minimum at positive vorticity centers. When low and high level vorticity maxima are induced by Coriolis forces, as in Equation (5), the effect is to reduce the updraft, essentially by transforming some of its energy into that of rotary motion. When midlevel vorticity is produced by the tilting terms, however, as in Equations (6)-(7), then the updrafts are accelerated under the vorticity maxima having the same sign as the Coriolis acceleration and decelerated in the opposite case. This explains why the right-moving half of the split pair is amplified by the Coriolis term in the original Klemp-Wilhelmson (1978b) model.

The effect of the mean shear on the pressure field depends strongly on the curvature of the hodograph, as does its effect on vorticity generation. If the shear is rectilinear, then it produces a negative pressure deviation at midlevels on the downshear side of the updraft maximum and a positive deviation upshear. Since no effect is produced at low or high levels where the updraft vanishes, the resulting vertical pressure force tends to tilt the updraft upstream, partially counteracting the downstream tilt produced by the shear on the updraft and on the buoyancy. No lateral propagation tendency is present. For the case of a rotating hodograph the local pressure minimum is still on the downshear side of the updraft maximum. The location of this minimum is, however, also rotating with height. Thus at a level 1/4 of a hodograph circle below the updraft maximum the pressure deviation is lateral to the mean shear direction and 180° out of phase with that 1/4 of a circle above the maximum. This leads to an updraft acceleration toward the center of the hodograph circle (see Figure 5). This effect tends to propagate a storm laterally to the hodograph and thus to sustain the generation of vorticity. Rotunno and Klemp have evaluated simulation model

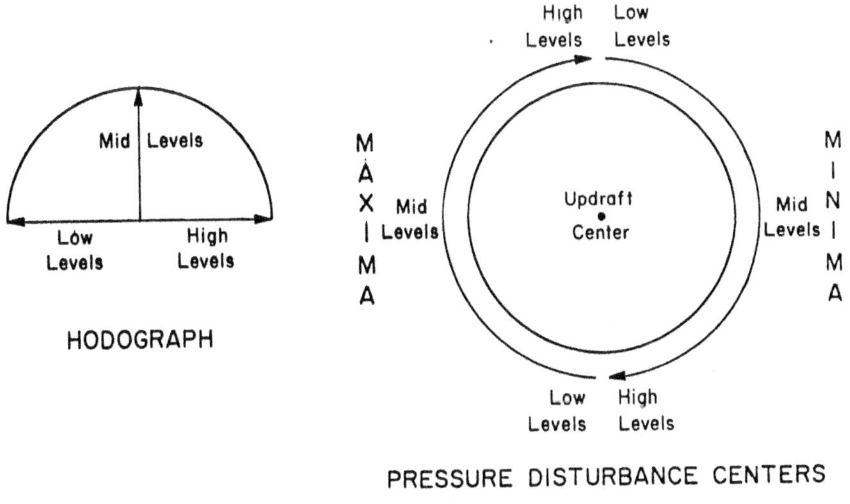

Fig. 5. Schematic illustration of the pressure disturbances produced around the edge of an updraft in a clockwise rotating hodograph. The pressure field contributes to updraft acceleration to the south at midlevels, counteracting the northward advection by the mean flow

results term by term for a convective updraft developing in a clockwise rotating hodograph. With the vorticity generation mostly controlled by transient terms, through Equation (6), the cyclonic side of the updraft is enhanced by a favorable vertical pressure gradient. This bias is attributed to the above effect.

An alternative version of the above analysis for a rotating hodograph can be obtained from the Beltrami flow mode. For Beltrami flow the pressure is determined completely from a Bernoulli expression,

$$p/\bar{\rho} + V^2/2 = \text{constant}$$

Use of this relation plus the solution form indicated in Equations (8)-(9) yields the pressure directly. The Beltrami solution requires that the amplitude of the rotational flow be greater than that of either the divergent or the vertical flow. Then the dominant asymmetric part of $V^2/2$ is twice the dot product of the mean and rotational flow vectors. When this is positive, as on the right side (looking downstream) of a cyclonically rotating vortex, then the pressure is low. This leads to results similar to the Rotunno-Klemp analysis above, but extensible to nonlinear amplitudes as well. It should be noted that this effect is completely different from the Magnus effect, earlier regarded as a strong candidate for explaining cyclonic movement of rotating storms. Although the pressure reasoning is superficially similar, the Magnus assumption is that a storm can be moved laterally as a solid body by horizontal pressure gradients to dynamically propagate the updraft. The Magnus prediction of rightward acceleration of a cyclonically rotating storm would be confined to the upper region, where the mean flow is faster than the storm motion. At low levels the Magnus predicted propagation would be in the opposite direction.

The effects of nonlinear terms can also be evaluated by their influence on generation of updrafts and downdrafts. From the analysis of Rotunno and Klemp, the nonlinear terms of the equations of motion appear in the Poisson equation for pressure in Cartesian coordinates as

$$\nabla^2(p/\bar{\rho}) = -\left(\frac{\partial u}{\partial x}\right)^2 - \left(\frac{\partial v}{\partial y}\right)^2 - \left(\frac{\partial w}{\partial z}\right)^2 - 2\,\frac{\partial v}{\partial x}\frac{\partial u}{\partial y} - 2\,\frac{\partial w}{\partial x}\frac{\partial u}{\partial z} - 2\,\frac{\partial w}{\partial y}\frac{\partial v}{\partial z}$$

The first three terms are the negatives of the squares of the components of divergence. In a buoyant cloud system they would be expected to be largest in the lower and upper levels, leading to higher pressure there. The last three terms are closely related to centrifugal force gradients in different planes and will tend to be positive near a vortex center. Both of the above effects can be interpreted alternatively by use of the equations of horizontal motion in two-dimensionally natural coordinates, which are in the downstream and cross-stream directions, respectively.

$$\frac{dV_H}{dt} + \frac{\partial(p/\bar{\rho})}{\partial s} = 0, \quad V_H\left(\frac{d\psi}{dt} + f\right) + \frac{\partial(p/\bar{\rho})}{\partial n} = 0$$

From the first equation, pressure increases as flow decelerates in a convergence region. From the second, the cross-stream pressure gradient equals the centrifugal force. Consistently with this, Rotunno and Klemp found that in the early stages of development of a convective element in shear, the updraft generation was enhanced under the growing vorticity center on either side of the original updraft maximum. They believe that this enhancement contributes to storm splitting.

Presumably the above process also contributes to updraft maintenance of a single rotating updraft. The details remain somewhat uncertain, however. As previously described, vorticity is generated ahead of the updraft center through the tilting terms but is moved back to that center by horizontal advection. If vertical motion is generated under the vortex center and is also advected back it will tend to lag the vortex center, which will itself then slow down to find the updraft, and the system will lose its motivation. Dissipative forces will presumably tend to keep the centers of vorticity and updraft closer to their generating regions and may then be responsible for maintaining motion. In any event the simulation results show that storms moving laterally to a rectilinear shear flow do so at a considerably slower rate than they can in a rotating mean hodograph under the same buoyancy conditions.

4 REFERENCES

Asai, T. 1970 Stability of a plane parallel flow with variable vertical shear and unstable stratification. J. Meteor. Soc. Japan, 48, 129-135

Asai, T. and I. Nakasuji 1973 On the stability of Ekman boundary layer flow with thermally unstable stratification. J. Meteor. Soc. Japan, 51, 29-42

Browning, K. A. 1964 Airflow and precipitation trajectories within severe local storms which travel to the right of the winds. J. Atmos. Sci., 21, 634-639

Fujita, T. and H. Grandoso, 1966 Split of a thunderstorm into anti-cyclonic and cyclonic storms and their motion as determined from numerical model experiments. J. Atmos. Sci., 25, 416-439

Hane, C. E. 1973 The squall line thunderstorm: numerical experimentation. J. Atmos. Sci., 30, 1672-1690

Klemp, J. B. and R. B. Wilhelmson, 1978a The simulation of three-dimensional convective storm dynamics. J. Atmos. Sci., 35, 1070-1096

Klemp, J. B. and R. B. Wilhelmson, 1978b Simulations of right and left moving storms produced through storm splitting. J. Atmos. Sci., 35, 1097-1110

Klemp, J. B., R. B. Wilhelmson and P. S. Ray 1981 Observed and numerically simulated structure of a mature supercell thunderstorm. J. Atmos. Sci., to appear, August issue

Lilly, D. K. 1979 The dynamical structure and evolution of thunderstorms and squall lines. Ann. Rev. Earth Planet. Sci., 7, 117-161

Marroquin, A. and D. J. Raymond 1981 A linearized convective overturning model for prediction of thunderstorm movement. Submitted to J. Atmos. Sci

Newton, C. W. and S. Katz 1958 Movement of large convective rainstorms in relation to winds aloft. Bull. Amer. Meteor. Soc., 39, 129-136

Orville, H. D. and F. J. Kopp 1977 Numerical simulation of the life history of a hailstorm. J. Atmos. Sci., 34, 1596-1618

Raymond, D. J. 1975 A model for predicting the movement of continuously propagating convective storms. J. Atmos. Sci., 32, 1308-1317

Raymond, D. J. 1976 Wave-CISK and convective mesosystems. J. Atmos. Sci, 33, 2392-2398

Rotunno, R. 1981 On the evolution of thunderstorm rotation. Mon. Wea. Rev., 109, 577-586

Rotunno, R. and J. B. Klemp 1981 The influence of the shear-induced pressure gradient on thunderstorm motion. Submitted to Mon. Wea. Rev

Schlesinger, R. E. 1978 A three-dimensional numerical model of an isolated thunderstorm: Part I. Comparative experiments for variable ambient wind shear. J. Atmos. Sci., 35, 690-713

Schlesinger, R. E. 1980 A three-dimensional numerical model of an isolated thunderstorm: Part II. Dynamics of updraft splitting and mesovortex couplet evolution. J. Atmos. Sci., 37, 395-420

Takeda, T. 1971 Numerical simulation of a precipitating convective cloud: the formation of a "long-lasting" cloud. J. Atmos. Sci., 28, 350-376

Weisman, M. A. and J. B. Klemp 1981 The dependence of numerically simulated convective storms on vertical wind shear and buoyancy. Submitted to Mon. Wea. Rev

Wilhelmson, R. B. and J. B. Klemp 1978 A numerical study of storm splitting that leads to long-lived storms. J. Atmos. Sci., 35, 1974-1986

Cumulus Rotation: Model and Observations of a Waterspout-bearing Cloud System

J. Simpson

1 Introduction, Motivation, and Background Research

Intense cumulus rotation has been found to precede severe tornadic storms. Substantial advances in documenting and understanding precursor conditions for destructive tornadic vortices have been made using Doppler radar observations. A typical tornado-associated "mesocyclone" has a diameter of 5-10 km, tangential wind velocities 15-25 ms^{-1} and vorticities in the neighborhood of 10^{-2} s^{-1}. Only about 60 percent of these strong mesocyclones produce funnels, however, suggesting that a special juxtaposition of a number of conditions is required.

Waterspouts appear to be similar, less intense atmospheric vortices, more accessible to measurement. This paper addressed the conditions required to produce a waterspout, using a three-dimensional cumulus model and observations made on a well-documented day (261) of the GATE[1] experiment of the Global Atmospheric Research Program. Primary emphasis is placed on the simulated development of cumulus-scale rotation. Second and more speculative emphasis is addressed to possible dynamic connections between the mesoscale rotation at intermediate cloud heights and the concentrated vortex funnel observed below cloud base.

Numerical model results (SCHLESINGER, 1978) and observations (HEYMSFIELD, 1981) show that rotation, usually in a cyclonic and anticyclonic vortex pair are the rule rather than the exception in cumuli growing in a windfield which varies with height. In particular, BLECHMAN's (1981) application of SCHLESINGER'S model, with no initial wind perturbation, shows mesovortex strength increases non-linearly with shear magnitude--rapidly at first and then levelling out at vorticity of about 10^{-2} s^{-1} when shear magnitudes reach about 1.5×10^{-3} sec^{-1} (in the layer 1-4 km with the same statically unstable thermodynamic sounding used by SCHLESINGER).

Two studies of the cumuli on day 261 of GATE have been performed with the same model using a precipitation scheme modified for tropical clouds. SIMPSON and VAN HELVOIRT (1980) simulated both the observed cumulonimbus and the cumulus congestus to investigate downdrafts, gust fronts and their impacts on the surface layer. The dependence of mesovortices on shear and their relation to organized entrainment was explored in further congestus model studies by SIMPSON, VAN HELVOIRT, and MCCUMBER (1981). The finding of mesovortices in all the model clouds,

[1] GATE stands for GARP Atlantic Tropical Experiment

Topics in Atmospheric and Oceanographic Sciences
Intense Atmospheric Vortices, ed. by Bengtsson/Lighthill

combined with recollection of an observed waterspout on GATE day 261 (one of two recorded on all 60 days of that field program) led to this effort to illuminate the combination of processes required.

2 Key Aspects of the Observations

The oceanic clouds and their surroundings in the study area on GATE day 261 were observed by five instrumented aircraft flying vertically stacked in a box pattern 150 km on a side. The northeast corner of the box, located at 10°N lat., 21°W long. contained a developing cumulonimbus system, which intensified while propagating slowly southwestward. Three of the GATE research ships were located just outside the box to the west and south, providing radar data, wind profiles (Fig. 2.1). thermodynamic soundings (Fig. 2.2) and boundary layer measurements.

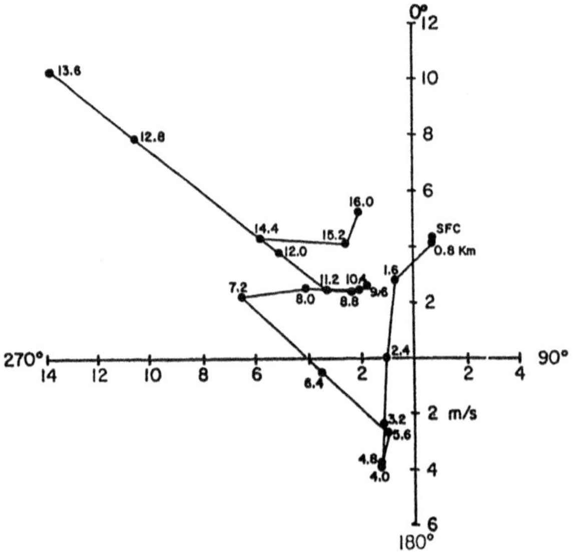

Fig. 2.1. Hodograph of observed wind profile near waterspout cloud on GATE day 261. Wind at any level is vector from center to level's image point

Fig. 2.3 shows the northeast corner of a detailed photogrammetric cloud map constructed from the aircraft films. Components of this map appear in extensive analyses of convection on this day (WARNER et al. 1979; 1980). The cloud heights (km) and locations agreed extremely well in comparison with both radar and satellite data.

The waterspout (Fig. 2.4) was captured on the side-looking cine camera of an Electra aircraft flying southward at 540 m elevation. The funnel was directly under the eastern edge of a congestus at the upshear end of a line flanking the cumulonimbus in Fig. 2.3. Restricted visibility prevented observation of the direction of funnel rotation. The congestus tower appeared stalled at about 7.8 km, as measured by a higher level aircraft (at 6.5 km elevation) just 2 min 50 sec before the funnel was

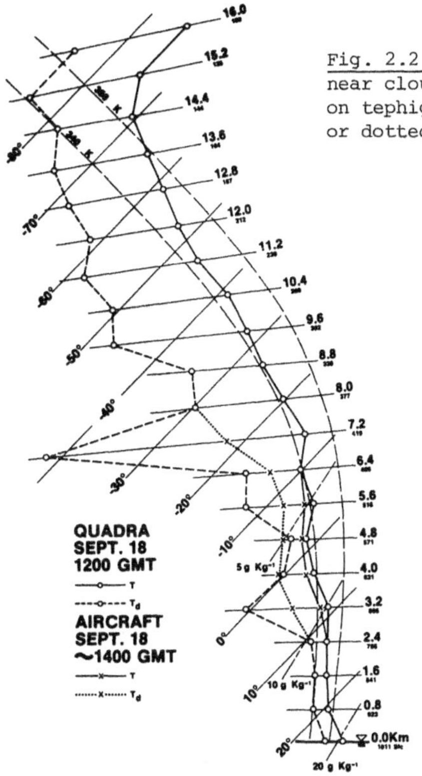

Fig. 2.2. Thermodynamic sounding at Ship QUADRA near cloud system at 1200 GMT day 261 plotted on tephigram. Solid line temperatures. Dashed or dotted dewpoint temperatures

Fig. 2.3. Portion of photogrammetrically constructed cloud map for cloud system near waterspout on GATE day 261. North upward. Four digit numbers times in GMT that C-130 (at 6.5 km) aircraft was at a location. Two digit numbers cloud heights in km

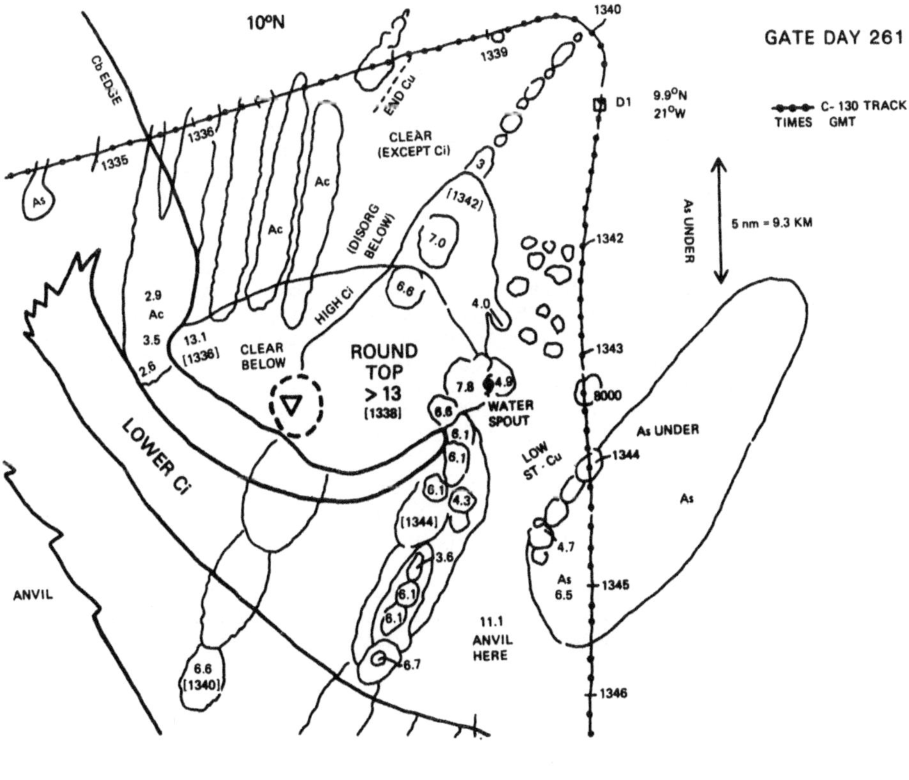

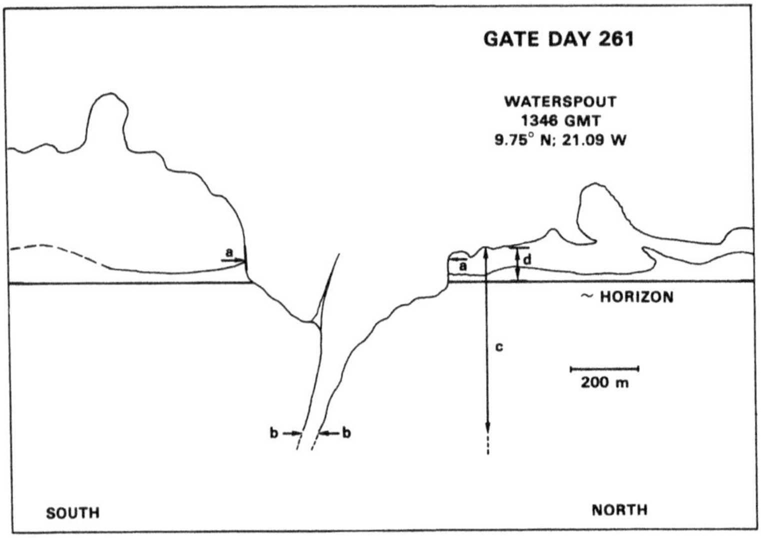

Fig. 2.4. Photogrammetrically constructed outline of waterspout from cine film on Electra aircraft flying southward at 540 m elevation. Waterspout diameter about 50 m

seen directly west of the Electra. The lower congestus towers at the upshear (northerly) end had a "boiling" appearance on the film, suggesting a more rapid rise rate than shown by members of the many congestus lines not associated with a cumulonimbus.

Typical significant features of the GATE cloud environment are 1) the stable dry lower cloud layer, which caused the rapid upward erosion of cumulus bases, and 2) the strong northerly wind shear between 0.8 and 4 km elevation (magnitude about 2.8×10^{-3} s^{-1}) which was previously shown to be essential for the mesovortex existence and orientation (SIMPSON et al. 1981).

Statistics on GATE cumulus updrafts (LEMONE and ZIPSER, 1980) reveal a sufficient condition for the rarity of waterspouts. Only 10 percent of the GATE updraft core sample exceeded 5 m s^{-1}, while Florida samples show updrafts more than a factor of two greater in the same percentile. Day 261 cumuli were active for GATE; both observations and models showed cores of about 6 m s^{-1} in the active stage of both typical congestus and cumulonimbus. Therefore it is likely that the flow force is ordinarily too weak. The gross flow force F is defined (MORTON, 1969) as $\int_S (\rho w^2 + p)\, dS$ where w is updraft speed, p is dynamic pressure and S is any surface on a circumscribing circuit. A buoyant term must also be incorporated for warm flows.

A main hypothesis is that the alteration of the ambient thermodynamic sounding by the cumulonimbus in Fig. 2.3 shown by the x's in Fig. 2.2 enabled the nearby congestus to develop an updraft strong enough to provide the necessary funnel-sustaining flow force, although this condition is not regarded as sufficient.

3 Model Cloud in the Modified Environment

The modified version of SCHLESINGER'S 3-dimensional cloud model was run with the x'ed sounding in Fig. 2.2 and the same initial cylindrical perturbation used in the other congestus cases of SIMPSON et al. (1981). Figs. 3.1 and 3.2 show profiles of the invigorated model cloud. The updraft peak ($\sim$10 m s^{-1}) is stronger than that of either the congesti or cumulonimbi modelled with the unmodified sounding (SIMPSON and VAN HELVOIRT, 1980). The model tower top ascends at about 4 m s^{-1} in agreement with observed relationships.

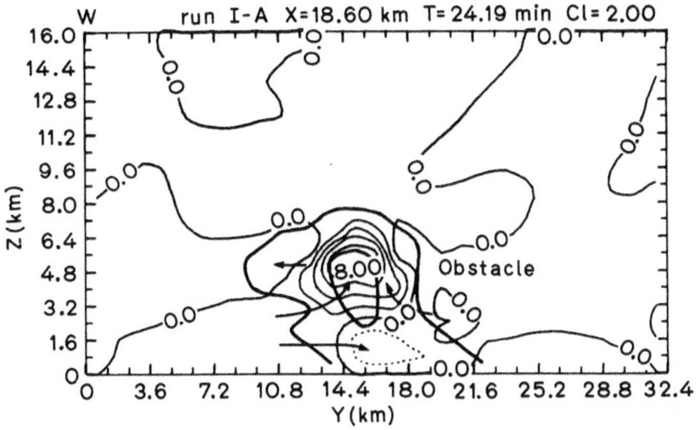

Fig. 3.1. Vertical profile through model waterspout-bearing congestus cloud in actively growing stage. Y-axis points north. Vertical velocities in m s^{-1}; each contour 2 m s^{-1}. Dashed line denotes downdraft. Heavy solid lines liquid water content, inner contour 3 g m^{-3}. Arrows indicate airflow relative to the cloud

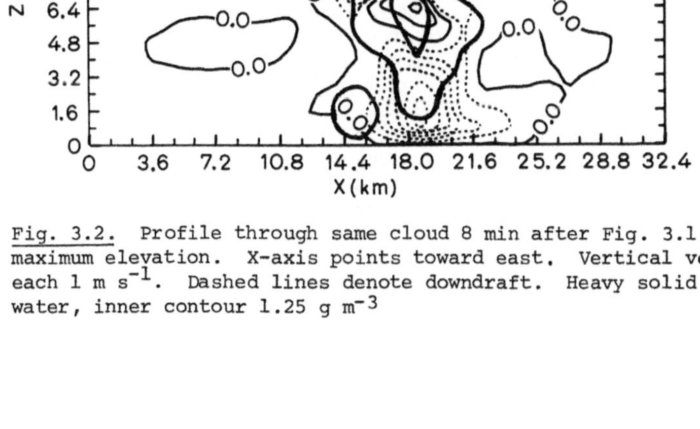

Fig. 3.2. Profile through same cloud 8 min after Fig. 3.1. Cloud top at maximum elevation. X-axis points toward east. Vertical velocity contours each 1 m s^{-1}. Dashed lines denote downdraft. Heavy solid lines liquid water, inner contour 1.25 g m^{-3}

In Fig. 3.1, if updraft velocity contours had been drawn at 25 cm s^{-1} instead of 2 m s^{-1}, the updraft extension to cloud base on the southeast side (rear, downshear flank) would have been apparent. Examination of the flow field relative to the cloud shows that its upper portions act as an obstacle toward relative flow from the north and east, while near cloud base the cloud acts as an obstacle to relative flow from southwest. Air is flowing into the downdraft (from the southwest) around the 2 km level, while it is entering the updraft (from both north and south) in the 3-4 km layer. By the time of Fig. 3.2 (8 min later) the updraft core has weakened to about 3 m s^{-1} and no longer has a downward connection to cloud base in the model. The lower portion of the model cloud is dominated by downdraft, with a core speed of 6 m s^{-1} at cloud base, which has eroded upward to about 1.2 km.

In the model cloud, precipitation reaching the surface begins at 20 minutes and terminates by 32 minutes. Rainfall is entirely confined to the northern half of the cloud and has shrunk to a tiny area at y=18 km by 24 minutes. On the aircraft film, no rain is seen falling from the base of the waterspout-producing cloud, although there are showers from the clouds to the south of it.

While the core updraft is only increased by one-third owing to the sounding modification, the flow force term containing the updraft is increased by 78 percent since it depends on the velocity squared.

The other postulated ingredients for waterspouts are examined next, before estimating the funnel time relative to the model cloud.

4 Cumulus Scale Rotation Related to Cloud Properties

Fig. 4.1 shows the vertical components of vorticity in the model cloud, as a function of time, at the level of strongest vorticities, namely 3.6 km. The peak value of 80 x 10^{-4} s^{-1} is in the range of mid latitude mesocyclones and two-thirds of the peak obtained by SCHLESINGER (1978) applying the model to mid-western severe storm conditions. A horizontal slice through the model cloud at 24 min (Fig. 4.2) shows the vortex pair at 3.6 km. All flow vectors are relative to the cloud motion, defined as the horizontal velocity of the center of gravity of the liquid water. The updraft core is split at this level with meso-anticyclone at the edge of the eastern (stronger) core and the mesocyclone collocated with the western (weaker) core. The meso-anticyclone tilts toward the south with height, but is continuous from cloud base to 6.8 km; the mesocyclone is continuous from 2 - 6.8 km, leaning more strongly to southeast. Both vorticity centers are within or at the edge of updraft from 2 - 6.8 km and one grid interval (1.2 km) or less away from updraft at cloud base. This is important because the funnel must be able to connect to the flow force (the rapidly rising tower). The vortex pair is strongest and most vertical in the layer 2 - 5.2 km, which is the layer of strong vertical windshear. It should be emphasized that the level of organization of vorticity in the cloud vortices reaches a maximum near the top of the shear layer extending from 0.8 to 4 km and then

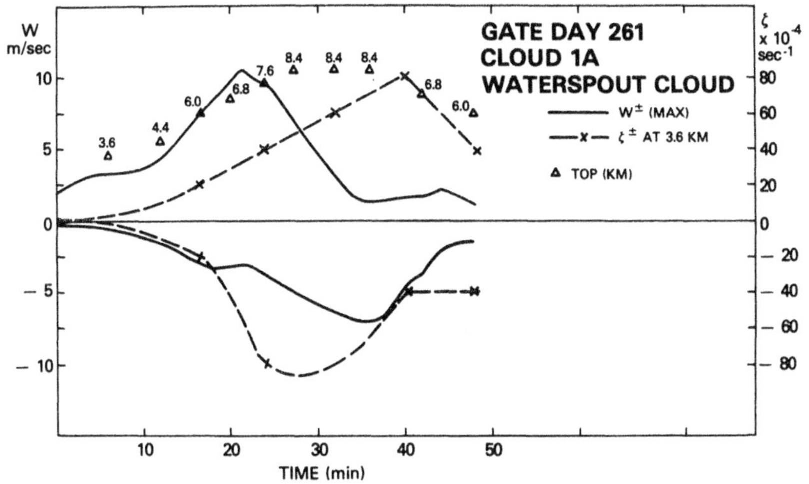

Fig. 4.1. Model cloud parameters as functions of time. Peak vertical velocities solid lines. Peak vorticities at the 3.6 km elevation (x 10^{-4} s^{-1}) dashed lines. Cloud top in km denoted by triangle

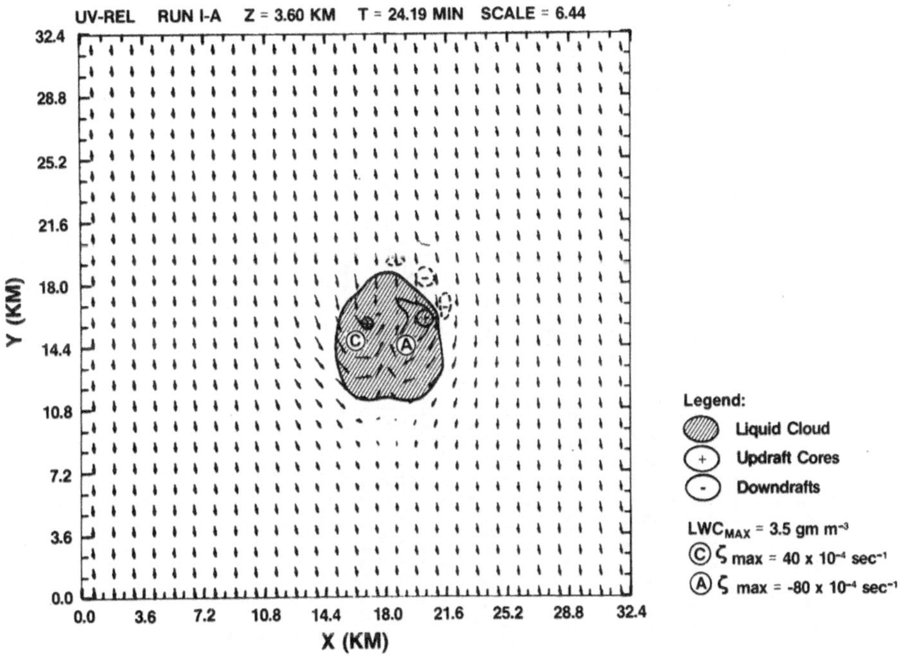

Fig. 4.2. Cloud-relative airflow at the 3.6 km elevation in active growth stage at 24 min. Longest arrow denotes relative velocity 6.44 m s^{-1}. C denotes center cyclonic mesovortex, A denotes center anticyclonic mesovortex. Plus signs show cores of split updraft (w>4 m s^{-1}); dashed lines containing minus signs denote downdrafts greater than 1 m s^{-1}. Hatched area liquid cloud

falls again as a consequence of diffusion without strong organization. The strong vortex layer also happens to coincide with the region of upward increasing updraft.

A horizontal slice through the cloud at 3.6 km, 8 minutes later than Fig. 4.2 (not reproduced) shows the cyclonic vorticity is stronger and the anticyclonic vorticity is about the same as previously, but the mesocyclone is within strong downdraft, while the meso-anticyclone is at the eastern edge of the downdraft, which now fills the model cloud up to 5 km (Fig. 3.2). Both vortices have little tilt at this time. The meso-anticyclone is in updraft at 5.2 km. The mesocyclone is within updraft at 6.8 km, but within downdraft at levels below.

At this time (32 min), the cloud tower has reached its maximum elevation, but has not yet started to descend. GOLDEN (1974) has observed waterspout funnels for a few minutes within obvious (precipitating) downdrafts. The mesocyclone peaks later and keeps extending further downward until after 40 minutes model time (Fig. 4.3). At the time of Fig. 4.3 (41 min) the cloud tower has just started its descent as the cyclonic vorticity peaks (Fig. 4.1). The mesocyclone is essentially vertical and is within weak updraft (1-2 m sec^{-1}) from cloud base level up to 5.2 km.

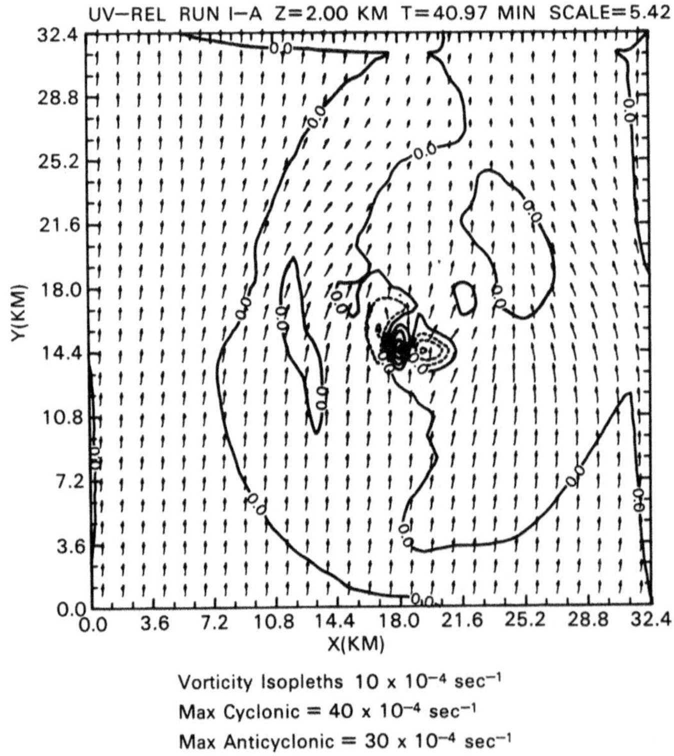

Fig. 4.3. Vorticity isopleths superimposed on cloud-relative airflow at 2 km elevation at 40.97 min model time. Longest vector is 5.42 m s^{-1}

A detailed examination of the terms in the equation for the vertical component of vorticity shows that the vortex pair is initiated by the tilting term. The vortices are intensified by the stretching or convergence term and advected vertically, upward at first and later down to cloud base as downdraft dominates the cloud increasingly after 32 minutes of model time. The downdraft also acts to intensify the vorticity by stretching.

Two factors are seen opposing a deeper, stronger, and less slanted vortex pair. The first is the rapid change in shear vector direction with height at 4.8 km, preventing the tilting term from helping the vortex pair develop at those levels. Second, the stable lower cloud layer typical of the GATE area causes the early appearance of downdraft in the lower cloud portions in both modelled and observed clouds. This causes the tilting term to oppose the vortex at low levels after a short time. In the Florida Keys, the lower cloud layer is more unstable, perhaps helping to account for more common and more intense waterspouts under similar or weaker windshear conditions. In the midwest, a very unstable lower cloud layer is coupled with shear, which only slowly veers in middle and upper levels providing the very strong mesocyclones found by Doppler radar.

5 Hypothesized Factors Related to Funnel Production

Florida waterspouts have measured circulation velocities of 4-35 m s^{-1} with diameters ranging from 7 - 150 m (GOLDEN, 1974; SCHWIESOW, 1981). Funnel vorticities are in the range 10^{-1} s^{-1} or greater. Highest vorticities found in midwestern mesocyclones are of the order 10^{-2} s^{-1} while the maximum in the model cloud was slightly less. The funnel on GATE day 261 measured about 50 m across so that even if its peripheral velocities were only 4-5 m s^{-1}, an order of magnitude larger vertical vorticity is still required than observed or modelled in cumulus-scale mesovortices.

There are no adequate mechanisms for spinning up a vertical component of vorticity of the order 10^{-1} sec^{-1} in the lifetime of a cumulus congestus cloud. We must, therefore, hypothesize that strong horizontal vortex tubes in the lowest boundary layer are suddenly tilted into the vertical. In neutral stability conditions over the oceans the logarithmic wind profile prevails, with shear vorticity (in the horizontal) of 10^{-1} s^{-1} or greater below about 10 m elevation and 1 s^{-1} or greater below about 1 m elevation. Adequate vorticity is clearly available.

On day 261, the boundary layer vorticity is associated with southerly flow increasing upward so that the vorticity vector points roughly westward.

The proposed tilting mechanism is the gust front from the neighboring cumulonimbus. Fig. 5.1 shows the lowest model level of the cumulonimbus, discussed in an earlier paper (SIMPSON and VAN HELVOIRT, 1980). The waterspout on the diagram is located by the observed distance between the waterspout and the cumulonimbus tower top. This estimates the waterspout location in the convergence zone and θ_E front associated with the model gust front.

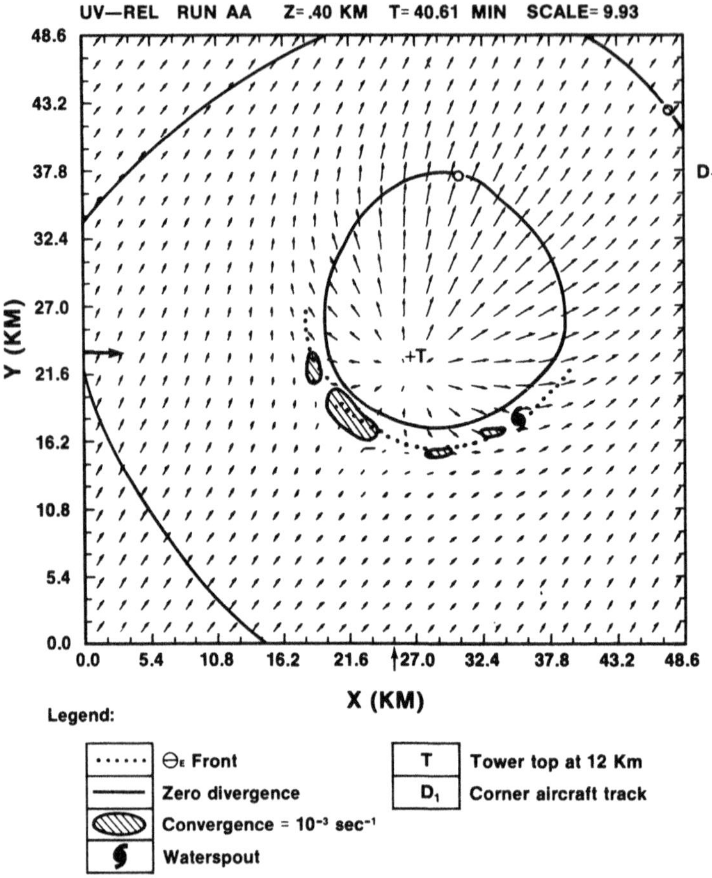

Fig. 5.1. Conditions below model cumulonimbus AA shown in Fig. 2.3 (further analyses in SIMPSON and VAN HELVOIRT, 1980) at mature stage. X-axis points east; Y-axis points north. Cloud relative airflow shown by vectors, largest 9.93 m s^{-1}. T denotes where observed cumulonimbus tower was located relative to observed waterspout location

The gust front is a wedge of cold dense air moving very slowly toward the waterspout location from about northwest. It can be envisaged as tipping bundles of the westward pointing vortex tubes upward into a vertical orientation so that the circulation around a tilted bundle is cyclonic (Fig. 5.2). GATE gust fronts on day 261 have been discussed by WARNER et al. (1981). Many (but not all) Florida waterspouts are observed on the upshear end of a congestus line flanking a cumulonimbus as in this case (GOLDEN , 1974).

The combination of the direction of boundary layer vorticity with the gust front location and motion appears to decide in favor of cyclonic rotation for the waterspout analyzed here.

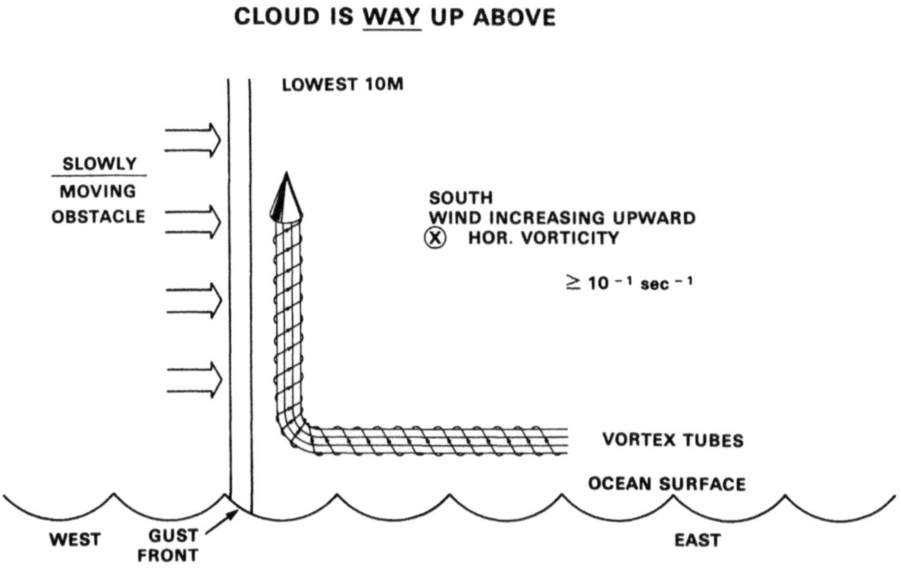

Fig. 5.2. Hypothesized upward tilting of vortex tubes in lowest boundary layer by gust front associated with cumulonimbus AA. Low-level vorticity associated with rapid upward increase of southerly wind just above ocean surface. Resulting cyclonic whirling postulated to connect with mesocyclone at cloud base producing visible funnel by rapid condensation

Other evidence favoring this hypothesis is that the waterspout-bearing cloud had "topped out" nearly 3 minutes prior to sighting the waterspout. Midwestern tornadoes are also found usually when the cloud tower is descending and when the mesocyclone has reached down to cloud base.

The above reasoning leads to estimating the time of the waterspout occurrence somewhere between 30-40 minutes cloud model time (see Fig. 4.1). An apparent problem is the weakened flow force and deep downdraft layer in the model cloud at 32 min; by 40 min the vertical velocity component of the flow is weaker still in the model cloud, although the mesocyclone is nearly vertical and in weak updraft up to 5.2 km. The difficulty here could be with the model, which was initiated with a non-repeating perturbation. Frequently, in real clouds a second thermal follows the first, or one may originate at mid-levels. It is also possible that the congestus just east of the waterspout rising through 4.9 km in Fig. 2.3 was providing some of the upward flow force at this time.

If the event sequence postulated here is on the right track, it becomes clearer why the dark spot is seen on the water first below the rotating cloud and the funnel is the last part of the system to appear. If the strong vorticity in the lowest 10 m or so is critical, intense whirling at the surface should appear before the funnel.

6 Concluding Remarks, Remaining Vital Questions and Recommendations for Further Work

It is concluded that most cumulus clouds in a shearing windfield will develop mesovortex pairs, hence areas of rotation, and that existing three-dimensional cumulus models are adequate to provide useful frameworks for synthesizing existing observations and for suggesting future observations. Improvements in the model ranging from microphysics, finer space resolution and different methods of initiation can make the simulations increasingly useful in examining important questions related to mesovortex evolution.

Many factors must coincide in space and time for an adequate mesovortex to exist at cloud base, connected upward to a convective flow force and downward to a suddenly tilted bundle of vortex tubes in the lower boundary layer. For the waterspout on GATE day 261, it seems that the nearby cumulonimbus was essential in two ways, first to strengthen the flow force and the mesocyclone via intensified updraft, and second to provide a steep density current (gust front) at just the time and place where the mesocyclone had reached cloud base. With the short active lifetime of GATE clouds relative to those in the Florida area, owing to the more stable lower cloud layer, it is not surprising that the "window" for coincidence of the necessary factors is short, and so waterspouts are quite rare in the GATE area.

A cloud which has a longer active lifetime of its updraft could use its own gust front to "scoop up" surface vorticity. This must be the case with those isolated waterspout-bearing cumuli off the Florida Keys; evidence (LEMON and DOSWELL, 1979) is similar for tornadic storms where updrafts and downdrafts may coexist side-by-side.

Major questions not addressed in this paper are 1) how the "scooped up" vortex tube bundle suddenly connects upward with the mesocyclone (and thence the flow force) which must involve pressure forces, and 2) what factors determine the scale of the funnel. These problems are probably better addressed in laboratory experiments, theoretical analyses and model simulations (e.g., SMITH and LESLIE, 1979).

Further observations are, however, required of waterspout-bearing cumulus clouds. The highest frequency in an accessible area is undoubtedly the Florida Keys where GOLDEN'S and SCHWIESOW'S find work should be followed up with the improved measuring systems now available. In particular, the in-cloud motion field needs to be measured as a function of time to relate updraft/downdraft structure to rotation. Fortunately, maritime cumuli in the Floirda area usually have plentiful precipitation particles inside them by the time they have grown to 3-4 km elevation so that Doppler radar could be used, which would be particularly valuable if airborne and so enabled to "chase" clouds.

A lower-flying aircraft carrying the Doppler lidar effectively used by SCHWIESOW (1981) could relate the waterspout properties to the associated cloud mesovortices above. Ambient wind and thermodynamic profiles might be obtained by aircraft dropwindsondes or boundary layer profiling systems as in the GATE. If resources were very limited, winds obtained by a network of pilot balloon releases together with time-lapse camera sites could provide the minimum additional essential information.

7 Acknowledgements

This research was done in collaboration with several colleagues, in particular B. R. Morton who will be co-author on the next stage of the paper.

8 References

BLECHMAN, J. B., 1981: Vortex generation in a numerical thunderstorm model. Mon. Wea. Rev., 109, in press

GOLDEN, J. H., 1974: Scale interaction implications for the waterspout life cycle. II. J. Appl. Meteor., 13, 696-709

HEYMSFIELD, G., 1981: Evolution of downdrafts and rotation in an Illinois thunderstorm. Mon. Wea. Rev., 109, in press

LEMON, L. R. and C. A. DOSWELL, 1979: Severe thunderstorm evolution and mesocyclone structure as related to tornadogenesis. Mon. Wea. Rev., 107, 1184-1197

LEMONE, M. A. and E. J. ZIPSER, 1980: Cumulonimbus vertical velocity events in GATE. Part I: Diameter, intensity, and mass flux. J. Atmos. Sci., 37, 2444-2457

MORTON, B. R., 1969: The strength of vortex and swirling core flows. J. Fluid Mech., 38, 315-333

SCHLESINGER, R. E., 1978: A three-dimensional model of an isolated thunderstorm. Part I. Comparative experiments for variable ambient wind shear. J. Atmos. Sci., 35, 690-713

SCHWIESOW, R. L., 1981: Waterspout velocity measurements by airborne Doppler lidar. J. Appl. Meteor., 20, in press

SIMPSON, J. and G. VAN HELVOIRT, 1980: GATE cloud-subcloud interactions using a three-dimensional cumulus model. Cont. to Atmos. Phys., 53, 106-134

SIMPSON, J., G. VAN HELVOIRT, and M. MCCUMBER, 1981: Three dimensional simulation of cumulus congestus clouds on GATE day 261. J. Atmos. Sci., 38, under revision

SMITH, R. K. and L. M. LESLIE, 1979: A numerical study of tornadogenesis in a rotating thunderstorm. Quart. J. Roy. Meteor. Soc., 105, 107-127

WARNER, C., J. SIMPSON, D. W. MARTIN, D. SUCHMAN, F. R. MOSHER, and R. F. REINKING, 1979: Shallow convection on day 261 of GATE: Mesoscale arcs. Mon. Wea. Rev., 107, 1617-1635

WARNER, C., J. SIMPSON, G. VAN HELVOIRT, D. W. MARTIN, D. SUCHMAN, and G. L. AUSTIN, 1980: Deep convection on day 261 of GATE. Mon. Wea. Rev., 108, 169-194

Observational and Theoretical Aspects of Tornadogenesis

R. P. Davies-Jones

1 Introduction

The tornado is the most intense atmospheric vortex, and the most mysterious. Many questions still remain about its formation. The field of tornado dynamics has been reviewed recently by DAVIES-JONES (1981). In theoretical work it is generally assumed that the tornado is a vertical vortex, but it is often inclined appreciably. The direction and degree of tilt offers insights, hitherto unexplored into the character of the velocity field in which the tornado is embedded.

This paper departs from previous research in that it deals with the vector form of the vorticity equation, written in terms of cartesian tensors, instead of the vertical component form. The traditional vertical vorticity equation hides the "physics" of the tilting term because of cancelled terms, and this has led to misconceptions. It is shown that the convergence and tilting terms are associated with the diagonal and nondiagonal elements, respectively, of the rate of strain tensor, and that both stretch and reorient vortex tubes. Shearing deformation plays the same role in the tilting term, as convergence in the convergence term. In general, the intensifying vortex is tilted toward a principal axis (the one of maximum stretching) of the rate of strain tensor.

Although both the solenoidal and eddy diffusion terms have been proposed as "generators" of tornadic vorticity (LEMON and DOSWELL, 1979; STARR, 1974), it is generally believed that the tornado acquires its tremendous vorticity through amplification of pre-existing vorticity by the combined actions of the convergence and tilting terms. The solenoidal term is only of secondary importance in mesocyclones[1] (2 to 5 km radius) (RAY, 1976) and, in the immediate vicinity of the tornado, the intense deformation due to differential rotation should inhibit the formation of strong solenoids. As explained below, the observed θ_W-gradients across tornado cyclones may be a consequence of the velocity field, rather than a direct cause of it. The eddy diffusion term is also not believed to be an important mechanism for concentrating vorticity in the tornado (LEWELLEN, 1976). Thus, the solenoidal and diffusion terms have been omitted in this paper.

The approach taken here is to treat the kinematics of vorticity generation in much the same way as PETTERSSEN (1956) approached the

[1]Solenoids generate toroidal vorticity near the top of rising thermals but this source of vorticity should have little direct effect on low-level motions.

Topics in Atmospheric and Oceanographic Sciences
Intense Atmospheric Vortices, ed. by Bengtsson/Lighthill

kinematics of frontogenesis. Because vorticity and thermodynamic properties are strongly linked (DUTTON, 1976, p. 383), the frontogenesis and "vortexgenesis" problems have both been solved as initial value problems under the restriction of linear fields. (The frontogenetic solutions are not presented in this paper because of space limitations.) The results are used to explain certain observed facets of tornadoes.

2 Mathematical Derivations

In a cartesian coordinate system rotating with angular velocity $\boldsymbol{\Omega}$, the vector vorticity equation is

$$\frac{d\boldsymbol{\omega}_a}{dt} = (\boldsymbol{\omega}_a\cdot\nabla)\,\mathbf{v} - (\nabla\cdot\mathbf{v})\,\omega_a - \nabla\alpha\times\nabla p + \nu\nabla^2\omega \quad , \qquad [1]$$

where the absolute vorticity, $\omega_a=\omega+2\boldsymbol{\Omega}$, the relative vorticity $\boldsymbol{\omega}=\nabla\times\mathbf{v}$ and the other symbols have their conventional meanings. On the right hand side, the first term, when expanded out in component form, contains a part which cancels with a part of the second term. The remainders of the first and second terms are the twisting (or tilting) and convergence terms, while the third and fourth terms on the right are the solenoidal and diffusion terms, respectively.

Equation [1] can be combined with the continuity equation,

$$\frac{d\alpha}{dt} = \alpha\nabla\cdot\mathbf{v} \qquad [2]$$

to obtain the Beltrami vorticity equation (DUTTON, 1976, p. 382)

$$\frac{d}{dt}(\alpha\boldsymbol{\omega}_a) = (\alpha\boldsymbol{\omega}_a\cdot\nabla)\,\mathbf{v} \qquad [3]$$

where the twisting and convergence terms have now been combined into one, and for reasons cited above, the solenoidal and diffusion terms have been omitted.

In the absence of diabatic heating and diffusion, wet-bulb potential temperature is conservative, i.e.,

$$\frac{d\theta_w}{dt} = 0 \quad . \qquad [4]$$

Applying the gradient operator to [4] yields

$$0 = \nabla\left(\frac{d\theta_w}{dt}\right) = \frac{d}{dt}(\nabla\theta_w) + (\nabla\theta_w\cdot\nabla)\,\mathbf{v} + \nabla\theta_w\times\omega \quad . \qquad [5]$$

From [3] and [5], it follows that

$$\frac{d}{dt}(\alpha\omega_a\cdot\nabla\theta_w) = 0 \quad , \qquad [6]$$

which is a statement of the conservation of potential vorticity (DUTTON, 1976, p. 383).

Further insight into Eq. [3] and [4] is gained by rewriting them in terms of Cartesian tensors. We define a right-handed orthogonal coordinate system with the x_1- and x_2-axes horizontal (their exact orientation is specified case by case) and the x_3-axis pointing upwards. The components of velocity, absolute vorticity and relative vorticity are (v_1,v_2,v_3), $(\omega_1,\omega_2,\omega_3)$ and $(\zeta_1,\zeta_2,\zeta_3)$, respectively, where $\omega_k=\zeta_k+f_k$ and $2\Omega=(f_1,f_2,f_3)$. After some algebraic manipulations, [3] and [5] can be written in tensor notation as

$$\frac{d}{dt}(\alpha\omega_k) = (\alpha\omega_j)D_{jk} \quad , \qquad [7]$$

$$\frac{d}{dt}\left(\frac{\partial\theta_w}{\partial x_k}\right) = -\frac{\partial\theta_w}{\partial x_j}\,G_{jk} \quad , \qquad [8]$$

where the indices run over 1,2,3 and the Einstein summation convention (that indices appearing twice in a term imply summation over that index) is in effect. The quantity $D_{jk} = (\frac{\partial v_j}{\partial x_k} + \frac{\partial v_k}{\partial x_j})/2$ is the rate of strain tensor, and the vorticity vector $\zeta_k = e_{ijk}\partial v_j/\partial x_i$ in terms of the alternator tensor, e_{ijk}. The velocity gradient tensor, $G_{jk} = \partial v_j/\partial x_k$, appearing in [8] can be written as the sum of the (symmetric) rate of strain tensor and the (anti-symmetric) rotation tensor

$$\Omega_{jk} = \frac{1}{2}\left(\frac{\partial v_j}{\partial x_k} - \frac{\partial v_k}{\partial x_j}\right) = \frac{1}{2}\begin{bmatrix} 0 & -\zeta_3 & \zeta_2 \\ \zeta_3 & 0 & -\zeta_1 \\ -\zeta_2 & \zeta_1 & 0 \end{bmatrix} \qquad [9]$$

in matrix representation. [Note that [7] and [8] and many of the following equations can be thought of as matrix equations with vectors represented as row (column) vectors for left (right) multiplication.]

Equation [7], which appears in TENNEKES and LUMLEY (1972, p. 82) and other texts, must be examined in component form to reveal how it contains implicitly the convergence and tilting terms. The vertical component of [7] is

$$\frac{1}{\alpha}\frac{d}{dt}(\alpha\omega_3) = \frac{1}{2}\left(\frac{\partial v_3}{\partial x_2} - \frac{\partial v_2}{\partial x_3}\right)\left(\frac{\partial v_1}{\partial x_3} + \frac{\partial v_3}{\partial x_1}\right) + \frac{1}{2}\left(\frac{\partial v_1}{\partial x_3} - \frac{\partial v_3}{\partial x_1}\right)\left(\frac{\partial v_2}{\partial x_3} + \frac{\partial v_3}{\partial x_2}\right)$$
$$+ \omega_3\frac{\partial v_3}{\partial x_3} = \frac{\partial v_1}{\partial x_3}\frac{\partial v_3}{\partial x_2} - \frac{\partial v_2}{\partial x_3}\frac{\partial v_3}{\partial x_1} + \omega_3\frac{\partial v_3}{\partial x_3} \quad , \qquad [10]$$

so the off-diagonal elements of D_{jk} generate the twisting term, while the diagonal elements, combined with $-\omega_k\frac{d\ell n\alpha}{dt}$ from the left side, give rise to the convergence term. Note that the twisting term in [10] represents the generation of vertical vorticity in regions characterized by both shearing deformation and non-zero initial horizontal vorticity in the (x_1,x_3) plane, and in similar regions in the (x_2,x_3) plane. By considering the case where $\omega_1=0$, ω_2 and ω_3 are positive initially and D_{23} is also positive, it is clear from the symmetry of D_{jk} that the twisting term also stretches vortex tubes as well as reorientating them, since the twisting term acts to increase both ω_2 and ω_3. The intensification of vorticity due to stretching, obtained by operating on both sides of [7] with the scalar product of ω, is described by

$$\frac{1}{\alpha|\omega|}\frac{d}{dt}(\alpha\omega_k\omega_k/2) = \frac{\omega_j D_{jk}\omega_k}{|\omega|} \quad . \qquad [11]$$

In severe storm research, there has been a recent trend toward identifying the convergence term as "the stretching term", but this practice is obviously misleading.

Generalizing PETTERSSEN's (1956) approach, a three-dimensional "frontogenesis" function can be defined for any conservative quantity, such as θ_w, by

$$F = \frac{d|\nabla\theta_w|}{dt} = \frac{\nabla\theta_w}{|\nabla\theta_w|} \cdot \frac{d}{dt}(\nabla\theta_w) \qquad [12]$$

(Miller, 1948). From [12] and [8]

$$F = -\frac{1}{|\nabla\theta_w|}\frac{\partial\theta_w}{\partial x_j} G_{jk} \frac{\partial\theta_w}{\partial x_k}$$

$$= -\frac{1}{|\nabla\theta_w|}\frac{\partial\theta_w}{\partial x_j} D_{jk} \frac{\partial\theta_w}{\partial x_k} \quad , \qquad [13]$$

since the antisymmetric part, Ω_{jk}, of G_{jk} makes no contribution to F. Equations [11] and [13] are strikingly similar; vorticity amplification and the intensity of frontogenesis are controlled by the same tensor, the rate of strain tensor. There is also a resemblance between Eqs. [7] and [8], which yield the material rates of change of individual vorticity and temperature gradient components, respectively. This likeness becomes apparent by noting that

$$(\alpha\omega_j)\Omega_{jk} = 0 \quad , \qquad [14]$$

so that adding [7] and [14] together yields

$$\frac{d}{dt}(\alpha\omega_k) = (\alpha\omega_j)G_{jk} \quad . \qquad [15]$$

Equation [15] has an exactly similar form to [8] apart from the minus sign, which is necessary for the conservation of potential vorticity. The main difference between [7] and [8], apart from sign, is the effect of the rotation tensor on the θ_w-field and its zero influence on the vorticity field. The rotation tensor causes the isentropes (but not the vortex lines) to seek a different direction from that of the principal axis of D_{jk} (PETTERSSEN, 1956, p. 204), and so affects the frontogenetic rate (but not the vorticity amplification rate) at times subsequent to the initial time (t=0). At t=0, F is dependent only on the observed temperature field and the irrotational part of the wind field (see Eq. [13]).

In the neighborhood of a point, the velocity field can be approximated by a linear field of motion, i.e.,

$$v_i = (v_i)_o + \frac{\partial v_i}{\partial x_j} dx_j \quad . \qquad [16]$$

where $(v_i)_o$ is the velocity at the point. PETTERSSEN (1956) used a two-dimensional linear velocity field to study the kinematics of frontogenesis. In two-dimensions the linear field of motion can be resolved into constant translation, divergence, vorticity, shearing deformation and stretching deformation (Fig. 1), which are mutually independent because there are four elements of G_{jk}. This approach can be generalized to investigate "vortexgenesis" in three dimensions. The local velocity field is completely specified by translation and eight quantities which are shear dependent, namely the three vorticity components, ζ_1, ζ_2 and ζ_3, the three shearing deformations $2d_{12}$, $2d_{23}$, $2d_{12}$, the divergence arising from flow parallel to the x_3-axis, d_{33}, and the stretching deformation in the x_1, x_2 plane, $\varepsilon = d_{11} - d_{22}$. In the original coordinate system defined earlier, d_{33} is the

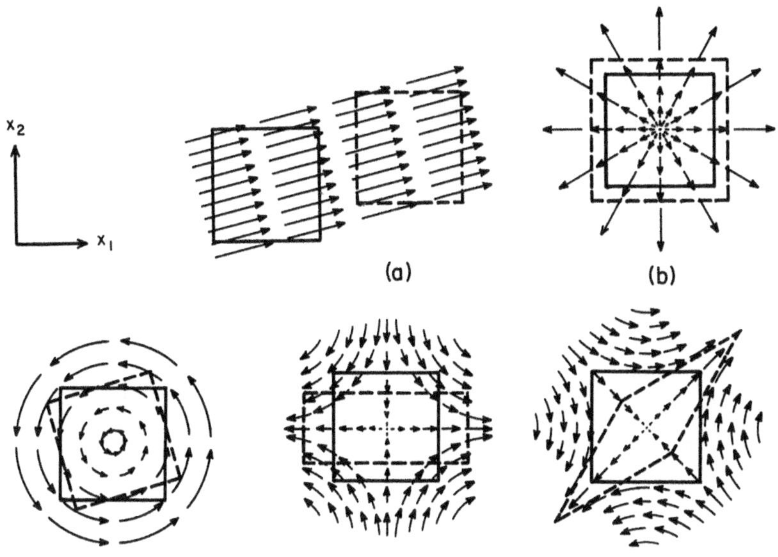

Figure 1. Pure types of planar motion at a point. (a) Translation; (b) divergence; (c) vorticity (positive); (d) stretching deformation; (e) shearing deformation. Arrow lengths are proportional to speed. Also shown in each case is the deformation (dashed) of an initial square (solid) of fluid

"vertical divergence" and $\hat{\varepsilon}$ is the horizontal stretching deformation, but for the moment consider that the orientation of the cartesian coordinate axes as unspecified. Under rotation of the coordinate axes, none of the eight quantities are invariant. In fact, the only invariants with obvious physical interpretations are the cubical dilatation and the magnitude of the vorticity vector. The continuity equation ($\nabla\cdot\mathbf{v}=d\ell n\alpha/dt$) is a constraint on the diagonal elements, and so all the components, g_{jk}, of G_{jk} are uniquely determined. Note that the trace of $[G_{jk}]$ is the cubical dilatation, $\frac{d\ell n\alpha}{dt}$. Thus,

$$G_{jk} = \begin{bmatrix} \frac{\delta+\hat{\varepsilon}}{2} & d_{12}-\frac{1}{2}\zeta_3 & d_{13}+\frac{1}{2}\zeta_2 \\ d_{12}+\frac{1}{2}\zeta_3 & \frac{\delta-\hat{\varepsilon}}{2} & d_{23}-\frac{1}{2}\zeta_1 \\ d_{13}-\frac{1}{2}\zeta_2 & d_{23}+\frac{1}{2}\zeta_1 & d_{33} \end{bmatrix} \qquad [17]$$

where $\delta=d_{11}+d_{22}$ is the horizontal divergence. The changes in shape of an initial cube of fluid induced by the various types of motions are shown in Fig. 2.

Linearizing the velocity field reduces the elements of the matrix $[D_{jk}]$ (denoted henceforth by $\mathbf{D}$) to constants when $\mathbf{D}$ is considered to be time independent (as in the present kinematical approach). Since $\mathbf{D}$ is symmetric, its eigenvalues, $\lambda^{(r)}$, are real, and if all the $\lambda^{(r)}$ are distinct, then its eigenvectors, $\begin{bmatrix} n_1^{(r)} \\ n_2^{(r)} \\ n_3^{(r)} \end{bmatrix}$ are mutually orthogonal and **D** is diagonalizable (KOLMAN, 1977). Since eigenvectors are determined only up to scalar multiples, they can be normalized by imposing the condition $n_i^{(r)} n_i^{(r)} = 1$. The normalized eigenvectors of **D** form an orthonormal basis over three-dimensional space, called the principal directions or principal axes of **D** (Fig. 3; DUTTON, 1976, p. 157). Since **D** is diagonalizable, there exists an orthogonal matrix, **P** , such that $\mathbf{P}^T\mathbf{D}\mathbf{P} = \mathbf{D}'$, a diagonal matrix (where $\mathbf{P}^{-1} = \mathbf{P}^T$, the transpose of **P**, defines orthogonality for a square matrix). If e_1, e_2, e_3 denotes the natural orthonormal basis for the original coordinate axes, and e_1', e_2', e_3' are the orthonormal set of basis vectors which define the principal axes, let $\mathbf{u} = \mathbf{P}\mathbf{u}'$ where u_i and u_i' are the ith coordinates of the same arbitrary vector in the two coordinate systems. It is easily shown that $\mathbf{D} = \lambda \mathbf{v}$ implies $\mathbf{D}'\mathbf{v}' = \lambda \mathbf{v}'$ so **D** and **D**´ have the same eigenvalues and characteristic polynomials. The coefficients of λ^2, λ and λ^0, namely the trace of **D**, $(d_{ii}d_{jj} - d_{ij}d_{ji})/2$ and the determinant of **D**, are invariant under rigid rotations of the coordinate axes.[2] Furthermore, since the linear transformation **P** takes the unit vectors in the prime coordinate system to the normalized eigenvectors of **D**, the columns of **P** are the eigenvectors of **D**, i.e.,

$$p_{ij} = n_i^{(j)} \ .$$

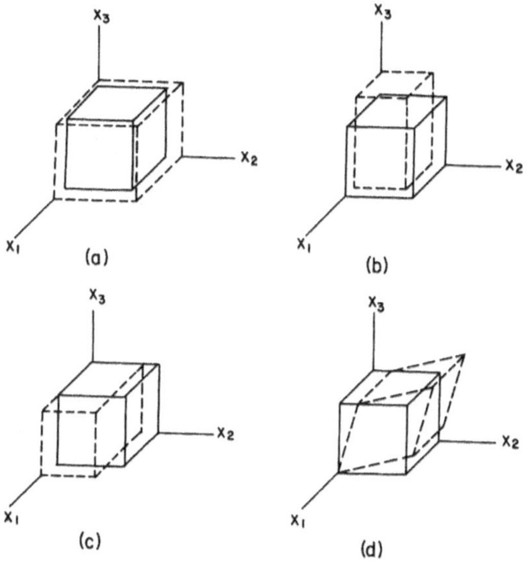

Figure 2. Deformation of a cube due to (a) cubical dilatation, $d\ln\alpha/dt$; (b) horizontal convergence, $-\delta$; (c) horizontal stretching deformation, $\hat{\varepsilon}$; (d) vertical stretching deformation, d_{23}. The effect of vorticity (not shown) is to revolve the cube around the rotation axis as a solid body. If the isentropic surfaces are initially horizontal, then the effects of these flows on the θ_w-surfaces is apparent from the displacements of the top and bottom faces of the cube. The amplification of the three vorticity components is visualized by the changes in length and orientation of the edges. The vertex at the origin is held fixed in all cases

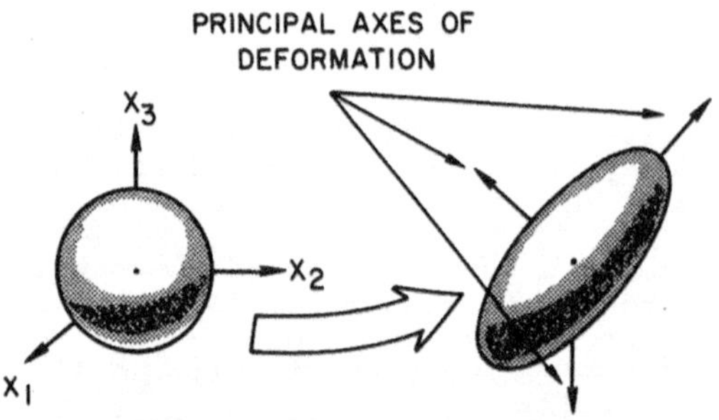

Figure 3. Deformation of a sphere into an ellipsoid in a three-dimensional linear velocity field

[2] The nth invariant is the sum of the nxn principal minors (for n=1,2,3).

Also, the inner product $(e_i, e_j') = p_{ij}$, so p_{ij} (or $n_i^{(j)}$) is simply the direction cosine of the jth eigenvector of **D** with the ith axis of the original coordinate system.

Equation [7] can be rewritten in terms of column vectors by taking the transpose of both sides, yielding

$$\frac{d}{dt}\left(\alpha \begin{bmatrix} \omega_1 \\ \omega_2 \\ \omega_3 \end{bmatrix}\right) = \mathbf{D}\alpha \begin{bmatrix} \omega_1 \\ \omega_2 \\ \omega_3 \end{bmatrix} \qquad [18]$$

Left multiplying both sides by **P** gives

$$\frac{d}{dt}\left(\alpha \begin{bmatrix} \omega_1' \\ \omega_2' \\ \omega_3' \end{bmatrix}\right) = \mathbf{D}'\alpha \begin{bmatrix} \omega_1' \\ \omega_2' \\ \omega_3' \end{bmatrix} \qquad [19]$$

where **D**′ is the diagonal matrix $\mathbf{P}^{-1}\mathbf{D}\mathbf{P}$ with the eigenvalues of **D** as the diagonal elements. Equation [19] represents a system of uncoupled linear differential equations in the Lagrangian formulation, and the solutions are

$$\begin{bmatrix} \omega_1'(t) \\ \omega_2'(t) \\ \omega_3'(t) \end{bmatrix} = \begin{bmatrix} \omega_1'(o)\exp(\lambda^{(1)}t) \\ \omega_2'(o)\exp(\lambda^{(2)}t) \\ \omega_3'(o)\exp(\lambda^{(3)}t) \end{bmatrix} \qquad [20]$$

for distinct $\lambda^{(r)}$. Left multiplying both sides by **P** yields

$$\begin{bmatrix} \omega_1(t) \\ \omega_2(t) \\ \omega_3(t) \end{bmatrix} = \sum_{r=1}^{3} \begin{bmatrix} n_1^{(r)} \\ n_2^{(r)} \\ n_3^{(r)} \end{bmatrix} \omega_r'(o)\exp(\lambda^{(r)}t) \quad . \qquad [21]$$

Solutions to the transpose of [8] can be obtained in exactly the same fashion if the matrix $[-G^T_{jk}]$ has real and distinct eigenvalues. Note that this matrix is symmetric only when the wind field is irrotational.

The fact that **D** can be diagonalized has great physical significance. With a linear velocity field, an appropriate rotation of the coordinate axes causes the tilting terms to vanish completely. Conversely, when the fluid is incompressible (i.e., $d_{11}+d_{22}+d_{33}=0$), the coordinate axes can be rotated so that in the new system the convergence terms vanish (i.e., the new rate of strain tensor has zero diagonal components). The implication here is that the convergence and twisting terms in the vertical vorticity equation should be interpreted physically with great care. The usual analogy that the convergence term amplifies vertical vorticity through stretching of vortex tubes and that the tilting term tips horizontal vorticity into the vertical without stretching is not an accurate picture. Both terms stretch and reorient vortex tubes, and it is shown in Section 3 that the tilting terms by themselves can not tilt initially horizontal vortex tubes past an inclination of 45° to the horizontal. The latter effect is due to the symmetry of **D**; both horizontal and vertical vortex tubes are tilted toward the dilatation axis of the vertical shearing deformation field.

3 Vortexgenesis

To study the kinematics of "vortexgenesis" in a linear velocity field, we hold the irrotational part of the field constant in time and obtain solutions of [7] which describe the time evolution of the rotational component of the field. Obviously, there comes a time when the vorticity becomes sufficiently large to strongly modify the irrotational flow component in reality, but this approach is enlightening and certainly no more limited than ROTUNNO's (1981) work, in which he solved the vertical vorticity equation for ζ, given a prescribed vertical velocity field and a linearized tilting term.

For two-dimensional (horizontal), incompressible, inviscid flow, Eq. [7] states that $d\omega_3/dt=0$, i.e., that the vertical vorticity of each parcel is conserved. Thus, vertical motions are prerequisite for tornado formation, a fact which has been recognized for decades.

The effect of cubical dilatation on a parcel's vorticity is seen readily from [7] by setting the right side equal to zero. Then $\alpha\omega_k$ is constant, so that an expanding (contracting) air parcel loses (gains) vorticity in accord with the conservation of angular momentum. A parcel descending from 150 mb to 900 mb only experiences about a five-fold increase in vorticity due to this effect.

The BURGERS (1948)-ROTT (1958) and KUO (1966) one-cell vortex steady state solutions of the equations of motion are well known (see DAVIES-JONES and KESSLER, 1974, for discussion). Because the horizontal flow is height-independent, the meridional and swirling flows are uncoupled, and the gain of vorticity of a parcel as it moves inward toward the rotation axis is described approximately by Eq. [7] with

$$\mathbf{D} = \begin{bmatrix} \delta/2 & 0 & 0 \\ 0 & \delta/2 & 0 \\ 0 & 0 & -\delta \end{bmatrix}, \quad (\alpha\dot{\boldsymbol{\omega}}_a)(0) = \begin{bmatrix} 0 \\ 0 \\ (\alpha\omega_3)(0) \end{bmatrix} . \qquad [22]$$

The loss of vorticity due to outward diffusion, which locally balances the vortex amplification due to convergence (stretching), is not represented in [22], nor is the horizontally varying horizontal shearing deformation, which is present in the steady state solutions. However, the latter has no effect on the solution of [7] since the tilting terms are zero, and ω_1, ω_2 remain zero for all time. Thus, the rotational and irrotational flow components are essentially uncoupled and **D** is, for all practical purposes, time invariant for the BURGERS-ROTT and KUO vortices.

The solution to [22] is

$$\omega_1(t) = \omega_2(t) = 0, \quad (\alpha\omega_3)(t) = (\alpha\omega_3)(0)\ \exp\ (-\delta t) \quad . \qquad [23]$$

Thus, in the absence of diffusion, the vorticity of a parcel increases exponentially with a growth rate equal to the horizontal convergence of the velocity field in which it is embedded. By Helmholtz's laws, the vorticity flux along a vortex tube is constant along its length, and in an inviscid, barotropic flow vortex tubes are "frozen" to the fluid. Therefore, the vortex tube which defines the tornado core (radius r_c) at time $t=\tau=-\ln(\omega_3(\tau)/\omega_3(0))/\delta$ (the time required for vorticity amplification to tornadic values from the initial background vorticity) has an initial radius $r_o=r_c(\omega_3(\tau)/\omega_3(0))^{1/2}$. For $r_c=50$ m, $\omega_3(\tau)=4$ s^{-1}, $\delta=-2\times10^{-3}$ s^{-1} and $\omega_3(0)=10^{-4}$ s^{-1}, corresponding to a powerful tornado forming in low-level convergent flow from synoptic-scale background vorticity, τ is 88 min and r_o is 10 km (neglecting changes in α). Note that when $\delta<0$ (as above) horizontal vorticity, if nonzero initially, would decay exponentially with time.

Next, we isolate the effects of non-horizontal twisting terms by considering

$$\mathbf{D} = \begin{bmatrix} 0 & 0 & d_{13} \\ 0 & 0 & d_{23} \\ d_{13} & d_{23} & 0 \end{bmatrix} \qquad [24]$$

with vertical vorticity initially zero (i.e., $\omega_3(0)=0$). By rotation of the coordinate system about a vertical axis, d_{13} can be made to vanish. Therefore, we choose the x_2-axis to lie in the plane of the vertical shearing deformation (so that $d_{13}=0$). This situation is applicable to the early stages of cloud growth in some of the early three-dimensional numerical simulations in which the environmental shear was linear with height and unidirectional, and coriolis effects were omitted (e.g., WILHELMSON and KLEMP, 1978). The solution of [7] with all $d_{ij}=0$ except d_{23} and d_{32} is

$$\begin{aligned}(\alpha\omega_1)(t) &= (\alpha\omega_1)(0) \\ (\alpha\omega_2)(t) &= (\alpha\omega_2)(0)\cosh d_{23}t + (\alpha\omega_3)(0)\sinh d_{23}t \\ (\alpha\omega_3)(t) &= (\alpha\omega_2)(0)\sinh d_{23}t + (\alpha\omega_3)(0)\cosh d_{23}t\end{aligned} \qquad [25]$$

Given sufficient time in the prescribed linear velocity field, the parcel's vorticity vector is rotated into the (x_2,x_3) plane and becomes inclined asymptotically (without overshoot) at 45° to the vertical, while the vorticity increases exponentially at a growth rate, d_{23}. In the special case when $\omega_3(0)=0$, the tilting term acts as a source of vertical vorticity and, in addition, it exponentially amplifies this source through vortex stretching. Acting alone, the tilting term can not tilt initially horizontal vortex tubes more than 45° from the horizontal. Only the source property of the tilting term is represented in ROTUNNO's (1981) semi-linear vertical vorticity equation, and the other properties generally have not been recognized in severe storms research. With $\omega_2(0)=2\times10^{-3}$ s^{-1}, a typical environmental value of vertical wind shear in supercell environments, and $d_{23}=10^{-2}$ s^{-1} (observed by Doppler radar in supercells), the time for vorticity to reach the same tornadic values as in the previous case (but now at 45° inclination) is $\tau=\cosh^{-1}[|\omega(\tau)|^2/(\omega_2(0))^2]/2d_{23}$ = 13 min (in the absence of diffusion). The geometry of this case is shown in Figs. 1(e) and 2(d). An initial square of fluid with sides parallel to the x_2- and x_3-axes is deformed into a rhombus of the same area but larger sides, and with the directions of the diagonals invariant. When the fluid is incompressible and the initial vorticity is horizontal or vertical, the increase in vorticity is simply proportional to the increase in the length of the sides. As the rhombus is deformed, the sides are turned toward the 45° inclination. Note that there is no strain component parallel to the x_1-axis in this case (i.e., the strain is planar). Prior to storm development, the isentropic (i.e., constant θ_w-) surfaces are typically horizontal with potential instability ($\partial\theta_\omega/\partial z<0$) from the surface to mid-levels (~600 mb). According to [8], the constant vertical shearing deformation field under consideration will tilt these isentropic surfaces toward the dilatation axis (45° elevation angle in the (x_2,x_3)-plane) and "sharpen" the θ_ω-gradients, while the intensifying vorticity field will rotate the surfaces around the direction of the vorticity vector. For $\omega_2(0)$ and $d_{23}>0$, the immediate consequence is the generation of a negative θ_ω-gradient in the x_1-direction and a positive one in the x_2-direction. By the conservation of potential vorticity (Eq. [6]), if $\omega_3(0)=0$ and $\nabla\theta_\omega$ is vertical initially, then the θ_ω-gradient along the vorticity vector is zero for all time.

Before treating the general case, note that the matrix **D** in [22] is similar to

$$\mathbf{B} = \frac{1}{2}\begin{bmatrix} 0 & \delta & \delta \\ \delta & 0 & \delta \\ \delta & \delta & 0 \end{bmatrix} \qquad [26]$$

since $\mathbf{D} = \mathbf{P}^T\mathbf{B}\mathbf{P}$ where the orthogonal matrix **P** is given by

$$\mathbf{P} = \begin{bmatrix} -1/\sqrt{2} & -1/\sqrt{6} & 1/\sqrt{3} \\ 1/\sqrt{2} & -1/\sqrt{6} & 1/\sqrt{3} \\ 0 & 2/\sqrt{6} & 1/\sqrt{3} \end{bmatrix} \qquad [27]$$

(see KOLMAN, 1977, p. 248). Thus, vortex lines will be stretched axisymmetrically about a direction $\cos^{-1}(1/\sqrt{3})=55°$ from the vertical in the same way as before (except for orientation), even though the convergence terms are zero in this case, and the tilting terms are zero in [22]. Note that the strain in this case is fully three-directional, in contrast to the previous solution [25] where it was planar and uncharacteristic of a tornadic flow field.

For the completely general linear velocity field, the mathematics is tractible only when referred to the prime (i.e., skewed) coordinate system in which the rate of strain tensor is diagonal. Let the x_3 -axis be the one that is most nearly vertical, and let δ' and ε' be the divergence and stretching deformation in the plane normal to this axis. Then the rate of strain tensor is given by

$$\mathbf{D}' = \begin{bmatrix} \frac{\delta'+\varepsilon'}{2} & 0 & 0 \\ 0 & \frac{\delta'-\varepsilon'}{2} & 0 \\ 0 & 0 & -\delta' \end{bmatrix} \qquad [28]$$

for an incompressible fluid. From [20] the solution of [19] in this case is

$$\begin{aligned} \omega_1'(t) &= \omega_1'(0)\ \exp\ [(\delta'+\varepsilon')t/2] \\ \omega_2'(t) &= \omega_2'(0)\ \exp\ [(\delta'-\varepsilon')t/2] \\ \omega_3'(t) &= \omega_3'(0)\ \exp\ [(-\delta't] \end{aligned} \qquad [29]$$

When $\delta'<0$ and $|\delta'|>|\varepsilon'|$, the x_3'-component of vorticity is amplified, provided that $\omega_3(0)\neq 0$, and the other two components reduced. Thus, in this example, the vortex tubes will seek the x_3'-direction, i.e., the principal direction of **D** which is closest to the vertical, and the sign of $\omega_3'(t)$ is dictated by the sign of $\omega_3'(0)$. When the above inequalities are not both satisfied, the irrotational part of the flow is no longer dominated by a "sink", and vorticity is amplified the most in one of the other principal directions.

The potential wet bulb gradient of a parcel and its vorticity are related through the conservation of potential vorticity (Eq. 6) and through Eq. [8]. Along the stretching axis, θ_w-gradients decrease with time, whereas the θ_w-gradients intensify along the contraction axes. In addition to becoming more perpendicular to the vorticity vector, the $\nabla\theta_w$-vector should rotate around it.

The irrotational part of a general three-dimensional linear velocity field has five independent components, too many to explore all possible variations. Therefore, we consider the case where the horizontal stretching and shearing deformation are both zero ($d_{11}-d_{22}=0$ and $d_{12}=0$) and choose the x_1- and x_2-axes so that $d_{13}=0$ also. If the incompressible continuity equation is approximately valid, the rate of strain tensor is then

$$\mathbf{D} = \begin{bmatrix} \delta/2 & 0 & 0 \\ 0 & \delta/2 & d_{23} \\ 0 & d_{23} & -\delta \end{bmatrix} \qquad [30]$$

Let μ equal $-2d_{23}/\delta$, i.e., the ratio of vertical shearing deformation to horizontal convergence. For $\mu=0$, 1 and ∞, the principal stretching axis lies in the (x_2,x_3) plane and is inclined to the vertical at angles of 0°, 17°, and 45°, respectively.

4 Application to a Tornado Case

The linear velocity field approximation is most valid in cases where the hodograph is straight (i.e., the environmental wind shear vector does not turn with height). In such conditions, severe storms tend to split with one part moving to the right of the shear vector, and a mirror image storm moving to the left (WILHELMSON and KLEMP, 1978). Choose the (horizontal) x_1- and x_2-axes to lie parallel and perpendicular to the shear vector. Prior to the storm, the vorticity vector points in the positive x_2-direction, and the $\nabla\theta_w$-vector is directed downward in the lower half of the troposphere. The storm in its early stages consists entirely of updraft, which draws up a loop of an initially horizontal vortex tube to form a vertical vorticity doublet, cyclonic on the right side of the updraft, anticyclonic on the left, in horizontal cross-sections (see Fig. 3 in ROTUNNO, 1981). The cyclonic and anticyclonic members of the doublet tilt toward each other with height and are most pronounced at mid-levels. The updraft splits, and the right (left)-moving updraft develops sufficient rotation at mid-levels to be classified as a mesocyclone (meso-anticyclone). Presumably the mesocirculations are centered close to the updraft cores and tilt only slightly with height at this stage because the convergent term (initially small because it does not appear in the linearized vorticity equation) becomes dominant over the tilting term (which is still important because it tilts the rotation axes of inflowing air parcels out of the horizontal). Slight asymmetries in the atmosphere usually cause the mesocyclone to be the more intense circulation and it may begin producing funnel clouds, and even small tornadoes if there is sufficient background vertical vorticity at ground level. The final phase appears to await the development of a significant downdraft in the mesocyclone. If there is no vertical vorticity in the flow prior to the storm as in some of the model simulations, then, prior to the downdraft reaching the surface, air at the ground in the mesocyclone is devoid of vertical vorticity because it has flowed into the storm along the surface and hence never undergone any tilting. Thus, if the initial vertical vorticity plays little role in the production of large tornadoes (as the observations seem to indicate), then the downdraft, although divergent, is vital to the lowering of devastating mesocyclonic and tornadic winds to the surface, because it transports to the ground air parcels with non-zero vertical vorticity which can subsequently be stretched in an adjacent updraft to build the vortex column down to the surface. Whether the downdraft is cool and saturated (as postulated by LEMON and DOSWELL, 1979) or warm and unsaturated is immaterial to this process. Note also that the outflow from the downdraft enhances low-level convergence in the vicinity of the updraft. During this stage, the tilting term becomes comparable to the convergence term, the circulation center migrates from near the updraft core to close to the updraft-downdraft interface and the mesocyclone tilts more with height. A major tornado is likely to form near the circulation center, and descending low θ_w-air on the right side of the mesocyclone forms a "clear slot" in the clouds behind the gust front (LEMON and DOSWELL, 1979). Sometimes, the clear slot is behind spiralling rain curtains (hook echo on radar) which wrap around the tornado from the rear. The tornado forms in gradients of both the vertical velocity and θ_w-fields (Fig. 4), but in rising air (RAY et al., 1976) presumably with high energy content (i.e., the air parcels which are stretched to form the tornado can not have originated from more than 1 to 2 km aloft). As the low θ_w-air spreads across the surface, it eventually occludes the mesocyclone, and chokes the tornado which tilts more and more as the flow around it becomes progressively more divergent. Typically, the tornado is deformed into a long, contorted, rope-like vortex, inclined roughly at 45°, and finally decays through the action of wave-type instabilities. Meanwhile, a new mesocyclone core may develop to the right and ahead of the old core near the point of occlusion in young updrafts above the bulging gust front. Tornadoes may occur repeatedly from the same storm in such reforming mesocyclone cores. The subsequent mesocyclone cores and tornadoes may build down to the ground more quickly and be more intense than the first ones because they have pre-existing ground-level vertical vorticity at their disposal.

The right-moving 24 May 1973 Union City, Oklahoma tornadic storm formed in the manner described above, and underwent a similar evolution. The environmental hodograph was practically straight, except for a slight clockwise turning with height at low levels which may explain why the right mover was more severe than the left mover (KLEMP and WILHELMSON, 1978). Doppler radar indicated that the tornado was inclined to the left of the storm motion vector with an average tilt from the surface to upper levels of 25° or 30° (BROWN et al., 1978). The tornado formed on the right rear flank of the updraft, and field observers witnessed the clear slot propagating cyclonically around the tornado from its right rear side. These observations, when viewed in the light of the kinematic insights

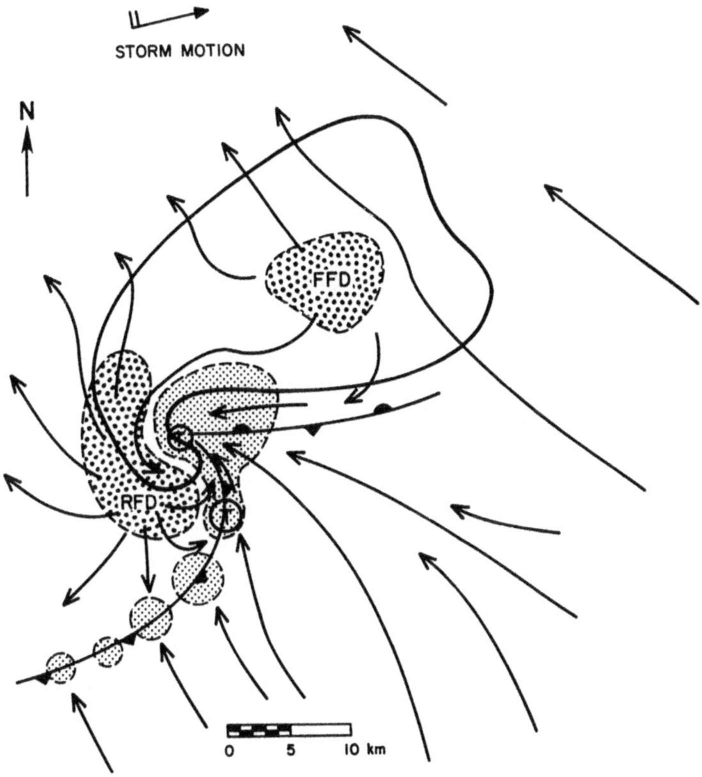

Figure 4. Plan view of an isolated right-moving supercell storm near the surface, showing positions of gust front (classical frontal symbols), updrafts (fine stippling), forward and rear flank downcrafts (FFD, RFD) radar echo (thick outline), storm-relative streamlines (arrows), and favored locations for tornadoes (T). Southern "T" also marks where a new mesocyclone core forms as old one occludes (adapted from Lemon and Doswell, 1979)

given in Section 3, suggest that the tilting term is of comparable importance to the convergent term in the production of large and devastating tornadoes, which should form along the line of greatest strain, and by the conservation of potential vorticity lie outside the undiluted updraft core.

If the convergent term were the dominant one acting throughout tornadogenesis, then tornadoes would form from background vertical vorticity (initially an order of magnitude weaker than background horizontal vorticity) in the core of the updraft, and would have no significant tilt. Experienced tornado field observers have gained the impression that only funnel clouds and narrow tornadoes are produced prior to downdraft air reaching the ground (e.g., the case studied by DAVIES-JONES et al., 1976). In a clockwise-turning hodograph example, RAY et al.'s (1976) data show that the tornado does not lie in the updraft core from the surface up to 7 km. The Union City tornado was tilted appreciably and was not located in the bounded weak echo region (indicative of maximum updraft). In summary, to explain the observed sequence of events which precede and accompany the formation of large tornadoes, it is hypothesized that the ultimate rotation source is ambient horizontal vorticity, and that the downdraft and tilting term are needed (in addition to the updraft and convergence term). Important forecasting parameters such as deep convective and conditional instability, and strong wind shear at low and mid-levels have obvious significance in the context of the tornado theory presented here.

5 Acknowledgments

Working at the National Severe Storms Laboratory has given me the opportunity to guide and participate in the Tornado Intercept Project, allowing me to observe tornadoes and tornadic storms firsthand, and also to exchange ideas with meteorologists who analyze Doppler radar data, especially, Mr. Don Burgess and Dr. Ken Johnson. The 1979 paper by Lemon and Doswell provided the initial "spark" for the development of the concepts discussed in this paper. The Tornado Intercept Project received funding between 1973 and 1979 from the Nuclear Regulatory Commission under Interagency Agreements AT(49-25)-1004 and RES-76-106. Ms. Sandy Mudd typed the manuscript and Ms. Joan Kimpel drafted the figures.

6 References

Brown, R. A., L. R. Lemon, and D. W. Burgess, 1978: Tornado detection by pulsed Doppler radar. Mon. Wea. Rev., 106, 29-38

Burgers, J. M., 1948: A mathematical model illustrating the theory of turbulence. Advan. Appl. Mechan., 1, 197-199

Davies-Jones, R. P., 1981: Tornado dynamics. Chapter X, Thunderstorms: A Social, Scientific and Technological Documentary: Part Two, E. Kessler (Ed.), U. S. Government Printing Office, Washington D.C., (to be published)

____________, D. W. Burgess, and L. R. Lemon, 1976: An atypical tornado producing cumulonimbus. Weather, 31, 336-347

____________ and E. Kessler, 1974: Tornadoes. Chapter 16 (pp. 552-595), Weather and Climate Modification, W. N. Hess (Ed.), John Wiley and Sons, New York, 842 pp

Dutton, J. A., 1976: The Ceaseless Wind. McGraw-Hill, New York, 579 pp

Klemp, J. B., and R. B. Wilhelmson, 1978: Simulations of right-and left-moving storms produced through storm splitting. J. Atmos. Sci., 35, 1097-1110

Kolman, B., 1977: Elementary Linear Algebra (2nd Edition). Macmillan Pub. Co., New York, 314 pp

Kuo, H. L., 1966: On the dynamics of convective atmospheric vortices. J. Atmos. Sci., 23, 25-42

Lemon, L. R., and C. A. Doswell, 1979: Severe thunderstorm evolution and mesocyclone structure as related to tornadogenesis. Mon. Wea. Rev., 107, 1184-1197

Lewellen, W. S., 1976: Theoretical models of the tornado vortex. Proceedings, Symposium on Tornadoes: Assessment of Knowledge and Implications for Man, Lubbock, Texas, Texas Tech University, 107-143

Miller, J. E., 1948: On the concept of frontogenesis. J. Meteor., 5, 169-171

Petterssen, S., 1956: Weather Analysis and Forecasting. Vol I (2nd Edition), McGraw-Hill, New York, 428 pp

Ray, P. S., 1976: Vorticity and divergence fields within tornadic storms from dual-Doppler observations. J. Appl. Meteor., 15, 879-890

___, C. E. Hane, R. P. Davies-Jones, and R. L. Alberty, 1976: Tornado-parent storm relationship deduced from a dual-Doppler radar analysis. Geophys. Res. Lett., 3, 721-723

Rott, N., 1958: On the viscous core of a line vortex. Z. Angew Math Physik, 96, 543-553

Rotunno, R., 1981: On the evolution of thunderstorm rotation. Mon. Wea. Rev., 109, 577-586

Starr, V. P., 1974: The tornado mechanism and its possible artificial duplication. Revista Italiana di Geofisica, 23, 267-271

Tennekes, H., and J. L. Lumley, 1972: A First Course in Turbulence. The MIT Press, Cambridge, Mass., 300 pp

Wilhelmson, R. B., and J. B. Klemp, 1978: A numerical study of storm splitting that leads to long-lived storms. J. Atmos. Sci., 35, 1974-1986

High Resolution Numerical Simulations of the Tornadic Region Within a Mature Thunderstorm

J. B. Klemp and R. Rotunno

1 INTRODUCTION

Recent multiple Doppler radar observations of supercell thunderstorms have documented that a particular dynamic structure consistently develops when a storm enters its tornadic phase. This structure is characterized by a downdraft outflow which wraps around the center of storm circulation at low levels and an associated distortion from a single cell to a two-cell updraft structure (Brandes, 1978; Lemon and Doswell, 1979). In this study we investigate the evolution of a storm's tornadic phase through high resolution numerical simulations initiated within the central portion of a 3D cloud model mature thunderstorm.

Tornadoes are most frequently observed to form within severe, long lived thunderstorms characterized by a single powerful updraft, the so-called supercell. Over the past twenty years the basic structure of the mature supercell storm has been revealed through analyses of observations based on visual, radar, aircraft and surface data. More recently, multiple Doppler radar observations have aided in evaluating conceptual models and in extending the understanding of these storms. Although these more detailed data on internal storm structure show numerous variations among individual storms, certain basic features are consistently observed which contribute to the intensity and long duration of supercell storms. Browning's (1964) description of the three-dimensional airflow within a severe right moving storm still provides a good qualitative picture of the dynamic structure of the supercell storm. Evolving in strong environmental wind shear, the convective cell is sustained by the continuous supply of moisture from low level inflow on one (front) side of the storm and the transport of precipitation out of the updraft and into the downdrafts on the other (back) side. These cold downdrafts, enhanced by evaporative cooling, spread out just above the ground, producing a gust front convergence line along the right flank of the storm which reinforces the lifting of moist inflow into the updraft. Right moving storms exhibit a mesocyclone, characterized by strong cyclonic rotation in the low to midlevels of updraft while their left moving counterparts display a mirror image configuration (Browning, 1968). The seemingly self-perpetuating mature structure is consistent with frequent observations of supercell storms persisting over periods of several hours.

[1]The National Center for Atmospheric Research is sponsored by the National Science Foundation.

Topics in Atmospheric and Oceanographic Sciences
Intense Atmospheric Vortices, ed. by Bengtsson/Lighthill

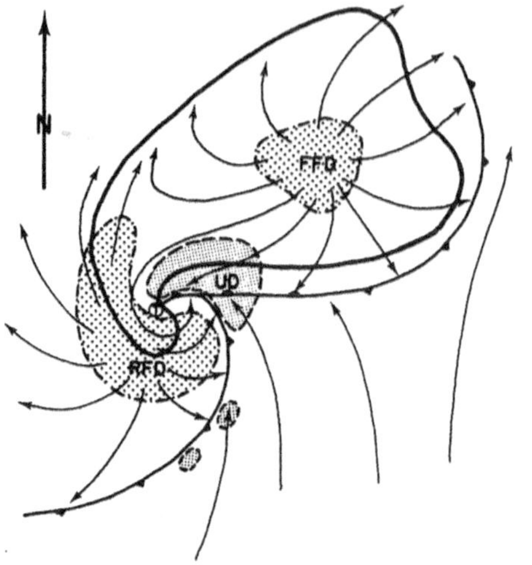

Fig. 1. Schematic plan view of a tornadic thunderstorm at the surface. Thick line encompasses radar echo. The thunderstorm "gust front" structure and "occluded" wave are also depicted using a solid line and frontal symbols. Surface positions of the updraft (UD) are finely stippled, forward flank downdraft (FFD) and rear flank downdraft (RFD) are coarsely stippled, along with associated streamlines (relative to the ground) are also shown. Tornado location is shown by an encircled T. (Taken from Lemon and Doswell, 1979)

As a storm enters its collapsing phase, however, significant changes in structure rapidly occur which create favorable conditions for tornado formation. Lemon and Doswell (1979) provide an excellent description of the observed transition in storm features during this phase which is partly summarized here for information purposes. Prominent features observed during the collapsing phase are the increase of reflectivity in the weak echo region associated with the storm's updraft core and the shifting of the midlevel mesocyclone center from this weak echo region to a location between the updraft and the downdraft along the rear flank. Strong horizontal shear is frequently detected near the mesocyclone center which characterizes the tornado vortex signature (TVS). As the updraft is declining in strength the downdrafts, and particularly the rear flank downdraft, are intensifying. The descent of the mesocyclone and the TVS to low levels occurs coincidentally with the development of the rear flank downdraft.

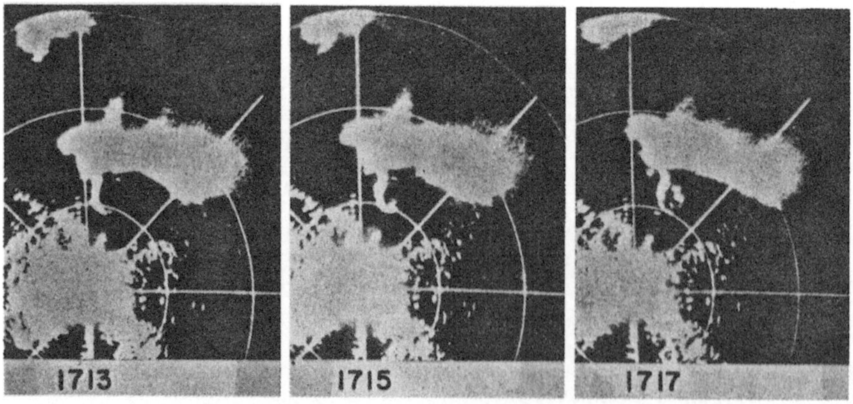

Fig. 2. Low level radar echo on 9 April 1953. Range markers are at 10 mile intervals. (Taken from Huff, Hiser, and Bigler, 1954)

As the low level downdraft is strengthening, downdraft outflow begins to progress cyclonically around the low level updraft in a manner which is visually similar to an occluding cold front in an extratropical cyclone (Danielson, 1975). This occluding structure is schematically illustrated by Lemon and Doswell in Fig. 1 which also shows the relative orientations of low level updrafts, downdrafts and precipitation. The tornado is typically located near the center of circulation at the tip of the occlusion with the tornado forming just on the updraft side of the updraft-downdraft interface. Secondary tornadoes, however, may develop along the trailing gust front. The hook echo forms behind the gust front within the rear flank downdraft (see Fig. 1) and as the occlusion progresses, the tip of the hook wraps up as illustrated in Fig. 2 from Huff, Hisler, and Bigler (1954). Notice the hook extending to the south from the southwest (right rear) quadrant of the main body of reflectivity associated with this storm. Rapidly following tornado touchdown at 1713 CST near Champaign, Illinois, the tip of the hook spirals around and fills with precipitation.

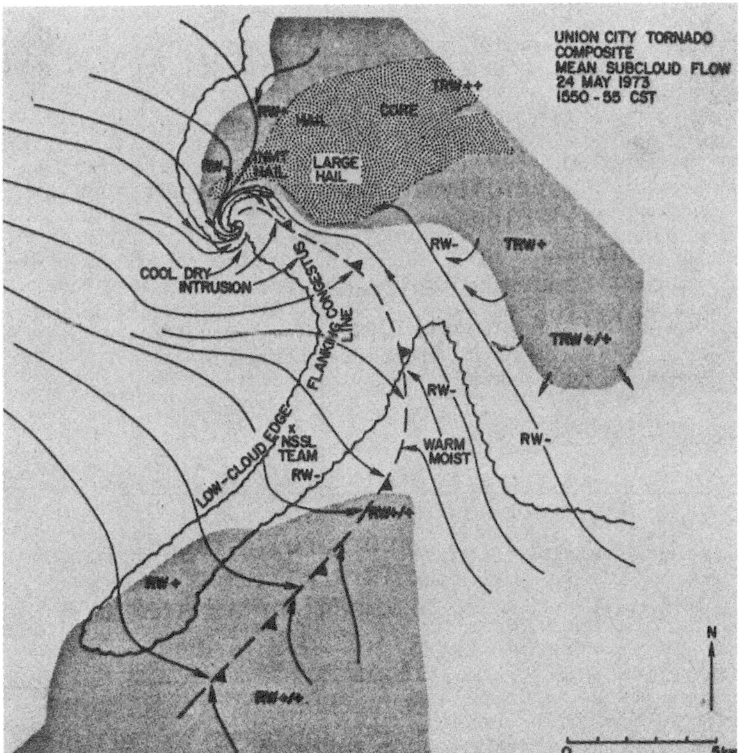

Fig. 3. Schematic plan view of composited mean subcloud mesoscale flow around Union City tornado (at circulation center) during mature state. Surface gust-front location is dashed, major low-level cloud boundaries, precipitation types and intensities are indicated. Regions of heavier precipitation are shaded. Note flanking line cloud base relative to surface gust-front, stream-flow, and tornado. (Taken from Golden, 1976)

The low level flow derived from surface observations during the Union City tornado also depicts this downdraft occlusion and is displayed in Fig. 3 from Golden (1976). The tornado near the circulation center descends to the ground near the right or rear periphery of a region of lowered cloud base called the wall cloud. Viewing the tornado from beneath the flanking line of clouds (see Fig. 3), relatively cloud free air is visible in front and to the left of the tornado (called the clear slot by Lemon and Doswell) which is associated with the unsaturated but precipitating rear flank downdraft. As the occluding gust front continues to advance, mixing of downdraft air into the updraft accelerates the collapse of the storm. For references to specific observations which document the events described above, the reader is referred to Lemon and Doswell (1979).

Three-dimensional numerical cloud models have successfully simulated important features of mature supercell thunderstorms and have complemented observational data in analyzing their dynamics (Klemp and Wilhelmson, 1978b; Schlesinger, 1978; Thorpe and Miller, 1978; Wilhelmson and Klemp, 1981). However, as a storm enters its tornadic phase, significant smaller scale features evolve which the limited grid structures of numerical models have been unable to resolve.

In the current investigation we seek to improve the understanding of a storm's tornadic phase through high resolution numerical simulations conducted within the central portion of a simulated mature supercell storm. This storm, the 20 May 1977 Del City storm, has been analyzed by Ray et al. (1981), Brandes (1981) and Klemp, Wilhelmson and Ray (1981). The new simulations suggest that the essential physics which promotes tornadogenesis is contained within the model equations and that with enhanced grid resolution the early development of these smaller scale processes may be captured. To clarify the nature of these simulations we shall describe in Section 2 the procedures used in restarting the numerical integrations on a finer mesh. In Section 3 numerical results with a 250 m horizontal grid are presented along with comparisons with observational data. Summary remarks are included in Section 4.

2 NUMERICAL APPROACH FOR HIGH RESOLUTION SIMULATIONS

In numerically simulating convective storms with finite computer resources, modelers must continually balance the desire for high grid resolution against the requirement of having a large enough integration domain to contain the storm plus a reasonable volume of its near environment. For the 20 May 1977 tornadic storm simulation discussed by Klemp, Wilhelmson, and Ray (1981) this compromise resulted in a $48 \times 48 \times 16$ km domain with 1 km and 500 m grid spacings in the horizontal and vertical directions, respectively. The numerical model used for this simulation is described by Klemp and Wilhelmson (1978a). Comparisons with analyzed multiple Doppler radar observations demonstrate the ability of the model to capture the important features of the mature supercell storm with the prescribed grid resolution. This mature storm structure is illustrated in Fig. 4 (taken from Klemp, Wilhelmson, and Ray, 1981) by the 1 and 4 km horizontal cross-sections for a 29×29 km portion of the modeled storm and

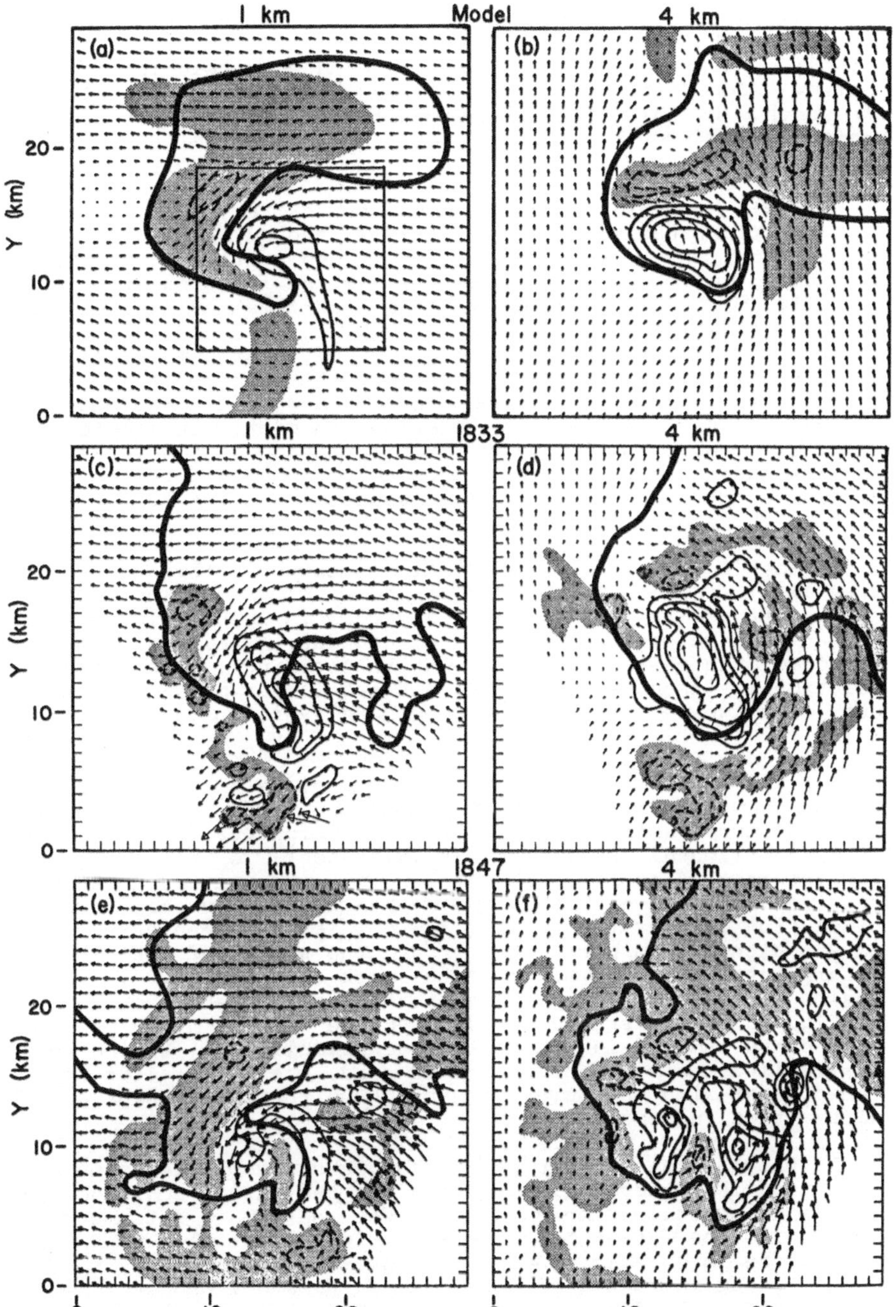

Fig. 4. Cross-sections at 1 and 4 km for model simulation and observations at 1833 and 1847. Vertical velocity is contoured in 5 m s^{-1} intervals except for 1847 which is in 10 m s^{-1} intervals. Shaded areas designate downdraft regions. The heavy solid line denotes the .5 g kg^{-1} rainwater contour in the model and the 30 dBZ reflectivity contour in the observed storm

the observed Del City storm at 1833. (All times referenced are in Central Standard Time.) A single powerful updraft is present with precipitation filled downdrafts producing cold low level outflow which reinforces the convergence line extending off to the south. By this time the mesocyclone center has already shifted away from the updraft center, although the rear flank downdraft is rather weak.

By 1847, however, the storm has entered its tornadic phase. The gust front has wrapped cyclonically around the low level circulation center (Fig. 4e) and a divided updraft structure is apparent at mid levels (Fig. 4f). These features are significantly smaller in horizontal scale than those evident at 1833 and do not evolve in the numerical model. This discrepancy could be due to inadequate physics in the model equations or to insufficient numerical resolution. Certainly the absence of surface drag and the simplification of microphysical and turbulence exchange processes through parameterization might contribute to these discrepancies. However, we hypothesize that numerical resolution is the primary limitation in simulating the occlusion processes occurring during tornadogenesis, subject to verification in these experiments.

In seeking to enhance numerical resolution within the storm, several alternative procedures can be considered. The most popular of these are the nested grid models in which solutions on a coarser mesh provide the lateral boundary conditions for an embedded fine mesh region (one-way interacting). The interior solution may then be used to adjust the larger mesh data to provide a two-way interacting procedure (Phillips and Shukla, 1973).

For the present study, our initial goal is to determine if the previously simulated mature supercell would contain smaller scale features associated with the occlusion phase of the storm if the horizontal resolution was increased. To this end a one-way interacting nested mesh model seems appropriate, at least for these initial experiments. We begin with the storm scale simulation illustrated in Fig. 4a, b which has been carried out to 2 h with a 1 km horizontal grid. The simulated data is then interpolated to a 250 m horizontal mesh within a 13.5×13.5 km region in the interior portion of this domain, shown by the box in Fig. 4a. The vertical grid structure with $\Delta z = 500$ m is maintained the same as in the storm scale simulation. Having reinitialized the model on the finer grid, the integration is resumed first with one Euler time step and then with leapfrog steps with a time increment of 1.5 s.

At the lateral boundaries of the higher resolution domain the situation is rather different than for the larger scale domain in the cloud scale simulations, since the domain size must be reduced to the extent that the storm is not entirely contained within the integration domain. In order to balance the need to maintain the influence of portions of the storm outside the domain against the desire to allow the boundaries to adjust to internal developments, the open boundary conditions described by Klemp and Wilhelmson (1978a) were modified to fix the scalar variables at the inflow boundaries. Testing these lateral boundary conditions (to be presented elsewhere) indicates that they do not produce any significant distortion in the simulations for the short time intervals of integration which are being considered.

The numerical results generated by these procedures must be interpreted with some caution. As smaller scale structure evolves on the finer grid there is a strong temptation to interpret the transient response as a physical evolution within the storm. Certainly the integrations are progressing forward in time while the transitions occur and a simulation time on the order of 10 min is a reasonable interval over which the occlusion processes might occur. However, the adjustment of the data to the higher resolution in the finite difference equations probably dominates the response and the transition during this adjustment may not have physical relevance. Consequently, we interpret the simulated fine mesh structure as being essentially present in the coarser grid data which had smoothed the small scale features through parameterized turbulence.

3 HIGH RESOLUTION NUMERICAL RESULTS

In conducting the high resolution integrations as described above, alterations in the flow structure are found to be most rapid in the vicinity of the low level convergence line where sharp gradients in the fields suggest that the numerical resolution is inadequate. For this reason we shall concentrate our attention on the low level storm structure in this simulation. Figure 5 illustrates the flow field and vorticity at z = 250 m from the cloud scale simulation at the time (2 h) when the higher resolution integrations are initiated. The domain displayed is just slightly smaller than the high resolution integration domain indicated in Fig. 4a. By this time the center of circulation has shifted away from the location of maximum updraft (Fig. 5a) and is located near the boundary between the storm relative easterly inflow and the cold downdraft outflow. The maximum vorticity (.023 s^{-1}) is nearly coincident with the circulation center and remains strongly positive along the convergence line which extends off to the southeast.

The production of vertical vorticity, ζ, can be analyzed by considering the forcing terms in the vertical vorticity equation:

$$\frac{d\zeta}{dt} = \left(\frac{\partial u}{\partial z}\frac{\partial w}{\partial y} - \frac{\partial v}{\partial z}\frac{\partial w}{\partial x}\right) + \left[\zeta \frac{\partial w}{\partial z}\right] + F \quad . \tag{1}$$

Here, (u, v, w) are the components of velocity in rectangular coordinates (x, y, z), d/dt is the Lagrangian time derivative and F represents dissipation terms. The terms in parentheses in Eq. (1) contribute to ζ through the tilting of horizontal vortex lines into the vertical while term in brackets alters the vertical vorticity through the stretching of vertical vortex tubes. Vorticity production through tilting in the storm scale simulation (Fig. 5c) is positive in the northern portion of the low level updraft and also along the inflow side of the convergence line. On the downdraft side of the convergence line negative vorticity production occurs through tilting of the horizontal vorticity embedded in the storm outflow. Vorticity production through the convergence of vertical vorticity (Fig. 5d) is strongly positive, as expected, along the convergence line in the region of strong cyclonic vertical vorticity.

By about 6 min of integration time on the refined mesh, the alterations in the model fields have slowed considerably, sug-

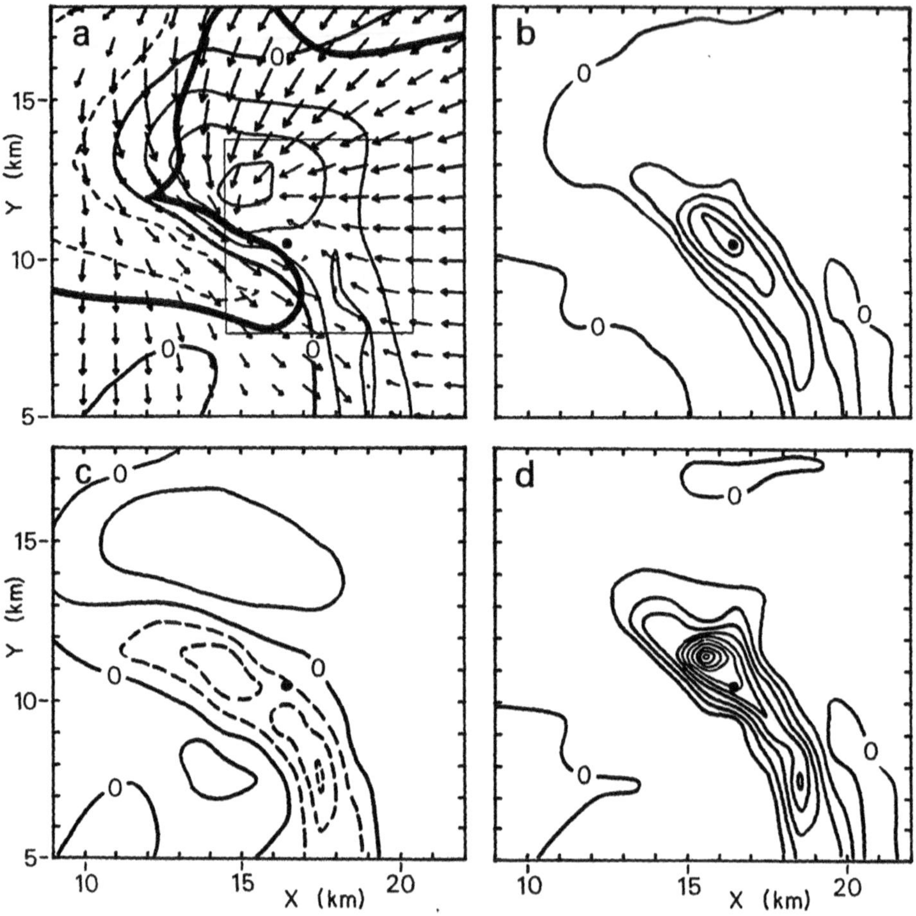

Fig. 5. Cross-sections at z = 250 m at 2 h in storm scale simulation of (a) flow field with horizontal vectors (one grid interval equals 20 m s^{-1}), vertical velocity (1 m s^{-1} contour interval) and .5 g kg^{-1} rainwater contour (heavy line), (b) vertical vorticity contoured at $.5\times10^{-2}s^{-1}$ intervals, (c) tilting term contoured at $1\times10^{-5}s^{-2}$ intervals, and (d) stretching term contoured at $2\times10^{-5}s^{-2}$ intervals. The black dot locates the vorticity maximum

gesting that the initial adjustments to the higher horizontal resolution are nearly complete. The corresponding z = 250 m flow field and vorticity characteristics are displayed at 6 min in Fig. 6. For clarity only the central portion of the integration domain is presented; the location of this 6×6 km region is indicated by the box in Fig. 5a.

By this time the downdraft outflow has progressed cyclonically around the center of circulation (Fig. 6a), with the gust front developing an occluded structure, similar to the observed low level structure at 1847 in Fig. 4e and also to the schematic illustration in Fig. 1. (For comparison Fig. 1 should be rotated counterclockwise by about 45°.) Two updraft maxima are evident at low levels; the one to the west is associated with the

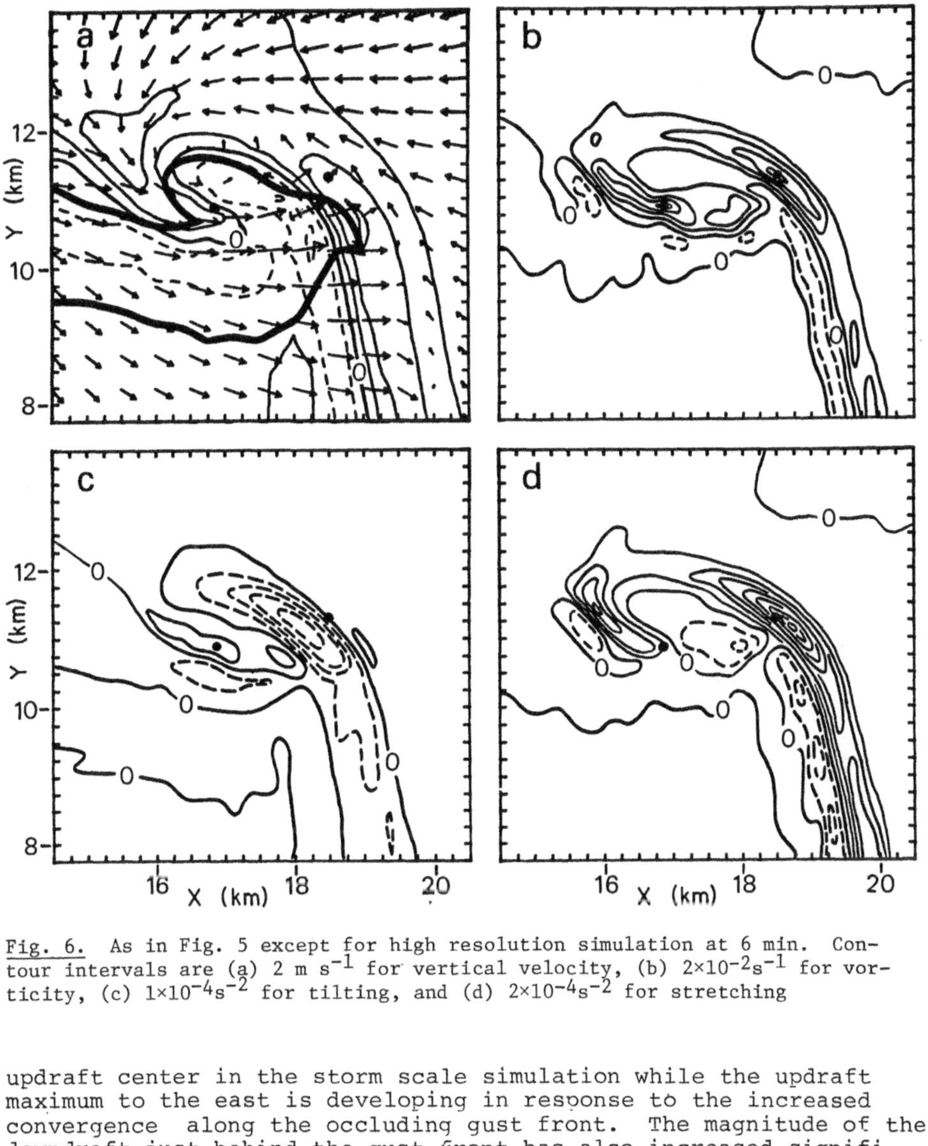

Fig. 6. As in Fig. 5 except for high resolution simulation at 6 min. Contour intervals are (a) 2 m s^{-1} for vertical velocity, (b) $2\times10^{-2}s^{-1}$ for vorticity, (c) $1\times10^{-4}s^{-2}$ for tilting, and (d) $2\times10^{-4}s^{-2}$ for stretching

updraft center in the storm scale simulation while the updraft maximum to the east is developing in response to the increased convergence along the occluding gust front. The magnitude of the downdraft just behind the gust front has also increased significantly to more than -4 m s^{-1}.

The rainwater contour in Fig. 6a depicts the finer structure of the tip of the low level hook (see Fig. 5a for reference). As the updraft and downdraft regions spiral around each other, the tip of the hook follows the advancing region of downdraft outflow and begins to wrap up as shown in Fig. 2. This feature is also resolved in the single radar reflectivity fields for the Del City storm presented by Brandes (1981) and Ray et al. (1981).

The development of fine scale structure in the low level vorticity field is particularly dramatic (Fig. 6b). While the storm scale low level vorticity maximum was coincident with the center

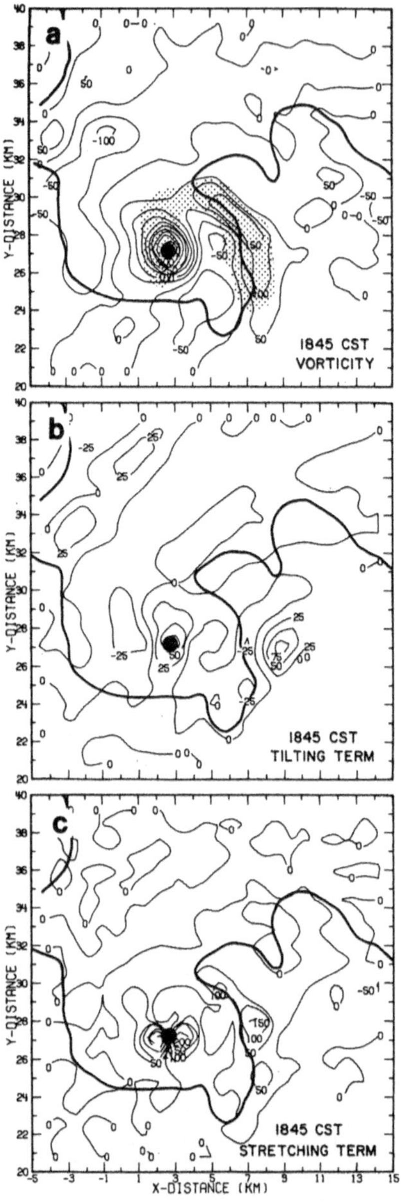

Fig. 7. Cross-sections at 1845 at z=1.4 km while tornado (black dot) is on the ground of (a) vorticity with contour intervals of $5\times10^{-3}s^{-1}$ and vorticity generation by (b) tilting and (c) stretching with contour intervals of 25 and $50\times10^{-6}s^{-2}$ respectively. Heavy line denotes the 30 dBz reflectivity contour and the updraft region is shaded

of circulation, the high resolution vorticity field forms a ring of strong cyclonic vorticity around the center of circulation. This ring of vorticity surrounding a downdraft core is qualitatively similar to the structure obtained by Rotunno (1981) in numerical simulations of multiple vortices produced in the laboratory. The maximum vorticity has increased by about a factor of four and occurs near the tip of the updraft region which is wrapping around the low level downdraft (Fig. 6a). This location, near the updraft-downdraft boundary, is consistent with the observed location of the tornado in the Del City storm (Ray et al., 1981; Brandes, 1981) as well as other tornadic storms (see Figs. 1 and 3). A second strong vorticity center is evident near the updraft center along the gust front which is also observed in the Del City storm as shown in Fig. 7a from Brandes (1981).

The production of vorticity through tilting and stretching processes display similarly enhanced small scale structure. Both fields exhibit maximum positive vorticity production just upwind of the two vorticity maxima based on the local flow directions within the updraft in the vicinity of these two locations (Fig.6a). Both fields also promote negative vorticity production along the downdraft outflow side of the convergence line. Similarities are apparent in comparing the distribution of the tilting and stretching terms with those computed by Brandes at 1.4 km shown in Fig.7b and c. At z=250 m in the model simulation, the contributions to positive vorticity are much greater from the stretching term than from tilting. Since stretching only amplifies existing vertical vorticity, the tilting of horizontal vorticity in the environmental shear into the vertical must provide the ultimate source of the strong vorticity apparent in Fig.6b. Although the positive tilting is small at low levels, vorticity produced through tilting may be exponentially amplified in time through stretching to produce the large vorticities observed in the model.

To illustrate the process by which rapid vorticity amplification might occur, consider the vorticity equation (1) for the situation in which the tilting term (deonoted here by T) is constant over a given portion of a parcel trajectory. We assume also that the divergence, $-w_z$, is constant over this interval and that mixing can be neglected (F=0). In this very simple situation Eq. (1) can be integrated following a parcel such that

$$\zeta = \frac{T}{w_z}\left\{e^{w_z t} - 1\right\} + \zeta_o e^{w_z t} \tag{2}$$

where ζ_o is the initial vertical vorticity of the parcel which we assume to be zero. Using appropriate estimates from the data in

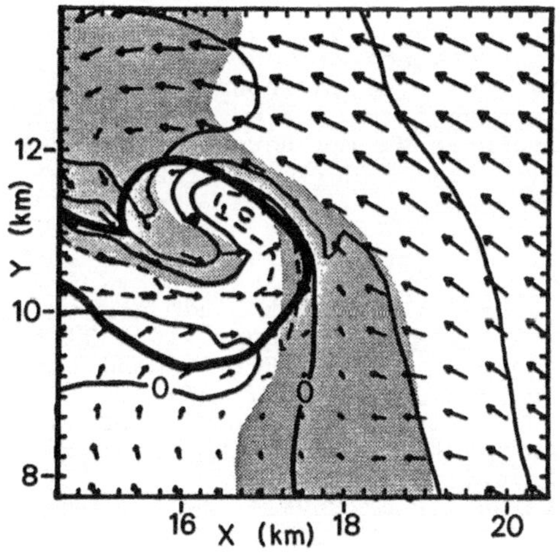

Fig. 8. Flow fields as in Fig. 6a except at z=1 km and vertical velocity is contoured in 5 m s^{-1} intervals. The shaded area denotes the region in which the cloud water exceeds .4 g kg^{-1}

Fig.6, let T=5 10^{-5} s^{-2} and w_z=1.6 10^{-2} s^{-1}. In this example vorticity produced initially through tilting will amplify according to Eq. (2) with an e folding time of one minute. The vertical vorticity will increase from 0 to 0.03 s^{-1} in 2.5 min, at which time the stretching term becomes 10 times greater than the tilting term. At a translation speed of 10 m s^{-1}, a parcel would travel only 1.5 km while this amplification occurred. Although this calculation is rather crude, it does suggest that inflow air can rapidly develop large vertical vorticity through tilting and strong stretching as it traverses the low level convergence zone.

At z=1 km (Fig. 8) the rear flank downdraft has increased in intensity to -10 m s^{-1} and the positive vorticity maximum of .05 s^{-1} is located at the center of circulation which is just on the updraft side of the updraft-downdraft interface. This vorticity maximum is less than at z=250 m since the strong low level adjustments in the flow have not yet been communicated significantly to higher levels. Notice also that the flanking line of cloud water extends to the south behind the gust front (evident in Fig. 6a) and that a clear slot has formed in the rear flank downdraft. These features are also similar to those in the Union City storm shown in Fig. 1.

4 SUMMARY

To investigate the dynamic processes by which tornadoes form within the parent storm, we have interpolated the cloud model fields within the central portion of a storm scale simulation and reinitiated the model within the mesocyclone of a mature supercell storm. Lateral boundary conditions are specified from the storm scale simulation which are found to produce little distortion during the 6 min integration time within the smaller domain. Increasing the grid resolution in this manner leads to strong enhancement of cyclonic vorticity along the gust front as the flow field adjusts to the finer mesh. Occlusion of the gust front occurs as the low level updraft and downdraft regions spiral cyclonically around the circulation center. At z=250 m a ring of high vorticity air surrounds the circulation center (which could be conducive to multiple vortex formation) while at 1 km the maximum vorticity is coincident with the center of circulation. Within the model, this strong positive vorticity is apparently produced by intense convergence of weaker vorticity generated through tilting of the low level environmental wind shear. Numerous aspects of this higher resolution structure bear resemblance to observed storm features during tornadic development.

5 REFERENCES

Brandes, E. A. 1977 Gust front evolution and tornado genesis as viewed by Doppler radar. J. Appl. Meteor., 16, 333-338
Brandes, E. A. 1981 Finestructure of the Del City-Edmond tornado mesocirculation. Mon. Wea. Rev., 109, 635-647
Browning, K. A. 1964 Airflow and precipitation trajectories within severe local storms which travel to the right of the winds. J. Atmos. Sci., 21, 634-639
Browning, K. A. 1968 The organization of severe local storms. Weather, 23, 429-434

Danielson, E. F. 1975 A conceptual theory of tornadogenesis based on macro, meso- and microscale processes. Preprints Ninth Conf. on Severe Local Storms. Norman, Amer. Meteor. Soc., 376-383

Golden, J. H. 1976 Comparison of Union City tornado life cycle with Florida keys waterspout life cycle. NOAA Tech Memo ERL NSSL-80, 135-140

Huff, F. A., Hiser, H. W., and Bigler, S. G. 1954 Study of an Illinois tornado using radar, synoptic weather and field survey data. State of Illinois Water Survey Div., 73 pp

Klemp, J. B. and Wilhelmson, R. B. 1978a The simulation of three-dimensional convective storm dynamics. J. Atmos. Sci., 35, 1070-1096

Klemp, J. B. and Wilhelmson, R. B. 1978b Simulations of right- and left-moving storms produced through storm splitting. J. Atmos. Sci., 35, 1097-1110

Klemp, J. B., Wilhelmson, R. B. and Ray, P. S. 1981 Observed and numerically simulated structure of a mature supercell thunderstorm. To appear in J. Atmos. Sci

Lemon, L. R. and Doswell, C. A., III 1979 Severe thunderstorm evolution and mesocyclone structure as related to tornado-genesis. Mon. Wea. Rev., 107, 1184-1197

Phillips, N. A. and Shukla J. 1973 On the strategy of combining coarse and fine grid meshes in numerical weather prediction. J. Appl. Meteor., 12, 763-770

Ray, P. S., Johnson, B. C., Johnson, K. W., Bradberry, J. S., Stephens, J. J., Wagner, K. K., Wilhelmson, R. B. and Klemp, J. B. 1981 The morphology of several tornadic storms on 20 May 1977. To appear in J. Atmos. Sci

Rotunno, R. 1981 Numerical simulation pertaining to the multiple vortex phenomenon. Proceedings of Joint IUTAM/IUGG Symposium on intense atmospheric vortices, Reading, England 14-17 July 1981

Schlesinger, R. E. 1978 a three-dimensional numerical model of an isolated thunderstorm: Part I. Comparative experiments for variable ambient wind shear. J. Atmos. Sci., 35, 691-713

Thorpe, A. J. and Miller, M. J. 1978 Numerical simulations showing the role of the downdraught in cumulonimbus motion and splitting. Quart. J. Roy. Meteor. Soc., 104, 873-893

Wilhelmson, R. B. and Klemp, J. B. 1978 A numerical study of storm splitting that leads to long-lived storms. J. Atmos. Sci., 35, 1975-1986

Numerical Studies of Tornado Structure and Genesis

L. M. Leslie and R. K. Smith

Abstract

Preliminary results from an axisymmetric numerical simulation of tornado growth in a mesocyclone updraught are described. Using an observed tornadic storm proximity sounding, the calculations show that the distribution and degree of buoyancy in a mesocyclone updraught can account for the generation and maintenance of an intense tornado when the initial level of storm rotation is within the observed range of values. The structure of the mature tornado is highlighted by a comparison of the vertical force balance in a rotating and non-rotating updraught simulation. The present paper extends our recent studies of tornadogenesis to include the effects of moisture diffusion and of negative buoyancy due to the water loading of the updraught.

Introduction

The extreme violence of tornadoes, coupled with their relatively short lifetime, small scale and comparative rarity, has hampered the collection of quantitative data on these vortices and only recently have their secrets begun to be exposed. This is due in no small measure to the development and deployment during the last decade of Doppler meteorological radars, two of which when used in conjunction enable the three dimensional airflow in severe thunderstorms to be determined (see e.g. RAY ET AL., 1980, BRANDES, 1981). It also follows the organization of tornado 'chase' teams from the National Severe Storms Laboratory and Oklahoma University (GOLDEN and MORGAN, 1972) which, guided by radar observers, have succeeded in adding visual documentations of tornado genesis to complement the radar data. The example *par excellence* of these combined observational efforts was the very detailed study of the Union City, Oklahoma tornado in 1973 (see the series of papers in the January 1978 issue of MONTHLY WEATHER REVIEW). The chase teams have also obtained high quality motion pictures which have been photogrammetrically analysed to provide valuable data on wind structure around the visible funnel cloud and debris cloud (GOLDEN, 1976). Although the tornado vortex is typically smaller in scale than radar sampling volumes, it appears to give a characteristic signature in single Doppler scans, showing up as a region of extreme shear between adjacent radar sampling volumes (BROWN ET AL, 1978).

Notwithstanding the technological advances of Doppler radar, which now enable tornado cyclone (or mesocyclone) circulations associated with severe thunderstorms to be identified and their evolution monitored, there is still a paucity of data on the wind and thermodynamic structure in a tornado, especially above cloud base, and many aspects of tornado dynamics remain uncertain or unexplained. In

Topics in Atmospheric and Oceanographic Sciences
Intense Atmospheric Vortices, ed. by Bengtsson/Lighthill

view of this, laboratory and numerical simulations have an important role in tornado research; in particular, numerical models provide a means of exploring the dynamical constraints on vortex motion, leading to a framework within which observational data may be interpreted.

Despite the fact that it is now widely believed that tornadogenesis is a result of intense buoyant convection in the presence of cloud scale rotation, there have been few studies of tornado dynamics which include the effect of latent heat release; indeed, the only numerical simulations that we are aware of, other than our own study (SMITH and LESLIE, 1979; henceforth referred to as SL) are those of PROCTOR (1978) and some very recent calculations by KLEMP and ROTUNNO in a companion paper in this Conference.

PROCTOR considers axi-symmetric flow in a cylindrical region of moist, conditionally unstable air, initially in uniform rotation at each height and increasing in strength with height. The flow domain, which is open at the radial and upper boundaries, is 6 km in diameter, but only 3 km in height, the grid resolution being 150 m in both radial and vertical directions. The application of a no slip surface boundary condition from the initial instant results in frictional convergence and ascent. This serves to trigger moist convection to maintain the meridional circulation, which itself leads to a concentration of the rotation. However, the radial boundary condition allows the rotation field to be steadily depleted and the ensuing vortex is transient. PROCTOR presents only a single calculation and does not relate the results to tornado observations; neither does he offer a view as to how the initial field of rotation comes about prior to convective activity. This detracts from his finding (p315) that an initial vortex "... whose intensity increases with height can trigger convection and generate a tornado".

Our own model (SL) has many similarities with PROCTOR'S, but the computational domain is much deeper (8 km) and the resolution much finer (25 m, both radially and vertically). Rotation is imposed by prescribing a swirling velocity component on air entering through the radial boundary, whereas the initial rotation in the flow domain is everywhere zero. The updraught is initiated by a 2 K temperature anomaly in the central region of the computational domain and, again, this triggers moist convection leading to continuing convergence and a consequent amplification of the rotation to form a vortex. Significantly, the evolution of the vortex core, which first develops in the lower part of the cloud, parallels that of the tornado vortex signature (TVS) observed in single Doppler radar scans. Furthermore, the model vortex extends to high levels in the cloud, but during its mature (steady state) stage, the maximum swirling velocities occur within the lowest kilometre or two above the ground, also in accordance with the radar data. The SL paper contains a detailed account of the dynamics of vortex intensification as revealed by the model.

A further result of our study is that for a fixed vertical distribution of ambient temperature and moisture, the vortex strength is sensitive to the imposed circulation; if the latter is too small, the cloud rotates weakly, but there is no evidence of a vortex or a funnel cloud; if it is too large, a vortex and its associated funnel cloud may form, but terminate aloft. This would explain why relatively few apparently suitable thunderstorms, even rotating ones, spawn damaging tornadoes.

Perhaps the most important conclusion of SL is the demonstration that the distribution of buoyancy in a mesocyclone updraught can account for the generation and maintenance of an intense tornado for a level of mesocyclone rotation within the observed range of values. However, it would appear that the realistic buoyancy fields were obtained fortuitously. The degree of conditional instability chosen for the ambient temperature and moisture sounding was relatively low, but this was offset by our neglect of the negative buoyancy due to liquid water loading of the updraught.

The model described here is broadly similar to that of SL, but uses an observed tornadic storm proximity sounding, the Dibble sounding, reported by DAVIES-JONES (1974). It also incorporates the effects of water loading and of the diffusion of moisture. Unlike the method used in SL, the present calculations employ a 'pseudo' vorticity-streamfunction approach preferred by a number of cloud modellers. This alternative was chosen for several reasons, the most relevant of which is that the diagnostic boundary value problem for the streamfunction is simpler to formulate than that for pressure. The model also uses a forward time differencing scheme instead of a leapfrog scheme, thus effecting a large saving in computer storage. At the time of writing, only two long time integrations of the new model have been performed. These include one tornado simulation together with a non-rotating updraught simulation for comparison. The broad features of the solutions are described below, with emphasis on some aspects which have not previously received detailed attention such as the role of liquid water. A revealing analysis of the axial fields in the steady state is also given for each simulation.

The Model

The computational domain is cylindrical with radius R(= 2 km) and height H (= 8 km). The continuity, momentum, thermodynamic, water substance and tangential vorticity equations are:

$$(\rho_a r u)_r + (\rho_a r w)_z = 0 \ , \tag{1}$$

$$u_t + u u_r + w u_z - v^2/r = - c_p \theta \pi'_r + D_r \ , \tag{2}$$

$$v_t + u v_r + w v_z + uv/r = D_\phi \ , \tag{3}$$

$$w_t + u w_r + w w_z = - c_p \theta \pi'_z + g(\theta'/\theta - q_l) + D_z \ , \tag{4}$$

$$\theta_t + u\theta_r + w\theta_z = \dot{h}/(c_p \pi) + D_\theta' \ , \tag{5}$$

$$q_{wt} + u q_{wr} + w q_{wz} = D_{qw} \ , \tag{6}$$

$$\eta_t + u\eta_r + w\eta_z = (2w\rho_{az}/\rho_a + u/r)(u\rho_{az} - \eta) + c_p(\rho_a\theta)_r \pi'_z + \\ + uw\,\rho_{azz} + (\rho_a v^2/r)_z - g\rho_a(\theta'/\theta - q_l)_r - c_p(\rho_a\theta)_z \pi'_r + D_\eta \tag{7}$$

where $\rho_a = \rho_a(z)$ is the ambient density (subscript 'a' denotes an ambient value); π is the Exner function of pressure; q_w and q_l are the mixing ratios for total water substance and liquid water, respectively; $\dot{h}$ is the rate of latent heat release and D_r, D_ϕ, ...

etc, represent the diffusion terms; otherwise the notation is standard (see e.g., SL). The expressions for the diffusion terms are the same as those used by SOONG and OGURA (1973), except that a uniform diffusivity of 10 m^2s^{-1} is used for all quantities. It is convenient to introduce a streamfunction ψ such that

$$\rho_a ru = \psi_z \quad , \quad \rho_a rw = -\psi_r \; , \tag{8}$$

and the 'pseudo' vorticity η, defined as $(\rho_a u)_z - (\rho_a w)_r$, is given by

$$\eta r = \psi_{zz} + r(r^{-1}\psi_r)_r \; . \tag{9}$$

Latent heat release is modelled exactly as in SL; briefly if q_v and q_{vs} $(= q_{vs}(\pi, T))$ denote the vapour and saturated vapour mixing ratios, respectively, we set $q_v = q_w$ and $q_l = 0$ if $q_w < q_{vs}$, and $q_v = q_{vs}$, $q_l = q_w - q_{vs}$ if $q_w > q_{vs}$. In this simple formulation, the rate of latent heat release is given by $\dot{h} = 0$ if $q_w < q_{vs}$, or $\dot{h} = -LDq_{vs}/Dt$ if $q_w > q_{vs}$, D/Dt being the substantive derivative, and the surface on which $q_w = q_{vs}$ defines the cloud boundary (see SL, p110, for further details).

The boundary conditions are:

On $z = 0$, $\psi = 0$, $u = 0$, $w = 0$, $v_z = 0$,

$\theta = \theta_a(0)$, $q_w = q_{va}(z)$, $\eta = r^{-1}\psi_{zz}$.

On $z = H$, $\psi_z = 0$, $u = 0$, $w_z = 0$, $v_z = 0$,

$\theta_z = 0$ if $q_l = 0$ or $-(L/c_p\pi)q_{vsz}$ if

$q_l \neq 0$, $q_{wz} = 0$, $\eta_z = 0$.

On $r = 0$, $\psi = 0$, $u = 0$, $w_r = 0$, $v = 0$,

$\theta_r = 0$, $q_{wr} = 0$, $\eta = 0$.

On $r = R$, $\psi_r = 0$, $(ru)_r = 0$, $w = 0$, and if

$u \lesssim 0$, $v = V_a$, $\theta = \theta_a(z)$, $q_w = q_{va}(z)$,

or, if $u > 0$, $(rv)_r = 0$, $(r\theta)_r = 0$ and

$(\eta/r)_r = 0$.

Here, V_a is a constant equal to 1 ms^{-1} and $\theta_a(z)$, $q_{va}(z)$ are based on the Dibble sounding. These conditions, which are not independent, are intended to permit free flow through the open boundaries and are broadly equivalent to those used in SL. Note, in this paper we assume free slip on the tangential velocity, but no slip on the radial velocity, at $z = 0$; this makes the conditions on v on $r = R$ and $z = 0$ compatible at the intersection of these boundaries.

Neither conditions are strictly appropriate for turbulent flow, but the calculations in SL suggest that the differences in model vortex behaviour with free slip or no slip on v are minimal; moreover such differences are easily anticipated (see e.g. BODE ET AL, 1975).

The Numerical Method

The numerical method is similar to that of SOONG and OGURA (1973), although we do not use a staggered grid. Essentially, v, θ, q_w and η are updated using (3), (5), (6) and (7), respectively, while ψ is obtained by solving (9) with the updated value of q. The meridional velocity components are then obtained using (8). The pressure field is found by integrating (2) radially, assuming $\pi'(R, H) = 0$, to give $\pi'(r, H)$; then (3) is integrated vertically to obtain $\pi'(r, z)$. The prognostic equations are integrated using a forward time, upstream space method until the flow attains a quasi-steady state as defined in SL. The grid size is 50 m in both directions, but in the vortex simulation, the calculations in the region $0 \lesssim r \lesssim 1$ km are repeated with a 25 m resolution, commencing from a time when the vortex begins to rapidly intensify, and using boundary conditions provided by the coarse mesh integration.

The Initial Conditions

The updraught is initiated by imposing a small temperature anomaly (+ 1K), distributed parabolically in the region $0 < r < 1$ km, $0 < z < 2$ km. The vapour mixing ratio is also increased in this region to maintain the ambient level of relative humidity. These anomalies are held fixed for 12 minutes, one third of the total integration time, and then phased out exponentially with an e-folding time of 1 minute. In the tornado simulation, an initial field of weak uniform rotation is also imposed, the tangential velocity being 1 ms^{-1} (= V_a) at radius R.

Results and Discussion

The non-rotating updraught simulation and tornado simulation are designated as experiments 1 and 2 respectively. The former is studied first.

(a) Non-rotating Updraught Simulation Figure 1 shows the radial profiles of vertical velocity at 1 km height intervals, the isentropic surfaces and the region of significant liquid water mixing ratio in the steady state of Expt. 1. The principal features are a narrow, intense updraught core, ½ km in diameter, surrounded by a weak annular downdraught. The flow is upwards on the axis, except below about 1 km, where there is slow downflow associated with a weak recirculation cell, the updraught being fed around this. The maximum updraught speed is 26.1 ms^{-1} at 8 km, a value which is less than the observed maximum of about 35 ms^{-1} at 5 km (DAVIES-JONES, 1974, p713), although the latter occurred as a sudden large fluctuation. The maximum temperature excess on the axis is about 7 K and occurs just below 8 km; this is 1-2 K lower than would be expected assuming pseudo-adiabatic ascent of surface air and is presumably a result of thermal diffusion. The annular downdraught, which attains a maximum speed of about 2 ms^{-1} at $z = 5$ km, is marked by an upward turning of the isentropes in Fig. 1b. The updraught is marked by a dip in the isentropes near the axis, as is the subsidence region below 1 km; the former is also indicated by the region of liquid water (Fig. 1c).

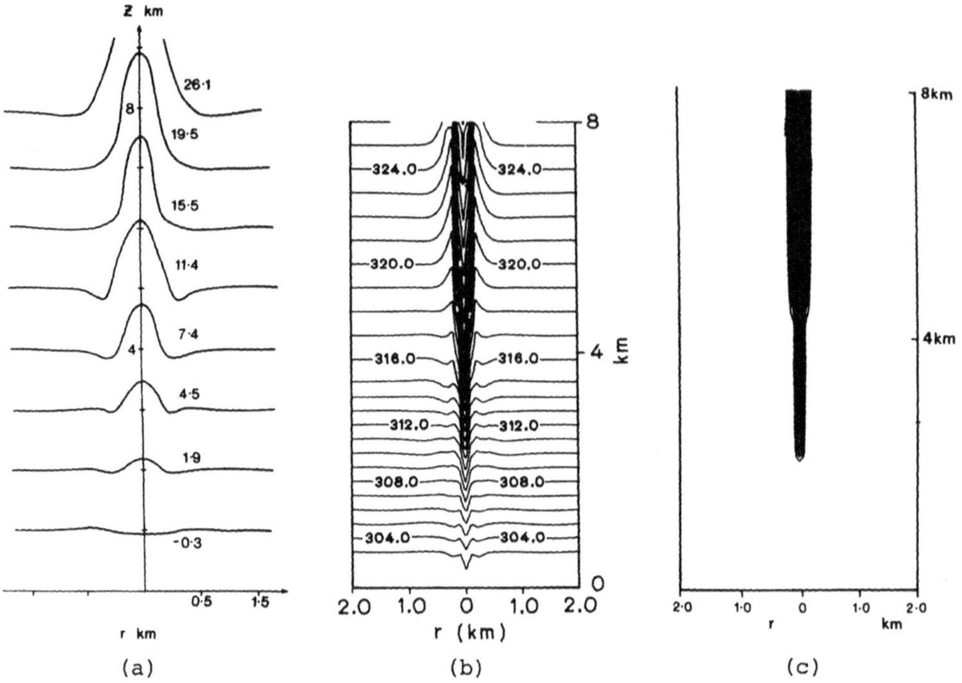

Fig. 1. (a) Radial profiles of vertical velocity in the steady state of Expt. 1; numbers alongside profiles label axial velocity in ms^{-1}. (b) isentropic surfaces; isopleths labelled in $^{\circ}K$. (c) region of significant liquid water (mixing ratio greater than 0.2 g kg^{-1})

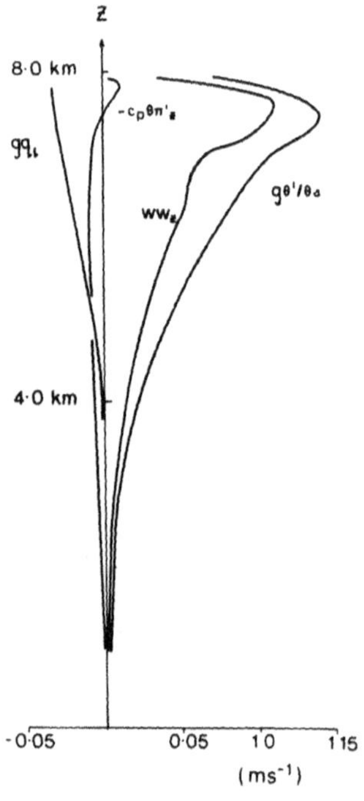

It is instructive to examine the steady state balance of forces in the vertical momentum equation near the axis to compare this with the vortex case. Figure 2 shows the principal forces at 50 m radius as functions of height for Expt 1. Above about 2 km, the most important of these is the 'pseudo' buoyancy force[1] $g\theta'/\theta_a$, with

[1] These are related to the actual buoyancy and pressure gradient terms by the formulae $g\theta'/\theta_a = - g\rho'/\rho - W$ and $- c_p\theta\pi'_z = - \rho^{-1}p_z + W$, where $W = (g\theta/\theta_a)[(\pi/\pi_a)^{1-(1/\kappa)} - 1]$; see Leslie and Smith (1978, p1286).

Fig. 2. Vertical profiles of the 'pseudo' perturbation pressure gradient, $- c_p\theta\pi'_z$; 'pseudo' buoyancy force, $g\theta'/\theta_a$; liquid water loading, $-gq_l$; and vertical acceleration at 50 m radius in the steady state of Expt. 1

the 'pseudo' perturbation pressure gradient[1] - $c_p\theta\pi'_z$ and the liquid water loading term $-gq_l$ making an important contribution also, especially at upper levels. The net force equals the vertical acceleration which is approximately ww_z near the axis. Note that the liquid water loading is everywhere a relatively small fraction of the 'pseudo' buoyancy force, showing that for the very unstable Dibble sounding, the water loading can be largely neglected.

The updraught width is rather less than that in SL and in this respect is less realistic[2], but the buoyancy levels are comparable with those observed in severe storms (DAVIES-JONES, 1974). The updraught is therefore suitable as a 'test-bed' for the tornado simulation described below.

(b) Tornado Vortex Simulation The growth of the vortex in Expt. 2 follows the pattern described in SL and is illustrated in Fig. 3 by the radial profiles of swirling velocity at selected heights after 20, 30 and 40 minutes of integration time. Note that the vortex intensifies first at intermediate levels (2 - 6 km; see Fig. 3a) and subsequently extends upwards and downwards. In the mature stage, after 40 minutes of integration, the maximum swirling velocity at the surface is 38.2 ms^{-1} at about 50 m radius, characteristic of a moderate intensity tornado.

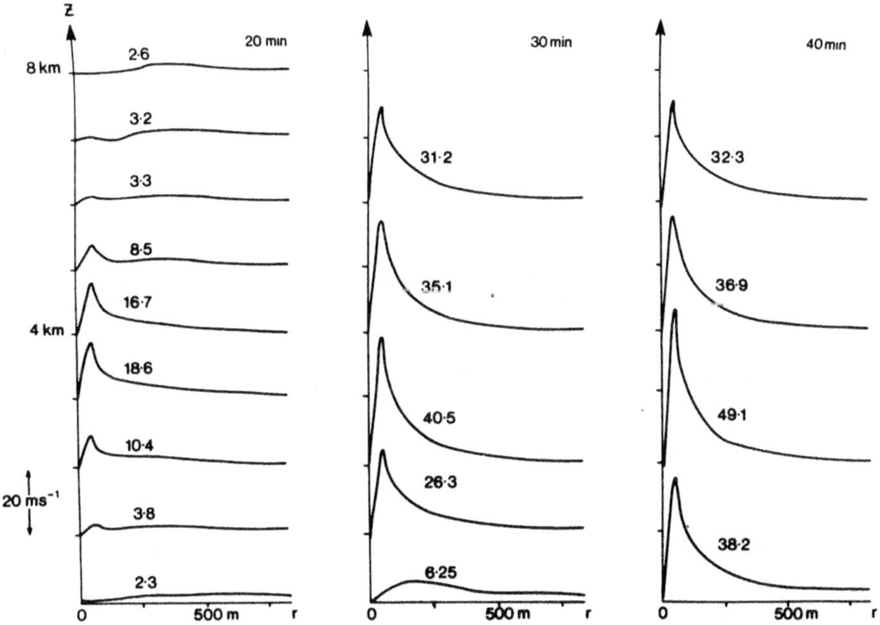

(a) (b) (c)

Fig. 3. Radial profiles of tangential velocity at selected height intervals in Expt. 2: (a) after 20 minutes; (b) after 30 minutes; and (c) after 40 minutes - the steady state situation. Numbers alongside curves indicate maximum velocities in ms^{-1}

---

[2] We expect that a broader initial temperature perturbation will lead to a broader updraught in future simulations.

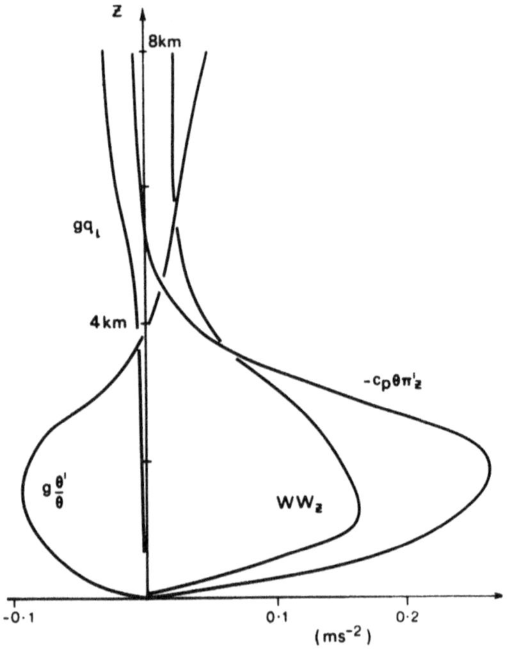

Fig. 4. Legend as for Fig. 2, except for Expt. 2 and profiles at 25 m radius

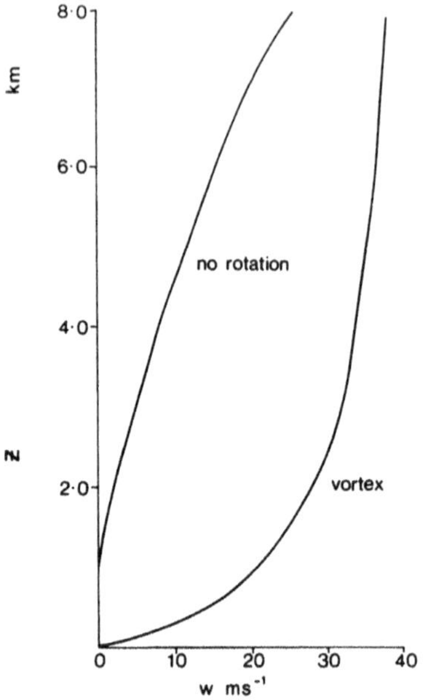

Fig. 5. Axial profiles of vertical velocity in the steady state for Expts. 1 and 2

The principal steady state balance of forces in the vertical momentum equation near the axis (r = 25 m) is shown in Fig. 4. The force fields are distinctly different from the profiles shown in Fig. 2. In the vortex case, the perturbation pressure gradient force is extremely large and positive in the lower part of the vortex, whereas the buoyancy force is negative below about 4 km where adiabatic cooling outweighs the condensational heating. However, the buoyancy force greatly exceeds the pressure gradient force at larger heights. Now liquid water is present at all heights above 600 m, about the lower end of the condensation funnel, but as in Expt. 1 it is never dominant.

Figure 5 compares the steady state axial profiles of vertical velocity between the non-rotating and vortex simulations. In the latter, the very large vertical accelerations at low levels contrast sharply with the non-rotating situation, and the large vertical velocities attained, especially at low levels, are quite consistent with photogrametrically deduced vertical wind speeds in tornadoes (GOLDEN, 1976).

Future Studies

The foregoing results are based on a preliminary analysis of the first two numerical experiments with our newly developed tornado model. Written initially on the basis of simplicity to reduce de-bugging time, the code is now being consolidated to make it more efficient. Other time integration schemes involving less implicit diffusion are also being considered. When this has been done, we propose to run further experiments, aiming at the simulation of a more realistic updraught width. It is also proposed to include a simulated tropopause at a higher level in the model, and to explore the energetics of vortex flows.

References

BODE, L., LESLIE, L.M. and SMITH, R.K. 1975. A numerical study of boundary effects on concentrated vortices with application to tornadoes and waterspouts. Quart. J.R. Met. Soc., 101, 313-324

BRANDES, E.A. 1981. Fine structure of the Del City-Edmund tornadic mesocyclone. Mon. Wea. Rev., 109

BROWN, R.A., LEMON, L.R. and BURGESS, D.W. 1978. Tornado detection by pulsed Doppler radar. ibid, 106, 29-38

DAVIES-JONES, R.P. 1974. Discussion of measurements of high-speed thunderstorm updrafts. J. Appl. Met., 13, 710-717

GOLDEN, J.H. 1976. An assessment of wind speeds of tornadoes. Proc. Symp. on Tornadoes: Assessment of Knowledge and Implications for Man, Texas Tech Univ., Lubbock, 5-42

GOLDEN, J.H. and MORGAN, B.J. 1972. The NSSL/Notre Dame tornado intercept program, spring, 1972. Bull. Amer. Met. Soc., 53, 1178-1179

LESLIE, L.M. and SMITH, R.K. 1978. The effect of vertical stability on tornadogenesis. J. Atmos. Sci., 35, 1281-1288

PROCTOR, F.H. 1977. Formation of a tornado from a rotating wind field: A numerical experiment. Proc. 10th AMS Conf. on Severe Storms, The Americal Meteorological Society, Boston, 315-321

SMITH, R.K. and LESLIE, L.M. 1979. A numerical study of tornado-genesis in a rotating thunderstorm. Quart. J. Roy. Met. Soc., 105, 107-127

SOONG, S.T. and OGURA, Y. A comparison between axi-symmetric and slab-symmetric cumulus cloud models. J. Atmos. Sci., 30, 879-893

A Numerical Simulation of Multiple Vortices

R. Rotunno

Introduction

It often occurs that tornadoes contain smaller subsidiary vortices which revolve about the tornado, which in turn, rotate in the same sense as the tornado (see FUJITA, 1970; FORBES, 1978). This phenomenon goes by various names ("suction vortices", "satellite vortices", and "secondary vortices"), however, for this work we follow CHURCH et al. (1979) and refer to the multiple vortex phenomenon (MVP). So, when we speak of the MVP we refer to that type of "suction vortex" which FUJITA (1976) has termed the "orbiting vortex". That the MVP may be more than a minor detail of the tornadic flow is suggested by damage surveys which indicate the most intense destruction of life and property is associated with cycloidal paths which FUJITA has termed "suction swaths". As with the tornado, very little is known about the internal circulation of the MVP. Photogrametric and ground survey data are inconclusive on such important questions as to i) what maximum wind speed is achieved by the tornado and ii) the relation between this and the MVP. MVs are clearly visible in photographs but it is extremely difficult to infer actual flow patterns.

WARD (1972) built a laboratory device intended to simulate tornadic flow. Among the facts which argue for the relevance of WARD's model is the result that for certain values of the swirl ratio (defined below), the single central vortex divides into two or more vortices. Until recently, the data from the WARD vortex chamber was, for the most part, from flow visualization techniques and surface pressure measurements. CHURCH et al. (1979) have reported on the internal flow, however, as of this writing have not measured the three velocity components separately as functions of space and time.

All analytical and/or numerical models of tornadoes to date are axisymmetric, i.e., with respect to a system of cylindrical coordinates, azimuthal variations are not allowed. Clearly, the physical system under discussion is essentially asymmetric. Simple analytical axisymmetric vortex models are of great utility for understanding some basic physical processes (see e.g., LEWELLEN, 1976) and provide a basic reference point for more complex numerical models which add certain refinements. However, for three-dimensional asymmetric vortices of the type discussed here even simple models are lacking. It is important to know how theoretical (based on axisymmetric models) estimates of the magnitude, location and duration of the maximum tornadic wind depend on the flow asymmetry.

Topics in Atmospheric and Oceanographic Sciences
Intense Atmospheric Vortices, ed. by Bengtsson/Lighthill

The aim of this work is to ascertain the general effect these flow asymmetries may have on the tornadic flow field. The principal research tool used is a fully three-dimensional numerical model in cylindrical coordinates; the model is based on WARD's physical model of a mesocyclone upon which WARD based his laboratory device. Principal findings are:

i) MVs arise as an instability of an axisymmetric flow as hypothesized by WARD (1972). Unstable waves grow to finite amplitude and become the MV. As an unstable wave grows, angular momentum is transported inward by the wave and the inner core is spun-up. The flow is thus stabilized and the large amplitude structures persist.

ii) The present model exhibits good agreement with certain features of the laboratory model, viz. the MV pattern progresses at approximately the local rotation rate of the flow at low levels. The local tangential velocity of the MV can be 20-30% larger than the azimuthal average at the radius of v_{max}, thus suggesting that MVs should indeed cause more local damage as has been observed. (Further, the 3-D model with MVs again has local tangential velocities between 20-30% higher than an equivalent axisymmetric model which uses the same external parameters.)

iii) The actual distribution of flow velocity within a MV is asymmetric with large positive vertical and tangential velocities on one side with reduced (or in some cases negative) vertical and tangential velocities on the other. This distribution is reasonable and can be explained as a natural consequence of the distribution of vorticity in the initially (unstable) axisymmetric flow. Other details of the flow are discussed.

Physical Model

BROOKS (1949) discovered the existence of a "wind and pressure system intermediate in size between the general parent low and the tornado funnel itself." The existence of a rotating updraft within severe storms was inferred by BROWNING (1964) from the hook-echo observed by radar.[1] The scale of the updraft diameter is typically O(5-12 km) (BROWNING,1977) which is at least an order of magnitude larger than a typical tornado funnel (O(100 m)). The inflow layer which feeds the updraft is O(1-2 km) in depth. The updraft radius, R, divided by the inflow layer depth h is termed the aspect ratio, h/R. WARD (1972) realized that all laboratory experiments purporting to model tornadoes used aspect ratios which were much too large, while the present discussion indicates h/R should be $\sim$.1-.5.

WARD's device is shown in Fig. 1a. There is a cylindrical chamber bounded above by a honeycomb grid over which suction

---

[1]That cyclonically curved echoes correspond to tornadoes was known from the mid-40's and documented by STOUT and HUFF (1953). BROWNING and LUDLAM (1962) first correctly identified the weak echo region as updraft; BROWNING (1965) later explained the hook as a consequence of updraft rotation.

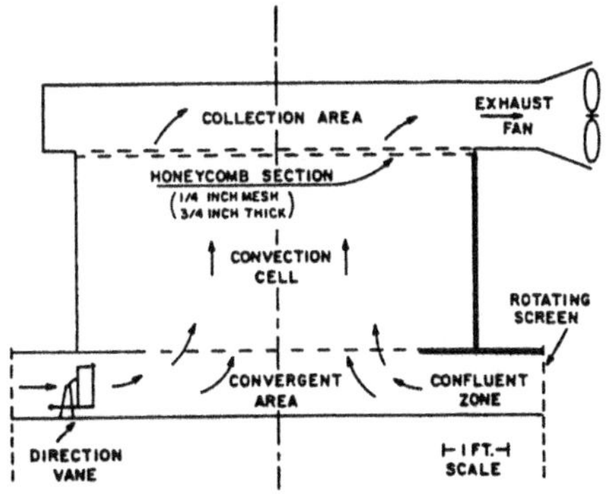

Fig. 1a. ↑ Schematic of WARD's (1972) laboratory device

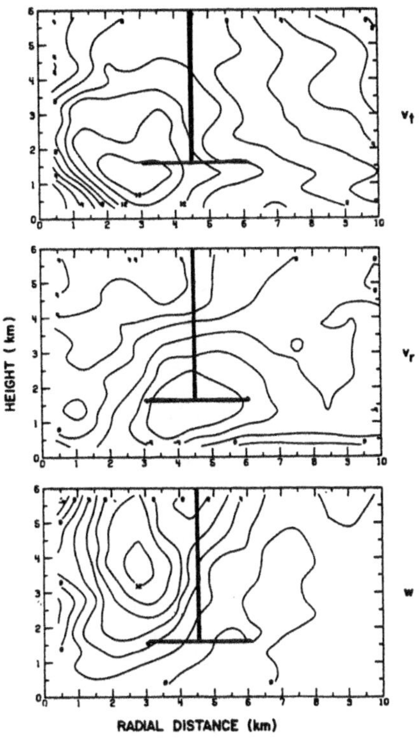

Fig. 1b. → Radial-height distributions of mean tangetial (v_t), radial (v_r), and vertical (w) wind (ms^{-1}) from dual-Doppler radar analysis. Radial distances are from the mesocirculation vertical axis. (After BRANDES, 1978)

is created by an exhaust fan. The rotating screen at the bottom imparts a specific amount of angular momentum to the in-rushing air.

Fig. 1b contains the azimuthally averaged (with respect to the apparent circulation center in the storm-relative frame of reference) tangential, radial and vertical velocities deduced from dual-Doppler radar coverage of a tornadic thunderstorm. So that the reader will have a clear idea of how we think the laboratory device relates to the data, we've sketched in, over the data in Fig. 1b, the basic outline of Ward's chamber. The numerical calculations reported here concern the flow within the "convergent area" and "convection cell" so marked in Fig. 1a, which would correspond to the flow below approximately 6 km altitude and within 3 km of the circulation center. It is important to note that within a domain of this size at least two important circulations are contained. The flow resolved by dual-Doppler radar in Fig. 1b is the mesocyclone; however, with the computing power now available, it is possible to resolve the mesocyclone together with more intense smaller scale vortex motion within a single numerical model. We will return to this point shortly.

Although no internal flow measurements were reported by WARD, a number of physical flow features could be ascertained from flow visualization techniques. Perhaps the most important piece of information obtained this way by WARD was that the core size of the small-scale vortex could be controlled by certain combinations of power-setting on the exhaust fan and rotation rate of the inflow screen. This combination is quantified by consideration of the swirl ratio, which is proportional to the ratio of the angular momentum supplied to the in-rushing air at the screen ($2\pi\Gamma_R$) to volume rate of flow (divided by the radius) through the chamber $2\pi\ Q/R$. Thus, the swirl ratio is defined

$$S \equiv \frac{R\Gamma_R}{2Q} \qquad [1]$$

We've yet to define what we mean by "core size"; in many studies "core" denotes the radius at which the maximum azimuthal velocity occurs. In the case of the familiar Rankine vortex (see e.g. LAMB, 1932) this is the radius at which the inner part, which exhibits solid body rotation, is separated from an outer irrotational flow. We have in the past differed with this definition. In comparing our results with the flow visualization experiments, we found (ROTUNNO, 1977) that the streamline which bounds the flow below separated from the surface at a certain radius and thereafter continues almost vertically out the top of the domain, thus, producing a pattern of flow where all of the swirling, in-rushing air is deflected upwards at r_c and thereafter exits through the "top" of the device. So as not to leave a vacuum in the region within r_c, air flows vertically downward through the top. Thus we may imagine a cylindrical boundary within the fluid where the radial velocity, u, is approximately zero. If, now, smoke (or any other tracer) is released within r_c and near the ground the smoke will fill this cylindrical region thus making visible a "core". This core size was found by Davies-Jones to depend on S. ROTUNNO (1977) who identified the radius of the separated boundary streamline with the size of the smoke-filled core of Ward's experiment found very good agreement between the laboratory core-size as a function of swirl ratio, $r_c(S)$ and that which resulted from the numerical integrations.

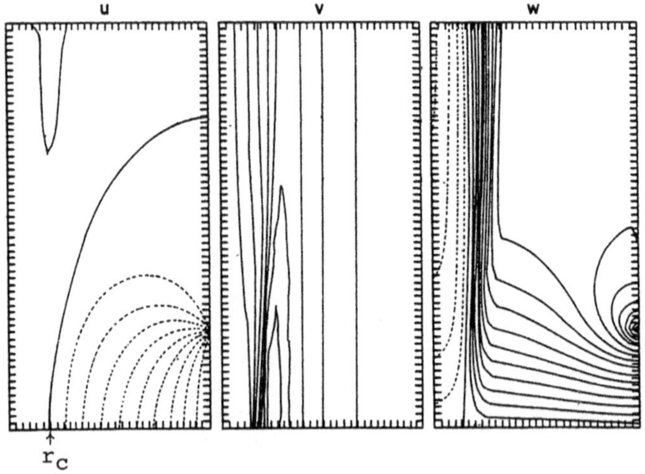

Fig. 2 . The steady-state patterns of u, v, and w for Re=1000 S=.4. The solid line cutting across the u contour plot is the zero line to the right of which the flow is inward and decelerating from the inlet value of -1.0 ft s^{-1}; the contour interval is 0.1 ft s^{-1}. The v field is contoured every 0.2 ft s^{-1}. The w field is contoured every 0.1 ft s^{-1}

Fig. 2 contains the steady-state radial, tangential and vertical velocities for a typical model experiment. The basic features of this solution compare favorably to the data displayed in Fig. 1b; radial inflow decreasing with decreasing radius, vertical velocity minimum (negative at some altitudes) on the central axis and increasing outwards and a tangential velocity maximum between the radius of maximum inflow and the radius where u vanishes are observed and simulated. So, we identify this flow with the mesocyclone. However, we do not wish to give the impression that in all situations there is a neat separation between tornadoes and mesocyclones. Some extremely large funnels (O(1 km)) may be regarded as small mesocyclones (AGEE et al., 1976). Multiple vortices are often observed in these large funnels (see e.g., FORBES, 1978).

Theoretical analysis of this type of flow are along the following lines: A non-rotating updraft may be thought of as being 'driven' by a pressure drop across the device, i.e., there is a conversion from potential to kinetic energy. Now, if angular momentum is conserved, the tangential velocity will increase with decreasing radius. This velocity cannot increase indefinitely since the kinetic energy will become much larger than the pressure drop available to drive the flow. Thus, it is hypothesized that the flow will not reach the center axis and separates at some finite radius r_c. This radius cannot be too large for if $r_c \approx R$, a too large pressure drop would be required to pass a specified volume flow through the thin annular ring. For a full discussion see LEWELLEN (1971) and DAVIES-JONES (1973).

The preceding discussion pertains to model solutions employing a free-slip (or zero stress) lower boundary condition. When no-slip conditions are employed the numerical solutions indicate an increased inward penetration of the flow near the bottom boundary (see ROTUNNO, 1979). Thus, the no-slip condition introduces another scale, i.e., the frictional boundary layer scale. The radius of v_{max} is much smaller at low levels than it is aloft where the flow is basically as it was in the free-slip case. So, as mentioned previously, there are two distinct types (and scales) of vortex motion within the domain of interest in an axisymmetric model.

If we relax the constraint of axisymmetry, there is another way that smaller, more intense vortices may form within the larger circulation. The core size (as defined above) increases with S; at a certain critical value the large vortex becomes unstable and smaller vortices form at r_c. Figure 3 is a tracing of the flow visualization experiment conducted by CHURCH et al. (1979, their figure 7). There are two intertwining helical smoke columns with apparent rotation of the same sign as that of the parent circulation ($r \lesssim r_c$). Note that the pitch of the helix which describes a smoke column is opposite in sign to the pitch of a (in general) helical streamline of flow round the multiple vortex system. Although there is no internal measurement, pressure measurements at the lower boundary indicate substantial pressure drops associated with the passage of each individual vortex (J.T. SNOW, private communication). It has been hypothesized by WARD (1972), DAVIES-JONES and KESSLER (1974), ROTUNNO (1978), and SNOW (1978) that these multiple vortices originate as an instability of the basic axisymmetric vortex. In this report, we examine the growth of small perturbations to the axisymmetric flow of fig. 2 using a three-dimensional numerical model. We are thus able to follow the development of perturbations as they grow to large amplitude. In the following section we briefly discuss the numerical analysis.

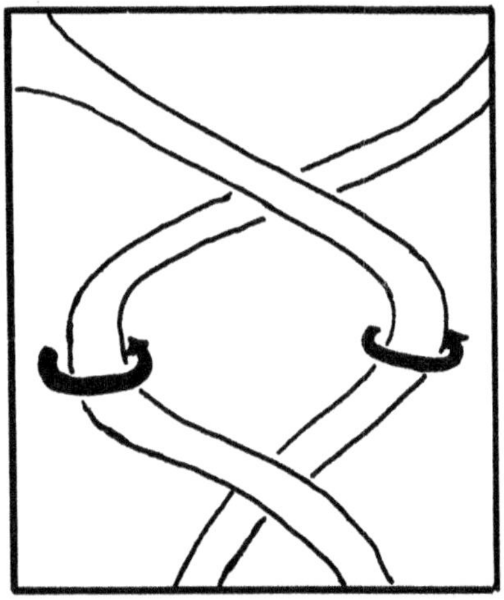

Fig. 3. A tracing of the flow visualization experiment conducted by CHURCH et al. (1979, their figure 7)

The Mathematical Model

The model developed here follows that of WILLIAMS (1969) closely. Williams developed a numerical model to study the flow in a rotating, differentially heated annulus; the Boussinesq equations are solved in cylindrical coordinates. Hence, much of WILLIAMS' work can be carried over directly. So, unless otherwise stated the model description in this section is identical, to WILLIAMS'. Our system is simpler, in one respect, because we have no thermal effects and need not solve a temperature equation, however, in another respect it is more problematic (i.e., fluid may pass into or out of the domain).

The governing equations for a constant density, constant viscosity fluid in cylindrical coordinates (r, Θ, z) are

$$\frac{Du}{Dt} - \frac{v^2}{r} = - P_r + Re^{-1}F(u) \tag{2}$$

$$\frac{Dv}{Dt} + \frac{uv}{r} = - \frac{P_\theta}{r} + Re^{-1}I(v) \tag{3}$$

$$\frac{Dw}{Dt} = - P_z + Re^{-1}H(w) \tag{4}$$

$$\frac{1}{r}(ru)_r + \frac{1}{r}v_\theta + w_z = 0 \tag{5}$$

where

$$(u,v,w) = \frac{1}{u_R}(u',v',w'); \qquad P = \frac{P'}{u_R^2}; \qquad (z,r) = \frac{1}{R}(z',r')$$

where u', v', w', are the dimensional components of the velocity vector in the r', Θ, z' directions, respectively; P' is the dimensional pressure divided by density and Re is the Reynolds number $u_R R/\nu$ based on the inflow speed u_R, chamber radius R and eddy viscosity ν. Subscripts denote partial differentiation. The functions F, I and H are the three components of $\nabla^2 V$.

In vector notation the system of equations is

$$\frac{\partial \mathbb{V}}{\partial t} + \mathbb{V} \cdot \nabla\mathbb{V} = - \nabla P - Re^{-1} \nabla x \omega \tag{6}$$

$$\nabla \cdot \mathbb{V} = 0 \tag{7}$$

where $\omega = \nabla \times \mathbb{V}$ is the vorticity vector. Equation (6) is prognostic for $\mathbb{V}$; to obtain P given $\mathbb{V}$, we take $\nabla \cdot$ (6) using (7) to obtain

$$\nabla^2 P = - \nabla \cdot \{\mathbb{V} \cdot \nabla\mathbb{V}\} \tag{8}$$

which is an elliptic equation for P. The four equations, (2) - (4) and (8), are solved numerically for the four unknowns u, v, w and P. The continuous equations are approximated by a second-order accurate finite-difference version on a staggered grid. Time integration is achieved by the 'leap-frog' method and the Poisson equation for P (Eq. 8) is solved by first taking the fast Fourier transform in the azimuthal direction, then employing a direct solver to obtain the solution for the Fourier components (functions of r and z). The time step is chosen to satisfy the CFL and diffusive stability criteria, viz. $\Delta t < \min (r\Delta\Theta/v_{max}, \Delta z/w_{max}, \Delta r/u_{max}$ and $\Delta t < \min Re/8 (r^2\Delta\Theta^2, \Delta z^2, \Delta r^2)$. For the parameters used the diffusive criterion was more strict. $\Delta t = .01$ and integrations were carried forward several thousand time steps. The number of grid points in the r, Θ and Z directions are, respectively, M=30, L=32 and N=30. Details of this technique may be found in WILLIAMS (1969).

The domain is a right circular cylinder which covers the central part of the vortex chamber shown in Fig. 4. Fluid enters through the sidewall for $o \leq z \leq h/R$ (=.5) and may leave or enter through the top ($z = H/R = 2$) for $a \leq r \leq 1$. A small center annulus (which

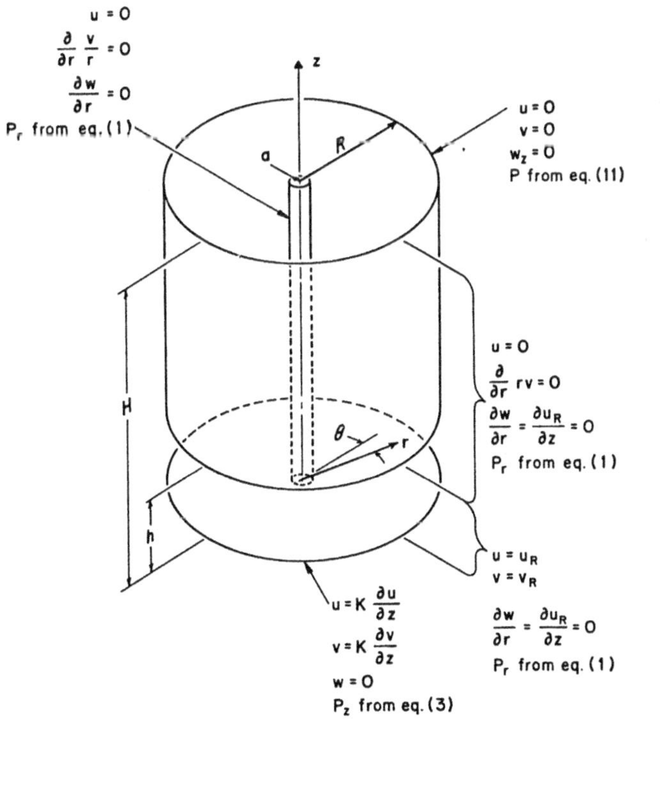

Fig. 4. Schematic of computational domain, boundary conditions are indicated

dows not exist in the laboratory or nature) is put into our model so as to avoid the severe time-step limitations imposed by the very small ($O(r\Delta\Theta)$) azimuthal grid lengths.

At $r = 1$, fluid enters the domain for $0 \leq z \leq .5$ with velocity $u = 1$ and with a specified azimuthal component $\bar{v} = v_R$. Since $\Gamma_R = v_R R$ and $Q = u_R Rh$, then from Eq. (1) $S = (R/2h)(v_R/u_R)$. For this study, $h/R = .5$. Thus, v_R/u_R is synonomous with the swirl ratio. For $.5 \leq z \leq 2$, $u = 0$ at a fictitious wall. At this wall, we require the vertical vorticity to be zero; $(rv)_r = 0$. Along the entire boundary ($0 \leq z \leq 2$) we require the azimuthal component of the vorticity vector $\eta = u_z - w_r = 0$. Thus $w_r = u_z$, which is prescribed because u is. Since u is prescribed as a time independent function of z at $r = 1$, we can use equation (2) to obtain

$$P_r\Big|_{r=1} = -\frac{D}{Dt}u + \frac{v^2}{r} + Re^{-1}F(u) \qquad (9)$$

At r=a, we set u=0 on the impermeable inner annulus, require zero stress ($(v/r)_r = 0$ and $w_r = 0$) and P_r is known at time τ from Eq.(2). At the lower boundary we require $w = 0$ at the impermeable surface and set $(u,v) = K(u,v)_z$, where K is a constant (in dimensional form, K has dimensions of length). When $K = 0$ the condition $u = v = 0$ (no-slip), while when $K \to \infty$ the condition becomes $u_z = v_z = 0$ (free-slip). G.I. TAYLOR (1915) first used this condition for a Ekman layer flow; later KUO (1971) also employed this condition. P_z is known at from Eq.(4). At the upper boundary, $z = 2$, we require the flow exit without radial or tangential velocity; $u = v = 0$. (As is done in the laboratory with a honey comb grid.) This is somewhat different than the conditions used by ROTUNNO (1977, 1979). In those studies $u = \partial v/\partial z = \partial\eta/\partial z = 0$ (η is azimuthal component of the vorticity vector). The reason for this is as follows. The axisymmetric flow could not have azimuthally travelling waves whereas the present three-dimensional calculation can and does; we found that, in addition to beeing more faithful to the laboratory apparatus, the present conditions vastly simplify the 'open' boundary condition problem. With $u = v = 0$. Equation (2) gives for P_r

$$P_r\Big|_{z=2}^{t=\tau} = -(wu_z)^{\tau} + \nu u_{zz}^{\tau-1} \qquad (10)$$

since the absolute value of P is irrelevant we choose P (r=a) = 0 and integrate (10) to obtain

$$P(r,\ \theta,\ z=2,\ t=\tau) = \int_a^r \{(-wu_z)^{\tau} + (\nu u_{zz})^{\tau-1}\}dr' \qquad (11)$$

So, we have pressure at the upper boundary.

Results and Analysis

We shall discuss one particular example ($S = 1$, $Re = 150$, free-slip lower boundary condition) in detail. For this particular set of parameters, the numerical solutions exhibit two intertwining helical vortices. Other combinations of the parameters (S, Re) lead to solutions exhibiting more or less multiple vortices, however, the basic physical structure of each individual vortex does not vary as a function of the number of vortices present. Hence, we shall concentrate on an example where only two multiple vortices form and analyze in some detail the associated three-dimensional flow pattern.

Initially, the flow has zero vorticity and v is set to zero everywhere. To obtain the initial u, w fields, we solve the following elliptic equation for u under the boundary conditions discussed.

$$\left(\frac{(ru)_r}{r}\right)_r + u_{zz} = 0$$

This procedure produces an irrotational convergent flow pattern.

Rotation is added for t > 0 at r = 1 by letting $v(r,\Theta,z,t) = v_R(1 - e^{-t/t_s})$ $(0 \leq z \leq .5)$ which is analogous to slowly (over a time period $t_s = 2$, in all cases) spinning up the rotating screen at the rim of the cylinder of the laboratory model. The swirling flow is then transported inward and upward by the initial u,w field. As explained previously, the swirling flow cannot reach a too small radius. Hence, the flow in the r-z plane develops a 'two-celled' structure. By the time t = 10, the flow reaches an axisymmetric steady-state configuration similar (except for the maximum observed in v near the top; this is due to the condition $v(^r,\Theta,2,t)=0$ used in this model) to that described by ROTUNNO (1977). Since the initial conditions are axisymmetric to within round-off error, the flow will remain axisymmetric unless it is unstable in which case small asymmetries could grow at an exponential rate. However, even at an exponential growth rate, it may take a long time for the instabilities to develop. Hence to conserve computer time, we add a small amount of random noise at a stage (t = 6) of flow development.

The first evidence of significant flow asymmetry in the horizontal contour plots is at approximately t = 22. Fig. 5 contains the azimuthally averaged u, v, w, P fields at t = 20. The flow is still essentially axisymmetric at this time. Figure 6 displays the same cross sections as shown in Fig. 5 a-d except at t = 60. The horizontal cross sections at $z = \Delta z$ are displayed in Fig. 7 . Considering the radial velocity together with the tangential velocity, clearly indicates smaller scale rotating wind field within a larger-scale rotating wind field. We identify these smaller scale vorticies as the multiple vortex phenomenon. Figure 8 is a composite of u, v, w, and P for one of the multiple vortices. Notice the horizontal circulation center nearly coincides with the pressure minimum gradient of w but in the updraft. The maximum azimuthal velocity is now approximately 2.1 as compared to the 1.6+ obtained in the axisymmetric flow shown in Fig. 6; this is an increase of approximately 30%. Just above this low level the pattern is essentially unchanged but shifts clockwise. The distance of the MV center from the central axis increases with height and the circulation pattern becomes more diffuse. The basic pattern shifts counterclockwise in time at a rate $\frac{\Delta\Theta_{pattern}}{\Delta t} \sim 2.36$ rad/s which may be compared with the mean rotation rate $\overline{V}(r,0)/r$ displayed in Figure 9.

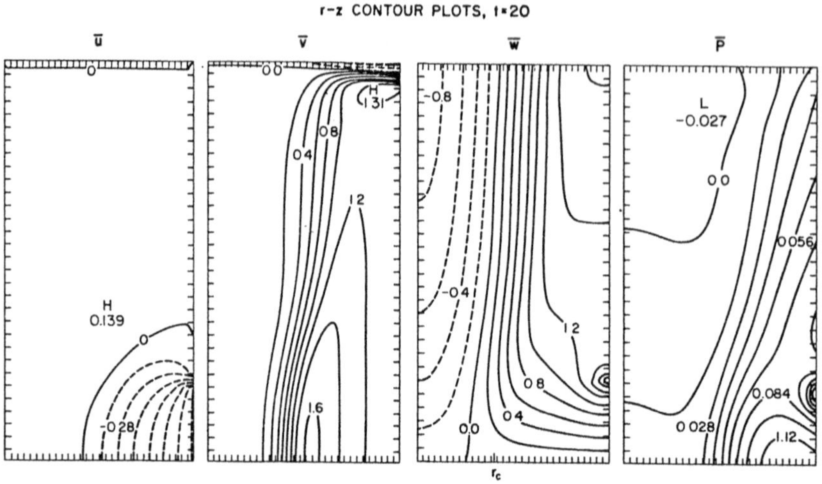

Fig. 5. Azimuthally averaged u, v, w, P fields at t = 20. The flow is still essentially axisymmetric at this time

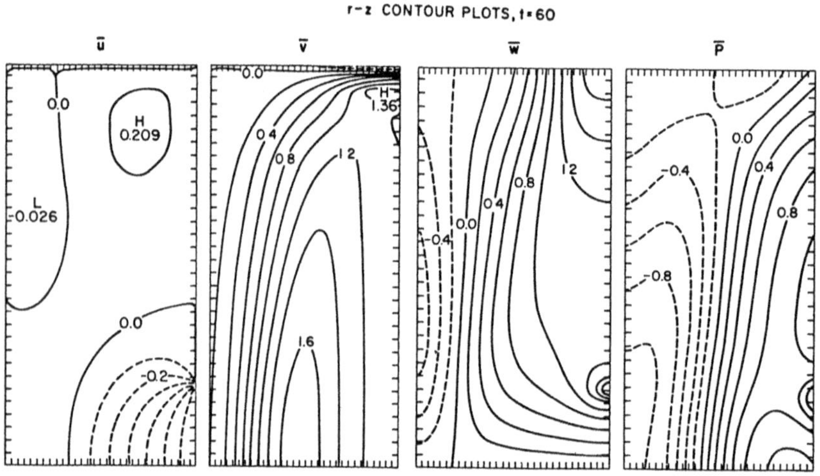

Fig. 6. Same as Fig. 5 except t = 60

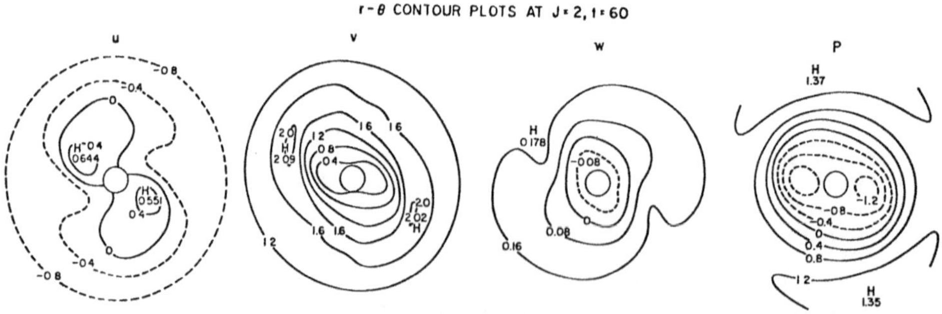

Fig. 7. Horizontal cross sections at z = z of u, v, w, P. Note the smaller scale rotating wind field within the larger scale rotating wind field

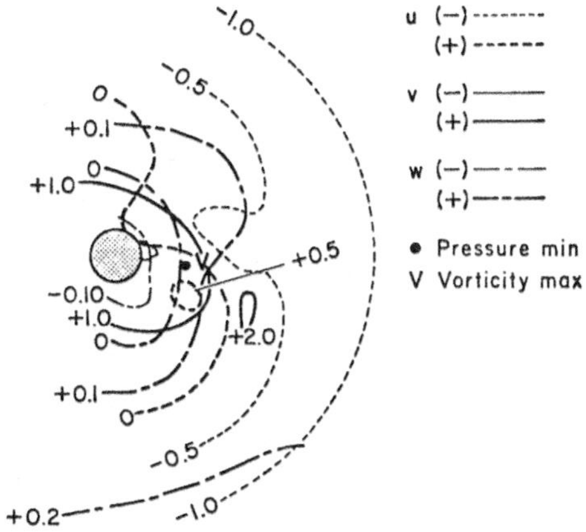

Fig. 8. A composite of u, v, w, and P for one of the multiple vortices

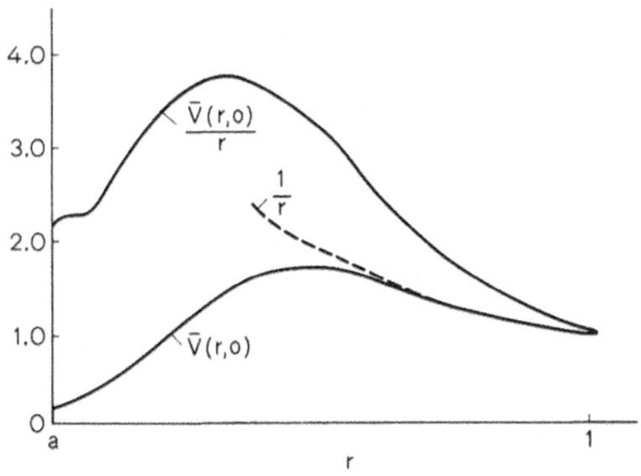

Fig. 9. Radial profiles of mean rotation rate and mean azimuthal velocity at z=0

Because the initial and boundary conditions of the model are axisymmetric, the multiple vortices must initially arise as an instabilily of this flow. What form does the instability take? Consider an idealization of the flow in Fig. 5; the flow for $r \lesssim r_c$ is downward and has very little tangential component. The flow for $r \gtrsim r_c$ is strongly swirling and rising. If the flow is nearly cylindrical (i.e., $\partial/\partial z = 0$), the mean vorticity at r_c is

$$\bar{\xi} = 0, \qquad \bar{\eta} = -\frac{\partial \bar{w}}{\partial r} < 0, \qquad \bar{\zeta} = \frac{1}{r}\frac{\partial \bar{\Gamma}}{\partial r} > 0$$

Thus, the vorticity vector points upward and in the negative azimuthal direction. If and when the flow becomes unstable, we expect the vorticity to accumulate along the helical vortex lines of the basic flow. The reason for the clockwise shift with height observed of the multiple vortex is apparent. Near the top of the domain the vortex lines are bent to the horizontal because u = v = 0, thus $\zeta = (rv)_r/r - u_\theta/r = 0$. This phenomenon is observed in the laboratory experiment described above (J.T. SNOW, private communication).

Finally, we analyse the increase in total energy which accompanies the transition to multiple vortices. Such an increase is, of course, not impossible since we have open system where energy may pass in and out. The energy equation is

$$\frac{\partial}{\partial t}\left(\frac{u^2+v^2+w^2}{2}\right) + \frac{1}{r}\frac{\partial}{\partial r}(ruH) + \frac{1}{r}\frac{\partial}{\partial \theta}(vH) + \frac{\partial}{\partial z} wH = Re^{-1}\, v \cdot \nabla^2 v \qquad (12)$$

where $H = P + \frac{u^2+v^2+w^2}{2}$, the total head. Integrating (12) over the domain gives

$$\frac{\partial}{\partial t} E = -\overset{I}{\iint} (uH)_{r=1} d\theta dz \; -\overset{O}{\iint} (wH)_{z=2} r dr dz \; + \overset{D}{\underset{\text{Volume}}{\iiint}} \text{Dissipation} \qquad (13)$$

H involves u, v, w and P and we specify conditions on the derivatives of w and P at inflow and outflow (so w and P may change).

Table I indicates that the presence of the multiple vortex does not allow the energy to pass out as quickly as it enters thus accumulation occurs until a balance between energy flux divergence and dissipation is reached.

Table I

T	E	I	O	D	I+O+D
0	2.25				
10	6.91	5.31	-4.31	-0.96	.04
20	6.90	5.20	-4.13	-0.94	.13
30	7.18	5.06	-3.77	-1.01	.28
40	8.16	5.27	-4.03	-1.22	.02
50	8.18	5.27	-4.04	-1.22	.01
60	8.17	5.28	-4.05	-1.22	.01

Due to limitations of computer time only a restricted combination of parameters could be studied; the cases studied are summarized in ROTUNNO & LILLY (1981).

Conclusions

We believe we've achieved the first successful simulation of the multiple vortex phenomenon. The highly asymmetric nature of the multiple vortex has precluded analytical descriptions heretofore; however, now a model based on solutions to the Navier-Stokes equations exists.

The model indicates that in a horizontal cross-section the circulation is centered on a strong gradient of the vertical velocity with the maximum tangential velocity on the side of maximum vertical velocity. The core of this asymmetric vortex shifts clockwise with height (the mean streamlines of the parent vortex are counterclockwise). The peak tangential velocity associated with the multiple vortices was found to be 20%-30% greater than either the azimuthally averaged larger scale flow or that obtained in an equivalent axisymmetric model. It is our guess that this amplification is probably underestimated by the present model due to the use of an eddy viscosity to parameterize turbulent transfer.

Acknowledgment

I would like to thank Prof. J. T. Snow (Purdue University) for a number of useful discussions. While preparing parts of this manuscript, the author was a visitor at Monash University where he had the benefit of several illuminating discussions with Profs. B. R. Morton and R. K. Smith.

The major portion of this work was funded by the U.S. Nuclear Regulatory Commission under contract no. 1531737.

References

AGEE, E. M., J. T. SNOW and P. R. CLARE 1976 Multiple vortex features in the tornado cyclone and the occurrence of tornado families. Mon. Wea. Rev., 104, 552-563

BROOKS, E. M. 1949 The tornado cyclone. Weatherwise, 2, 32-33

BROWNING, K. A. 1964 Airflow and precipitation trajectories within severe local storms which travel to the right of the mean wind. J. Atmos. Sci., 21, 634-639

_____ 1965 Some inferences about the updraft within a severe local storm. J. Atmos. Sci., 30, 533-545

_____ 1977 The structure and mechanisms of hailstorms. Meteor. Monogr., 20, 533-545

_____, and F. LUDLAM 1962 Airflow in convective storms. Quart. J. Roy. Meteor. Soc., 88, 117-135

BRANDES, E. A. 1978 Mesocyclone evolution and tornadogenesis: some observations. Mon. Wea. Rev., 106, 995-1011

CHURCH, C. R., J. T. SNOW, G. L. BAKER, and E. M. AGEE 1979 Characteristics of tornado-like vortices as a function of swirl ratio: a laboratory investigation. J. Atmos. Sci., 36, 1755-1776

DAVIES-JONES, R. P. 1973 The dependence of core radius on swirl ratio in a tornado simulation. J. Atmos. Sci., 30, 1427-1430

_____, and E. KESSLER, 1974: Tornadoes. Chapter 16, Weather and Climate Modification, ed. by W. N. HESS, John Wiley, N.Y., 552-595

FORBES, G. S. 1978 Three scales of motion associated with tornadoes. Final Report to the U.S. Nuclear Regulatory Commission (NUREG/CR-0363) RB, 359 pp

FUJITA, T. T. 1970 The Lubbock tornadoes: a study of suction spots. Weatherwise, 23, 160-173

_____ 1976 History of suction vortices. Proc. of the Symposium on Tornadoes, Texas Tech Univ., 28-88

KUO, H.-L. 1971 Axisymmetric flows in the boundary layer of a maintained vortex. J. Atmos. Sci., 28, 20-41

LAMB, H. 1932 Hydrodynamics, Dover Publications, 738 pp

LEWELLEN, W. S. 1971 A review of confined vortex flows. NASA Rep. CR-1772, 219 pp

_____ 1976 Theoretical models of the tornado vortex. Proc. of the Symposium on Tornadoes, Texas Tech Univ., 107-144

ROTUNNO, R. 1977 Numerical simulation of a laboratory vortex. J. Atmos. Sci., 34, 1942-1956

_____ 1978 A note on the stability of a cylindrical vortex sheet. J. Fluid Mech., 87, 761-771

_____ 1979 A study on tornado-like vortex dynamics. J. Atmos. Sci., 36, 140-155

_____ 1981 A numerical model pertaining to the multiple vortex phenomenon. Final Report to U.S. Nuclear Regulatory Commission (NUREG/CR-1840), 51 pp

SNOW, J. T. 1978 On inertial instability as related to the multiple vortex phenomenon. J. Atmos. Sci., 35, 1660-1671

STOUT, G. E., and F. A. HUFF 1953 Radar records Illinois tornadogenesis. Bull. Amer. Meteor. Soc., 34, 281-284

TAYLOR, G. I. 1915 Eddy motion in the atmosphere. Phil. Trans. Roy. Soc. London, A215, 1-120

WARD, N. B. 1972 The explanation of certain features of tornado dynamics using a laboratory model. J. Atmos. Sci., 49, 1194-1204

WILLIAMS, G. P. 1969 Numerical integration of the three-dimensional Navier-Stokes equations for incompressible flow. J. Fluid Mech., 37, 727-750

The Laboratory Modelling of Atmospheric Vortices: A Critical Review

T. Maxworthy

Abstract Recent experiments on concentrated vortices produced by swirling flow into a constricted, cylindrical tube have resulted, we believe, in a new understanding of vortex flows in a variety of similar devices, the main purposes of which have been to simulate atmospheric vortices. As a major part of this review we present a synopsis of these new results and show their connection to various laboratory models of atmospheric vortices. The utility of such models in the study of large scale atmospheric flows is also discussed.

I. Introduction In this review we concentrate on a description and interpretation of laboratory models which have been or could be used to reveal the characteristics of concentrated atmospheric vortices. In this we take a rather radical point of view since we plan to interpret this work using knowledge gained from a recent series of detailed experiments on vortices in a tube designed to simulate flows in certain components of industrially-important turbomachines. As we hope to show, the apparatus used in both cases, although different in scale and purpose, are actually quite similar and the flow fields created are closely related. It also happens that numerous other vortex experiments can be described and classified using these results as we will show by reference and actually describe in a few representative cases. We have found that most interest has been centred on tall, columnar vortices of the dust devil or tornado type with little emphasis on flat vortices of the hurricane type, of which there is only one example known to this author and which is discussed later.

Several interesting reviews on vortex motion have been published over the past several years. The first of importance is included in a general article on "Experimental Analogies to Atmospheric Motion" by FULTZ (1951) which is helpful because it summarizes succinctly most of the results of 150 years of work on the subject. FULTZ'S conclusion that very little useful quantitative work had been performed to that time has been recently criticized in PETERSON (1976) by PETERSON himself who quotes work by LETZMAN (193) and (1933) on detailed measurements in laboratory vortices. In Fig. 1, we show a picture from FULTZ (1951) of early observations by WILKE (1785) with the first indications of a vortex breakdown of which this author is aware. In all, FULTZ, in his usual thorough way, quotes some 32 older articles, the majority of which are qualitative in nature and serve to show that, if nothing else, columnar vortices are very easy to produce in a variety of different ways.

More recently MORTON (1966) has reviewed much of the observational background and discussed modelling parameters appropriate under a variety of different circumstances. This was followed by a few years ago by a series of articles in PETERSON (1976) by W.S. LEWELLEN, D.K. LILLY, and especially important from our present point-of-view, an article on "Laboratory Simulation of Tornadoes" by R.P. DAVIES-JONES. Also of great interest in this proceedings volume is much of the discussion between the various attendees which shows quite clearly the confusion over the role of laboratory models in the study of atmospheric phenom-

Topics in Atmospheric and Oceanographic Sciences
Intense Atmospheric Vortices, ed. by Bengtsson/Lighthill

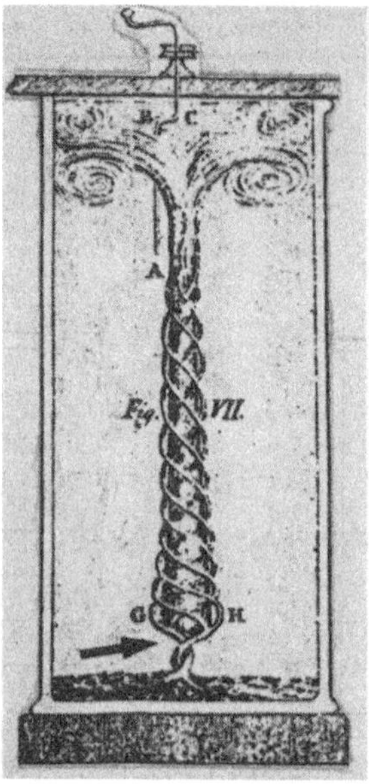

Figure 1. Drawing from FULTZ (1951) of an experimental vortex from WILKE (1785). A vortex breakdown at the wall is indicated by the arrow! The flow was produced by rapidly rotating a bent wire in the top of the tank

ena and a general lack of a consensus over the critical features to be incorporated in such models. Here it is not our intention to repeat the material presented by these authors, we wish to concentrate on those features which may be new or which have not presented forcefully enough previously.

From all this work it is clear that two essential features are required to produce some kind of vortex, one is a source of vertical flow force, and the other a source of angular momentum. We note that a primary source of vorticity is not required, although it may be present and important, since we will show this can be generated locally under suitable circumstances! Taking each of these in turn; the flow force has been shown to be generated in at least two ways; by buoyancy associated with either latent heat release or sensible heat in the column itself, is in hurricanes and probably tornadoes in the former case and dust devils in the latter, and/or by a buoyancy driven inertial updraught that is coherent over a "large" radial and vertical expanse. Often the buoyancy force in such cases acts locally e.g. aloft in the case of the "mother" cloud of a tornado spawning a cyclone. Only in the case of the dust-devil is the source clear and free of contention. Here the buoyancy force, due to the concentration of sensible heat from levels close to the ground into the rising column, is unquestionably the major force in operation, as has been successfully modelled in a number of recent experiments.

The source of vertical vorticity is an even more difficult problem with several mechanisms competing for attention and none unambiguously the favourite. LILLY in PETERSON(1976) spends considerable effort to show that a large scale ($\sim$ 5km radius) updraft can eventually concentrate sufficient planetary vorticity to initiate a tornado itself or at the very least a tornado cyclone. Baroclinic generation of vorticity, at fronts for example, is an important source of vorticity that can then be further amplified by convective activity to form concentrated vortices. In fact, numerous authors, (e.g. MORTON (1966), LILLY (1976), MAXWORTHY (1973)) have invoked this latter stretching of pre-existing vertical vorticity by convective activity as a method of generating a sufficient concentration to form a vortex column under many circumstances and this seems a very plausible mechanism for most tornado types of flow, for example. However, it has been suggested (MAXWORTHY 1973) that often such eddies are unlikely to be large enough to explain the long lives led by some types of vortices, dust devils especially. It appears that often some other source is necessary and it is possible that the boundary layer of the ambient wind could act as such a source since the vortex is able to tip this horizontally oriented vorticity into the vertical direction. MORTON (1966) suggests a similar mechanism associated with the randomly oriented ensemble of vortex elements associated with the atmospheric boundary layer. Here we extend this idea and suggest that the normal "bursting" phenomenon associated with the eruption of "viscous sub-layer" fluid, as found in all experiments on neutral turbulent boundary layers may be greatly modified by buoyancy and cause the elements of vorticity oriented in the direction of the free stream to rise very vigorously to form a small vortex. This vortex could then both gain further vorticity and buoyancy from the layer closest to the ground, i.e. we suggest that dust devils in particular, may start as buoyant sub-layer bursts!! Furthermore, vorticity generated by separation from topographic feature, either natural or man-made (e.g. buildings) may also be important in initiating vortex action.

Experimental models of these processes have, in all cases but two, concentrated on simulating a source of angular momentum and a source of vertical flow force in an axisymmetric geometry. They have, in general, avoided the problem of simulating the motion of a vortex in a mean wind moving over a surface. Clearly such modelling would be very difficult on any reasonable scale and it is unlikely that such an effort will be made in the foreseeable future. Based on results obtained up to now, it seems unlikely that the ommission of this feature has resulted in any serious loss of understanding over the dynamics of the main, upper, part of the vortex, but clearly features associated with such motion near the ground e.g. suction spots (FUGITA 1972) and details of vorticity transport (MAXWORTHY 1973) may not be faithfully reproduced. In what follows we concentrate on axisymmetric models, while discussing the others in

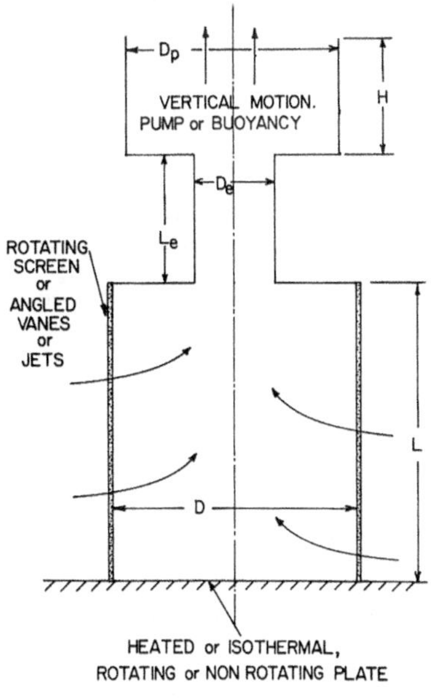

Figure 2. Canonical form of apparatus. The majority of known experiments fit this scheme

less depth; one (FUGITA 1972) because it does not appear to have the correct ingredients for an adequate model and the other (MAXWORTHY, private communication) because it is not well documented.

II. Experimental Models In reviewing all known experimental apparatus, it has become clear that they are nearly all variations of one general model that is shown in Fig. 2. This "canonical" form of apparatus consists of a bottom plate of diameter D that may or may not be rotating, surrounded by a cylindrical surface of height L at which angular momentum can be added to the inflowing fluid. This may consist of angled, closely-spaced vanes, a rotating screen or one or more tangential jets. At the upper end the chamber is terminated by a orifice of diameter D_e and length L_e followed by a second chamber of length H. Fluid is circulated through these chambers either by a means of a pump or by buoyancy forces. The latter, in turn, are generated either by heating the bottom plate (in which case the force acts over the whole length of the vortex column) or by buoyancy introduced near the exit (in which case it acts essentially as a pump). In some cases e.g. in the tubes used to study vortex breakdown, both the bottom surface and the entrance to the contraction section may be given smooth contours to promote and stabilise the breakdown at a given location within the tube L_e.

We have emphasised this interrelationship between all of the available models because we wish to show how they are connected to one particular apparatus that has been explored in great detail and which can be used to "explain" the features of less well documented experiments. The model has been built at the Brown-Boveri Co. (B.B.C.) Research Centre by a group headed by Dr. M. P. ESCUDIER in order to model flow in certain turbomachinery components. Dr. ESCUDIER'S discussion of his work is included as a separate paper in the proceedings of this meeting and the reader is referred to that paper for greater detail. It is also critical that the B.B.C. measurements were all performed with a non-intrusive laser Doppler anemometer which did not disturb the flow. The majority of the other experiments to be discussed measured velocities with pitot-tubes or hot wires which almost certainly affected the very flow they were supposed to measure. (A fluid dynamical version of Heisenberg's Principle!). Rotating flows are very sensitive to such perturbation and all intrusive measurements should be regarded with suspicion. The results of interest here will now be summarized.

A. Inertially Driven Models In all cases described here fluid is pumped mechanically through the system and the bottomboundary is nonrotating.

a). The B.B.B. Model* Here the tube diameter, D, is 55 mm (see Fig.2) while all of the other quantities L, D_e, L_e, D_p and H can be varied independently over wide ranges. The only flow variable is the flow rate Q, while angular momentum is added to the flow by means of a single tangential slit of width, t, at the cylindrical surface. The characteristics of the vortices produced in such a geometry, except when L is small of the same size at D, is shown simply by reference to figures 3 and 4. For a given geometry the photographs of figure 3 show three classes of flow, designated classes A, B and C that appear sequentially as the Reynolds number $R_e = Q/L\nu$, for example, is increased (see E2 and E4). In a Class A flow, a jet-like axial velocity profile (see figs. 3a and 4a) exists throughout the tube which breaks

*This model has been described in details in a series of articles by ESCUDIER, BORNSTEIN and ZEHNDER (1980) [E1]; ESCUDIER, BORNSTEIN, ZEHNDER & MAXWORTHY (1980) [E2]; ESCUDIER and ZEHNDER (1982) [E3]; ESCUDIER, BORNSTEIN, ZEHNDER & MAXWORTHY (1982) [E4], and ESCUDIER (1981) [E5]. The designations (E1 - E5) respectively are used in what follows to refe to these papers.

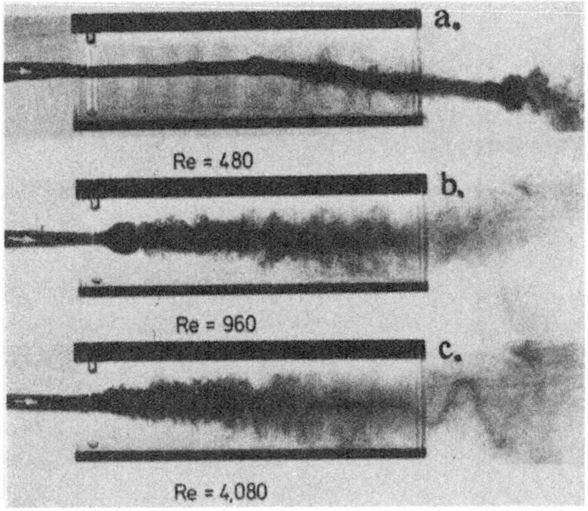

Figure 3. Photographs of the three major flow classes found in E2, 4. a). Class A, b). Class B, c). Class C. Flow from left to right

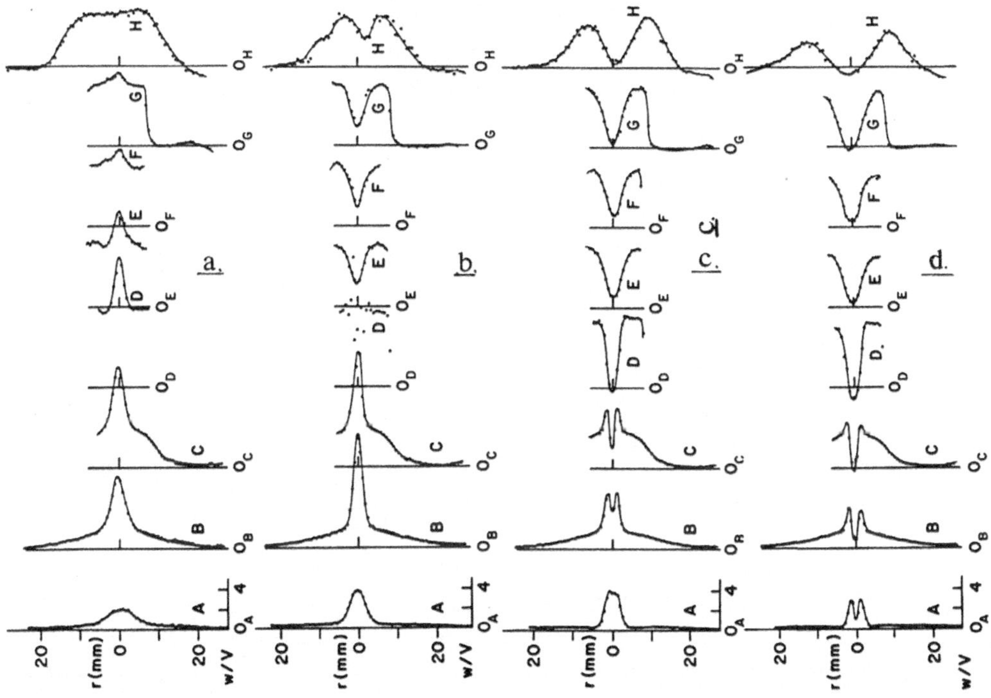

Figure 4. Axial velocity profile for each class a). Class A (Re = 480), b). Class B (Re = 960), c). Class C, d). Class C at higher Reynolds number (Re = 4080) [From E2,4]

down beyond the exit. As R_e is increased, this breakdown moves into the contraction tube to form a Class B flow with a jet-like axial profile upstream and a wake-like profile downstream. As shown in E3 and E4 by comparison to BENJAMIN'S (1965) theory of vortex breakdown and by analogy to open-channel flow (figure 5) this breakdown represents a transition from a supercritical to a subcritical state. In E4 we

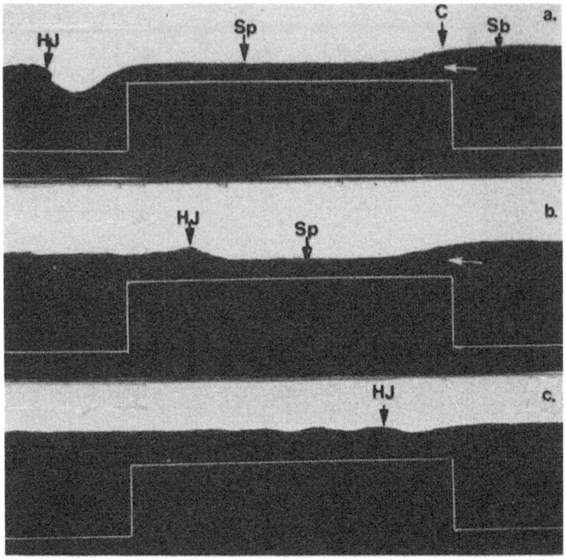

Figure 5. Open-channel analogous to a). Class A with a supercritical flow in the contraction undergoing a jump at the exit; b). Class B here the jump moves into the contraction; c). Approach to a Class C flow. Flow is almost subcritical everywhere. Jump is very weak. Actual Class C flow has an almost flat surface and is not worth showing; where C, S_b and S_p indicate critical, subcritical and supercritical flows respectively. HJ shows position of the hydraulic jump. Flow from right to left. (MAXWORTHY, private communication)

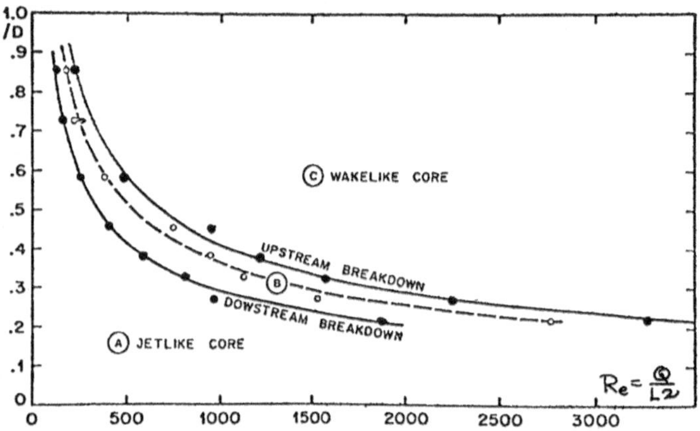

Figure 6. Regime diagram showing variation in Re for each flow class for fixed L, Le, t and variable De [From E2,4]

also discuss in great detail the connection of this vortex breakdown to the stability of the flow before and after the jump and are able to rationalize the appearance and characteristics of this largely misunderstood feature. Further increase in R_e results in the movement of the jump to the entrance to the contraction (figs. 3b and 4b Class B) at which point it becomes weaker and weaker to finally disappear as R_e is increased still further (figures 3c and 4c Class C). The axial velocity profiles show that the wake-like profiledownstream of the jump penetrates further into the main part of the tube and that the flow is now sub-critical everywhere (figure 4c and d). Experiments on a variety of different geometries show that this sequence is always repeated except that the A-B and B-C transitions occur at different values of R_e in each case (figure 6). We note here for future reference that as $L_e \to 0$ these two transition curves become one and no

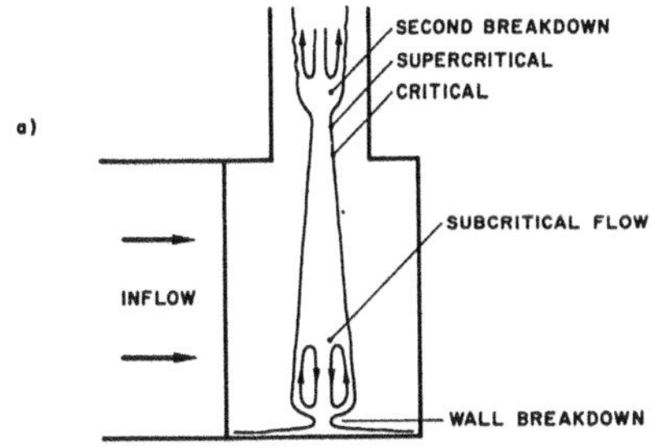

Figure 7. Diagrammatic view of the overall flow field found in a). A Vortex tube for which L >> De, D; showing the formation of a vortex jump at the ground plate. In c). L is small and the wall jump moves into the contraction. d). Increasing Re moves it back to the wall. b). A second contraction causes the subcritical flow to become supercritical with a second breakdown c.f. fig. 7a

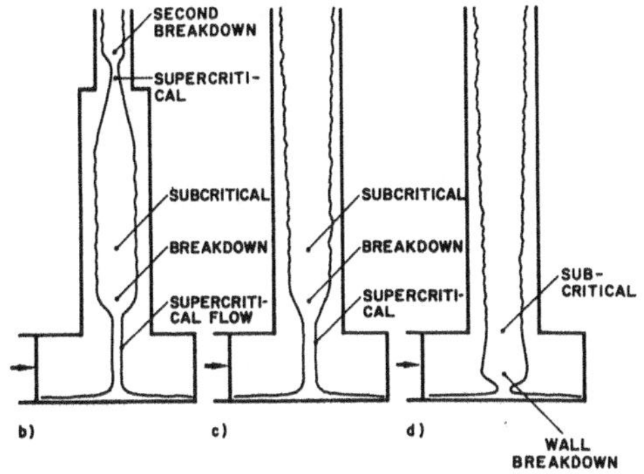

Class B then exists. These results have been rationalized in E3, in which the parametric dependency of R_e on the other appropriate dimensionless parameters is clearly revealed (see E3 and E5). In particular at large R_e, the Class C flow is invariable found unless De/D is quite small. When L is large, a vortex breakdown also occurs on the wall as a result of the supercritical eruption of boundary layer fluid (Fig. 7a) (See MAXWORTHY 1972 and section II Ad). When L is not large however, the interaction between the pressure field due to flow into the contraction and this wall breakdown can cause it to move into the contraction itself (fig. 7c) and only further geometrical changes (fig. 7b) or increases in R_e (fig. 7d) can cause either changes in the subcritical flows downstream or in the location of the jump. It is important to note that often the flow downstream of the vortex jump consisted of a spiral core or cores with a sense opposite to that of the basic rotation. (See fig. 3c) In E4 we relate this observation to the rapid spatial growth of spiral instabilities to finite amplitude in the wake-like axial velocity profile created by the vortex breakdown itself. Similar observations of multiple cores have also been seen in the wall breakdown (MAXWORTHY 1972).

Finally, a strong dependence of core diameter (d) on geometry (especially D_e) has been found with (d) also a function of R_e for the Class A jet-like profiles and independent of R_e for the Class C wake-like velocity profiles. As shown in E3 and E5, one possible set of scaling parameters is a Reynolds number $R_e' = W_e De/\nu$; a measure of the angular momentum imparted to the flow, a swirl parameter $\Omega = \Gamma/W_e D_e$ and the ratio of radial to tangential velocity at the inlet section i.e. the cotangent of the in flow angle $R = U/V$. Where $W_e = 4Q/\pi D_e^2$ $U = Q/\pi DL$ and $V = \Gamma/\pi D$ while Q is the volume flow rate and Γ the circulation at the inlet section. All other investigators have used similar quantities defined in slightly different but equivalent ways.

In all of the cases to be considered in the rest of this discussion, the geometries are fixed except in a few cases where minor changes can b made (e.g. WARD (1972), CHURCH & SNOW 1979). This in turn means that the only available variables to change the flow classes are related to the flow rate and the input of angular momentum at the cylindrical wall. This is equivalent in the B.B.C. experiment to having control over only the flow rate Q and the width of the inlet slit, t, restrictions which would have made interpretation of the data very hard. This probably accounts for the difficulty most investigators appear to have had in placing their flows into any sort of reasoned perspective, and makes the B.B.C. experiments particularily useful as a unifying concept.

b). The "Ward" Vortex Generator In 1972 the late NEIL WARD reported on a model based on what he believed to be the important flow processes occuring in a tornado. This apparatus has since been used by JISCKE and PARANG (1974), DAVIES-JONES (1973) and in slightly modified form by LESLIE (1977). A second version was built at Purdue University and several papers have resulted: SNOW, CHURCH, BAKER, and AGEE (1977) [P1] CHURCH, SNOW and AGEE (1977) [P2], BAKER (1978) [P3], CHURCH, SNOW, BAKER (1979) [P4], CHURCH and SNOW (1979) [P5], SNOW CHURCH and BARNHART (1979) [P6], BAKER and CHURCH (1979) [P7], BARNHART (1979) [P8], SNOW and PAULEY (198) [P9], SNOW and CHURCH (1981) [P10] We will use the designations [P1]-[P10] to refer to these papers in what follows. Both pieces of equipment fit our canonical classification scheme precisely, as is shown in the following table.

	L_e	D_e	L	D	D_p	H	Screen Rot^n Rate	Q
Ward simulator (Oklahoma)	0	0·3-1·22m	0·3-0·61m	2·44m	1·83m	0·91m	0·5-7p.p.m.	0·62-1·56m^3/
Ward simulator (Purdue)	0	0·4-1·22m	0·17-0·61m	3·04m	2·84m	1·66m	0-11r.p.m.	0·24-2·03m

Angular momentum was added to the flow at the rotating screen shown in figure 8. (From CHURCH and SNOW 1979 Contrary to the case where angular momentum is supplied by vanes or jets, it appears that a rotating screer

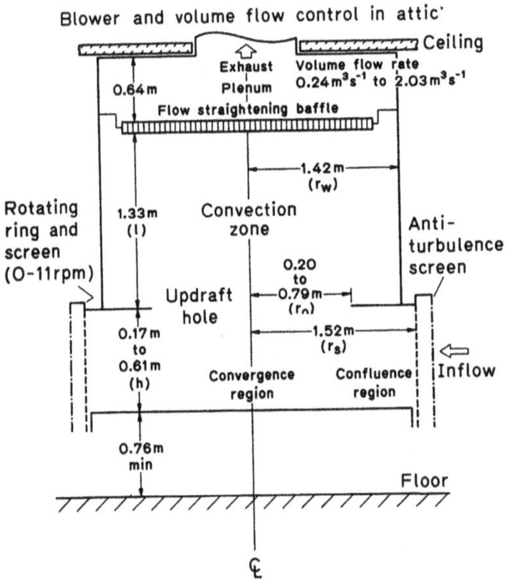

Figure 8. Diagram of Ward type of vortex generation from P5 c.f. figure 2

adds net vertical vorticity to the flow under some circumstances and this augments the central vorticity that comes from tipped boundary layer vorticity and that produced at the centre line.

Despite WARD'S (1972) reasoned justifications for the geometry of his tornado simulator and even though he gave the various regions of the device names suggesting they had meteorological significance (e.g. confluence zone, convergence area, convection cell) his device is no more no less than one more example of our canonical form and hence has the characteristics of such a device. The majority of the results of the Oklahoma investigators fit into the classification scheme developed in E2, E3, and E4, as we will show in some detail in what follows. This means, of course, that the vortex characteristics depend crucially on the geometry e.g. De and L in their case, which have no counterpart in natural flows. One must therefore view their results very critically before using them as models for such flows. In particular it has been claimed, because such devices produce multiple vortices and some natural storms have been observed to exhibit a similar feature, that they are good tornado simulators. However, it is clear from viewing such simulations that the diameter of the exit hole (De) acts as a definite constraint on the location of the vortices and must be important in determining their dynamics. One is immediately reminded of the results of HIDE and TITMAN (1968) on the stability of rotating shear layers created by differential rotation of a disc, as also pointed out by DAVIES-JONES (1977). Such vortices appear under circumstances where the central downflow has a large radius so that the upflow is forced into a relatively narrow annulus with the central hole as its outer diameter. The most reasonable explanation of the appearance of these vortices is that this annular shear layer becomes unstable at a wave number appropriate to the basic axial flow and swirl distributions. Calculations of such a flow have been carried out by LESSEN, SINGH and PAILLET (1974) and they have also been used in E4 to rationalise the appearance of multiple cores behind vortex breakdown far from a wall. It now appears that the vortex breakdown which exist at the wall itself, in the Ward simulator, falls into the same category and that one has a self consistent explanation for multiple cores under a variety of circumstances. A similar explanation has been presented by SNOW (1978). The other observations of the Oklahoma investigators also fall into the classification scheme we have already presented. The majority of the flow photographs (e.g. WARD'S fig. 4) are of the B.B.C. Class C flow, while the core diameter measurements of DAVIES-JONES (1972) and JISCHKE and PARANG (1974) show the type of dependence on swirl parameter and independence of Reynolds number typical of this flow class. The upstream movement of the wall vortex breakdown (see figs. 7c and d) found by these authors is also typical of generator for which De/L is of order one. In fact, this characteristic has been exploited by those interested in vortex breakdown per se to produce stable well defined flows of this type (see section II F).

DAVIES-JONES (1977) has presented a complete review of all of this work and it is recommended that the reader go there for more detailed comments.

Many of the features mentioned in the previous paragraphs are seen also in the results of the Purdue investigators. In the papers listed at the beginning of this section (P1 - P10), they have visualised and measured a number of different configurations as we will outline in what follows. In fact with a certain feeling of uneasiness we have added to some of their results in light of the B.B.C. experiments.

Initially we refer to P2, P4 and P5. Interpretation is difficult because they do not give enough information for us to be sure of the location of the various vortex jumps with respect to the exit hole (De), and they also create a certain amount of confusion by repeating the same results in a number of different publications. In figures 9, 10 and 11, we show the three conditions of importance. Here figure 9 clearly shows B.B.C. Class A (since no Class B flow exists in a case where $L_e = 0$), with the ground breakdown, acceleration to supercritical flow at the contraction

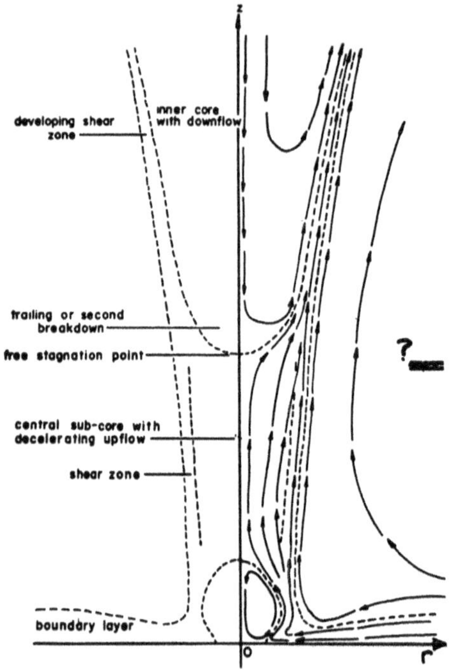

Figure 9. Sketch of Class A flow found in P5 with location of contraction (D_e) marked in approximately (c.f. fig. 4b)

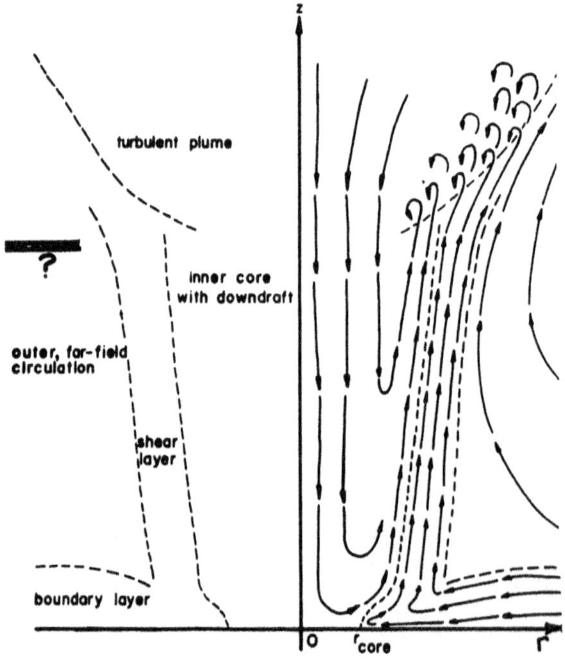

Figure 10. Class C flow found by Purdue investigators with wake-like flow penetration to the ground (c.f. fig. 4d)

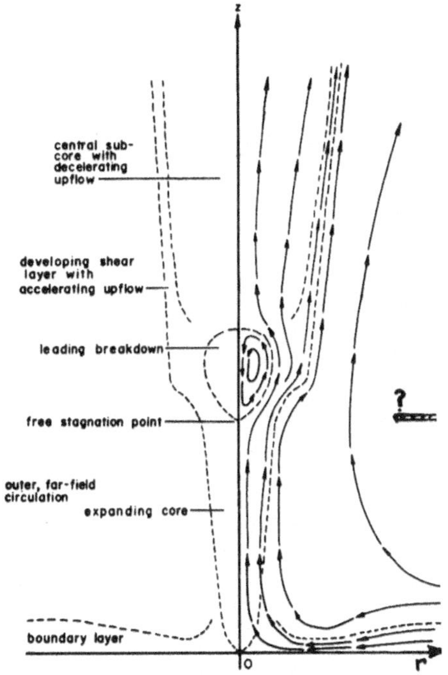

Figure 11. Here ground vortex jump has moved into the contraction (c.f. fig. 7c)

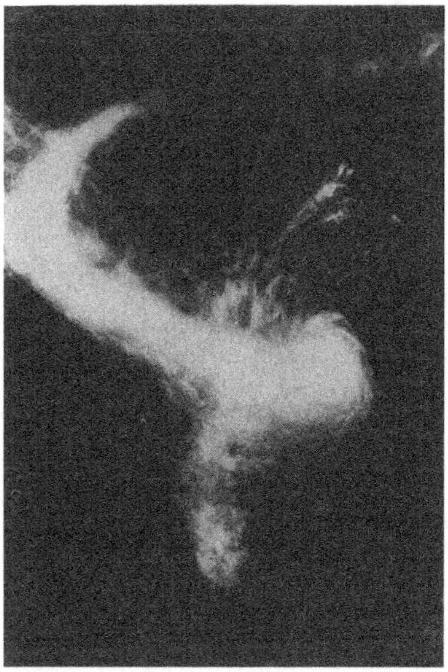

Figure 12. Photograph of spiral wave downstream of a concentration, from P5

(which we have sketched in tentatively based on the B.B.C. results) and a second breakdown followed by a more rapidly growing core. Figure 10 shows that their flow is of B.B.C. Class C at high Re when the wake-like profile initially behind the breakdown has penetrated the surface and the plume beyond the contraction is spreading rapidly and the flow is subcritical throughout. Their suggestion, in P5, that these flow are similar to the buoyancy driven cases of MULLEN and MAXWORTHY may have some validity with the contraction accelerating the flow to a supercritical state rather than the buoyancy forces. Only figure 11 has no immediate correspondence to the B.B.C. experiments because the latter have not considered small enough values of L/De, up to now. However, such flows are typical of apparatus designed to study vortex breakdown as shown in fig. 7c and section II F The other extreme flow which has not yet been achieved in the B.B.C. apparatus are those with multiple cores starting at the lower boundary.

However, several other Purdue observations also correspond to observations in the B.B.C.'s apparatus and in fact, in other experiments too. For example the large scale spiraling motions of the whole core found downstream of the contraction in P5 fig. 8, P4 fig. b, (See our fig 12), for example, are exactly like those found at a similar location in E3 and E4 (see fig. 3c) and were also observed by CHANAUD (1965); who related the observed whistling sound emanating from such a device to the rotation rate of this vortex spiral. Here again the appearance of such vortices depends crucially on the existence of the flow restriction, a circumstance which has no counterpart in natural flows. The appearance of similar spirals in the latter is therefore likely due to other causes and the flow simulator does not modelling them.

In P6 pressure measurements on the ground plane are presented. Here, as in MAXWORTHY (1972) (see section II Ad), these pressures relate directly to the presence of a vortex breakdown at the wall. However, the authors' claim that they reproduce the natural flow successfully must be considered skeptically by those wishing to use their results to understand the prototype.

c). The C.U.A. (Catholic University of America) Models CHANG (1969), YING and CHANG (1970) and WAN and CHANG (1972) report on two models with essentially the same geometry, differing only slightly in scale. In both, angular momentum was added to the flow by a rotating screen of diameter (D) 198 cms in the first case and 190 cms in the second. Fluid was withdrawn by a fan through ducts of diameter (De) 21.6 cms and 20.3 cms respectively. The dimension L is 157 cms in case one and 183 cms in the other. In both cases no attempt was made to view the flow within the exit duct, and the Reynolds numbers were large. In CHANG (1969) six flow régimes were identified. The second, the so called "funnelless tornado-like vortex", appears to be of B.B.C. Class A with a jet-like axial velocity profile but this identification is by no means certain. The next, the "single-funnel vortex", is of the B.B.C. Class C, existing for moderate values of the applied circulation (See Case 2 figure 13). As this circulation is increased still further the observations become dominated by the effects of rotation and the core size now correspond exactly to the size of the outlet duct, De. In this case, boundary layer fluid from the top surface penetrated downwards along the outer edge of the core, as is found in many experiments in which the whole tank is rotated (see section II C), and hence has no counterpart in natural flows although reversed flows do appear to occur under some circumstances. For this reason, the detailed measurements in TING and CHANG (1970) must be viewed with a certain amount of suspicion. In all cases a weak downward flow existed at a radius of 10-15 cms which is exactly the size of the exit duct. We suggest that a flow like that shown in Case 1 figure 13 actually existed and that their simulations while interesting are misleading as models for tornado-like flow, in particular one can see that the entrainment into the core is in the wrong direction and only enters the main upflow by a circuitous route. However, they do present a great deal of useful information on the structur of the boundary layer beneath such vortices and these are of conventiona type with radial inflow at the ground and radial outflow at higher levels.

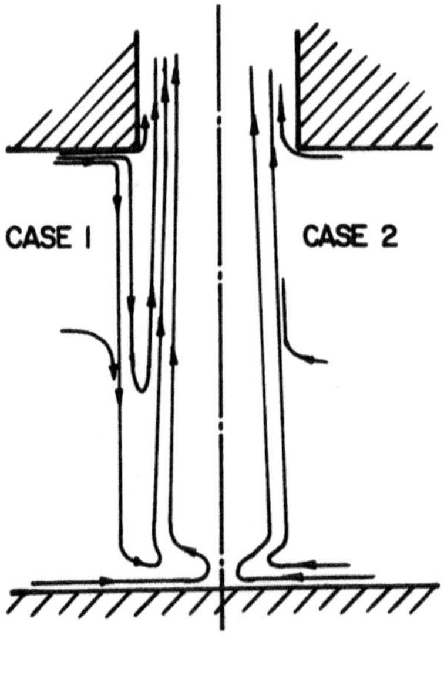

Figure 13.
Diagrammatic view of two flow types found in the C.U.A. vortex generator. In Case I the flow is in almost solid body rotation everywhere and boundary fluid from the upper wall penetrates downwards along the core boundary. In Case II the swirl is reduced and the flow corresponds to B.B.C. Class C

A more complete set of data was presented in WAN and CHANG (1972) although here again they only considered cases with an outer annular downflow at a radius corresponding to the edge of the exit duct. Both cases show characteristics of the B.B.C. Class C with wake-like axial velocity profile and in particular, reversed flow on the central axis in one case. DAVIES-JONES (1977) has reviewed the measurements with accuracy and in some detail, and we need not repeat them here.

d). MAXWORTHY'S (1972) Model This model has only a loose relationship to the canonical form shown in figure 2. Here the downstream contraction wassupplied by a shrouded propellor, but the exact relationship between the diameter of this device and D_e (fig. 2) is unclear. Angular momentum and vorticity were added by rotating both the side and top walls of the tank while the bottom surface remained fixed. The Reynolds numbers of operation were low and it now appears that all cases were within the Class A flow regime. As far as we are aware, this work was the first to note the dynamical significance of the vortex breakdown which invariably appeared beneath the vortex at its intersection with the bottom wall. Boundary layer fluid erupted at the centre line as a rapidly swirling, high velocity, supercritical jet which immediately underwent a jump transition to a subcritical state. Substantial axial pressure gradients were created by this feature so that the pressure distribution on the end wall and that in the main part of the swirling flow, above the jump, were quite different. This jump becomes the critical feature that allows the flow to adjust to changes in downstream conditions.

B. Convective Models BARCILON (1967), FITZGARRALD (1973), MULLEN and MAXWORTHY (1977). All of these models have the common characteristic that the vertical flow force was added to the vortex column by heating the bottom plate and allowing the radial inflowing boundary layer to transport the heated fluid into the vortex core. Bouyancy forces then acted over the whole height of this core to accelerate the fluid therein In all cases Dp = De = D (see fig. 2) and the vortex was allowed to rise into the atmosphere of the surrounding room. In BARCILON'S case, angula momentum was added by a rotating, porous screen while in the other two experiments closely-spaced, angled vanes were used. It is critical to note that in these latter cases no vertical vorticity was added to the flow once the starting transients had been convected out of the apparatus.

BARCILON'S (1967) observations were essentially qualitative and only served to show that indeed a vortex core could be produced by heating in a swirling environment. Viewing of his published photographs shows the presence of a vortex breakdown at the wall, while the core is long, thin and turbulent. Comparison to the results of FITZGARRALD (1973) suggests that these are of the latter's type IV with a large area of downflow at the core center. They correspond to B.B.C. Class C flows.

The next model in chronological order is due to FITZGARRALD (1973) who, following a suggestion of the present author, used a series of closely-spaced, angled vanes to generate a swirling inflow. As discussed in some detail in MULLEN and MAXWORTHY (1977) this generates no vertical vorticity directly at the vanes. The latter is formed by two mechanisms, by the tipping of horizontal boundary layer vorticity into the vertical direction and by direct generation at the vortex core as already described in Section A. FITZGARRALD'S measurements are, as he admits, limited since his instrumentation, a single channel laser Dopple anemometer and rake of thermistor probes, was placed at one location in the experimental tank and measurements were taken as the vortex drifted through the measuring volume. His measurements of maximum vertical velocity scale with a stability parameter $[T_p - T_\infty]/T_\infty \tan^2\theta$ and a velocity $V^2 = gR\,(T_p - T_\infty)/T_\infty$ where T_p and T_∞ are the plate and ambient temperatures respectively, θ is the angle of the inlet vanes and R the plate radius. He claims agreement with field measurements but there are so

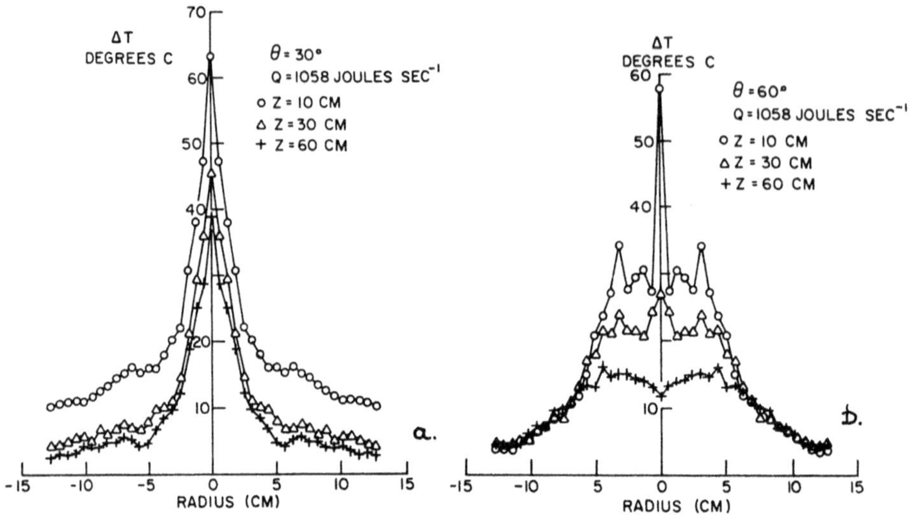

Figure 14. a) Radial temperature profile in a heated vortex showing no backflow and hence jet-like axial velocity profiles everywhere. b) Same as a) except for more intense swirl. The ance of cooler central temperatures indicates some central backflow at higher levels with more jet-like flow lower down equivalent to a B.B.C. Class C flow

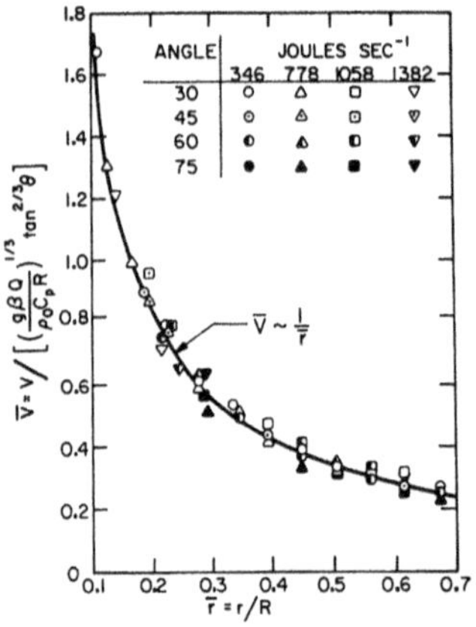

Figure 15. Outer flow potential vortex velocity distribution for the dust-devil types of MULLEN and MAXWORTHY (1977)

few points that the comparison is open to considerable doubt. Other measurements of the vertical distribution of vertical velocity and maxi mum temperature are open to the same criticism. In the latter case, a claimed Z^{-1} dependence with height (Z) is extracted from a line drawn through four scattered points. While in fact the actual slope as measured from his figure 8 is closer to 2/3!

More interesting is his classification of the vortex types with a supercritical core at low swirl angles with a vortex breakdown appearin a substantial distance above the plate and moving downwards as the swir is increased further. His type IV flow with the breakdown at the plate surface and reversed flow in the core corresponds to the Class C flow o the B.B.C. model and other flow types appear because buoyancy forces are able to accelerate the flow and allow the existence of a supercritical jet-like flow in regions where it would not exist in an isothermal vortex.

The most complete model of this type is due to MULLEN and MAXWORTHY (1977). They swept a long resistance wire through the vortex and then deconvolved the resulting signal using Abel's integral to obtain radial temperature distributions at various heights (See fig. 14). They also measured velocities by photographing marker soap bubbles filled with helium. Their results show quite clearly the transition from a jet-like flow to a wake-like reversed flow as the vane angle is increased and a vortex jump moves down the column. Unlike FITZGARRALD'S observations, they always found a vortex jump at the wall followed by a subcritical flow which accelerated to a supercritical state at low swirl angles unde the action of buoyancy forces. It was this supercritical flow which underwent a transition to a subcritical state at lower and lower levels as the swirl angle was increased. Furthermore, the vertical distributio of temperature showed clearly that the upper part of the core after breakdown was well described as a non-rotating turbulent jet with $\Delta T \sim Z^{-5/3}$ in disagreement with FITZGARRALD'S results.
Core diameters scaled with both Reynold's number and boundary layer thickness and the outer flow swirl velocity profile was close to that of a potential vortex (fig. 15), scaling with a reference velocity $[g\beta Q/\rho_o C_p R]^{1/3} \tan^{2/3}\theta$, where g is acceleration due to gravity, β the volume coefficient of expansion, Q power to the vortex, ρ_o the density and C_p the specific heat of fluid, R the radius of the apparatus and θ the vane angle.

It is our opinion that this type of convective model is the most realistic representation of natural flows being less influenced by the geometry of the devices. The core size, velocity distribution etc. are determined without imposing a specific upflow area. The scaling relationship are well established and can be used to predict natural flow.

C. Rotating Tank Models. In these models (MORTON 1963, LONG 1961, TURNER and LILLY 1963, TURNER 1966 and GLUCK 1972) the whole tank rotate with constant angular velocity and had a constant diameter D. Vortex motion was instigated by either withdrawing fluid from a small central hole or using buoyancy forces to create vertical motion at the centre line. In general, the effects of the overall rotation was so dominant that their value as models of atmospheric vortices is limited. In particular, in the majority of such flows, the interaction of the vortex with the boundary layer on a solid wall is relatively simple and both can be matched through the Ekman layer compatability relationships (see TURNER 1966 for example). As the tank rotation rate is reduced, inertia effects become large and a wall vortex breakdown forms and Ekman layer dynamics can no longer be used so that the boundary layer and vortex can now only be coupled through the jump conditions across the vortex breakdown, while the pressure field of the breakdown is at the same time impressed on the inflowing boundary layer. A solution to this complex coupled flow has eluded analysis up to now.

D. Miscellaneous Models. i). Geophysically Motivated Experiments
These models can be mostly dismissed as being inadequate and/or irrelevant in one or more ways.

In HSU and FATTAHI (1975) a rapidly rotating jet was discharged towards a stationary plate. As a model of vortex interaction with ground it suffered from several disadvantages. There is no angular momentum (circulation) at a distance from the centre line and hence no inflowing boundary layer. From all that is known about atmospheric vortices such an inflow plus subsequence eruption at or near the centre line is essential to any realistic model of such flows. Any vortices produced in a flow configuration like this are due to local concentrati of the original jet vorticity, but the overall flow has no relevance to atmospheric vortices.

FUGITA'S (1972) apparatus consisted of a number of rotating cups belo a series of exhaust holes located above a liquid surface. The entire cup-suction mechanism could be translated over this surface for a distance of 15 feet in an attempt to simulate the interaction of a moving vortex with the ground. The most intense vortices were produced when the exhaust holes were closed. DAVIES-JONES (1977) criticism of this apparatus is still valid; we note additionally that it is virtually impossible to assign quantitative parameters to this system since the relationship between cup rotation rate, updraught and circulation is complex, depending on such quantities as room size and shape, exact cup con figuration etc. Furthermore, it is well known that sufficient suction above a stationary plate can induce vortex flow from the reorientation of the vortex lines of the inflowing, non-rotating boundary layer.

DESSEN'S (1972) apparatus consisted of a stationary cylindrical tank with a baffled, straight-radial-bladed fan at the upper end. He was mainly interested in the effects of lower boundary roughness on the vortex core. His two experiments showed an increase in core diameter while his vertical velocity profiles showed a change from a wake-like to a jet-like configuration when the boundary was rough. With no further information it is impossible to know exactly how the overall flow field looked in his two cases and hence make too much sense of his results Like Fugita, the relationships between fan speed, updraught velocity and circulation are obscure, although under slightly better control than in this latter case. As a qualitative experiment it at least showed the modifying effect of wall roughness and pointed the way to a further possible area of study.

MAXWORTHY (private communication) has created a moving vortex over a heated plate in a wind tunnel (1 ft. x 1 ft. cross section) by suddenly changing the angle of attack of a vertically oriented, two dimensional wing placed at the entrance to the test section. Unfortunately, the scale of the experiment was so small that no meaningful results were obtained but the technique may be more useful in a larger facility.

ii). Experiments Motivated by Application to Engineering Problems Since most engineering models are strictly not within the boundaries of the present review we only note here that there is a considerable literature on the subject, all of which is useful in trying to understand vor tices in general and vortices produced in laboratory simulation in parti cular. In E4 we note the relationships between the various models as presented in the literature in CHANAUD (1965), CASSIDY and FALVEY (1970), GARTSHORE (1962), GORE and RANZ (1964), LUGT (1959, 1962), among others.

E. MAXWORTHY'S (1972) Hurricane Model The apparatus shown in fig. 16 is the only attempt to model a hurricane flow known to this author. The annular suction region is particularly crucial because it forces the for mation of a central, quiescent "eye". The boundary layer fluid erupts

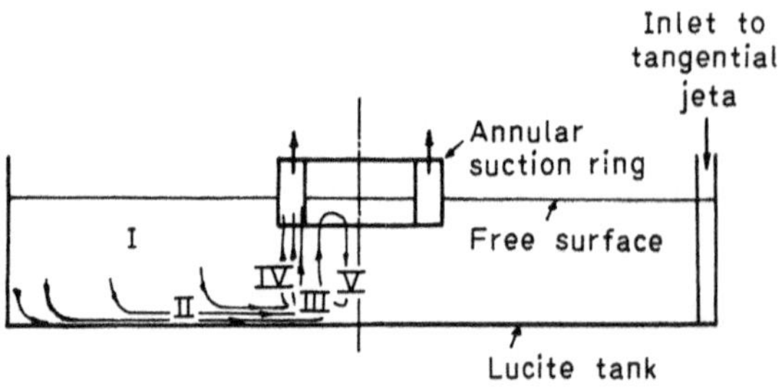

Figure 16. Hurricane model of MAXWORTHY (1972), showing annular suction ring and form of meridional flow at the wall

around a ring and the meridional flow is forced to reverse its path as shown in the figure. The dynamical significance of this "annular vortex jump" is discussed in detail in MAXWORTHY (1973)(reviewed by DAVIES-JONES 1977) where it is shown that the rapid change in direction at the ground is required to support the radial pressure gradient applied by the outer swirling flow.

F. Vortex Breakdown Models Equipment designed to study the phenomena of vortex breakdown itself is mostly of the type shown in fig. 7c. In general L is of the same order or smaller than D and De < D. Under such circumstances and at moderate Reynolds numbers, the wall vortex breakdown moves into the exit tube, which is usually tapered slightly to provide stabilization of breakdown location. The detailed results obtained by HARVEY (1962), GARG and LEIBOVICH (1979), FALER and LEIBOVICH (1977) and SARPKAYA (1971) have had only a peripheral impact on the study of atmospheric vortices because of their specialized nature. As our knowledge of all types of vortex flows increases, there is no doubt that these results and those that follow in the future will occupy an important position in our understanding of the details of atmospheric vortex flows.

REFERENCES

BAKER, G. L., 1978, M. S. Thesis, Purdue Univ., W. Lafayette In
BAKER, G. L. and CHURCH, C. R., 1979, J. Atmos. Sci., 36, 2413
BARCILON, A., 1967, J. Atmos. Sci, 24, 453
BARNHART, B. J., 1979, M. S. Thesis, Purdue Univ. W. Lafayette In
BENJAMIN, T. B., 1962, J. Fluid Mechs., 14, 593
CASSIDY, J. J. and FALVEY, H. T., 1970, J. Fluid Mechs., 41, 727
CHANAUD, R. C., 1965, J. Fluid Mechs., 21, 111
CHANG, C. C., 1969, Proc. Sixth Conf. on Severe Local Storms. A.M.S
CHURCH, C. R. and SNOW, J. T., 1979, J. Rech. Atmos., 13, 111
CHURCH, C. R., SNOW, J. T. and AGEE, E. M., 1979, Bull. A.M.S., 58, 900
CHURCH, C. R., SNOW, J. T., BAKER, G. L. and AGEE, E. M., 1979, J. Atmos. Sci., 36, 17
DAVIES JONES, R. P., 1973, J. Atmos. Sci., 30, 1427
DAVIES JONES, R. P., 1977, See Peterson (1977) p. 151
DESSEN, S. J., 1972, J. Appl. Met., 11, 72
ESCUDIER, M. P., 1981, See Proc. IUTAM/IUGG Symp. on Intense Atmos. Vortices
ESCUDIER, M. P., BORNSTEIN, J. and ZEHNDER, N., 1980, J. Fluid Mechs., 98, 49
ESCUDIER, M. P. and ZEHNDER, N., 1982, J. Fluid Mechs., In press
ESCUDIER, M. P., BORNSTEIN, J., ZEHNDER, N. and MAXWORTHY, T., 1980, ASME Symp. on Vortex Flows, Chicago, p. 15
ESCUDIER, M. P., BORNSTEIN, J., ZEHNDER, N. and MAXWORTHY, T., 1982, In prep. for subm to J. Fluid Mechs
FALER, J. H. and LEIBOVICH, S., 1977, Phys. Fl., 20, 1385
FITZGARRALD, D. E., 1973, J. Atmos. Sci., 30, 894
FUGITA, T. E., 1972, SMRP Res. Paper., 97, U. of Chicago
FULTZ, D. 1951, "Compendium of Meteorology", p. 1235, A.M.S. Boston
GARG, A. K. and LEIBOVICH, S., 1979, Phys. Fl., 22, 2053
GARTSHORE, I. S., 1962, N.R.C. Canada, Rept. LR-343, N.R.C. No. 6968
GLUCK, D. F., 1972, Ph.D. Thesis, U.S.C., Also N. A. Rockwell Rept. SD72-SA-0108
GORE, R. W. and RANZ, W. E., 1964, A.I.Ch.E. Jour., 10, 83
HARVEY, J. K., 1962, J. Fluid Mechs., 14, 589
HIDE, R. and TITMAN, C. W., 1968, J. Fluid Mechs., 29, 39
HSU, C. T. and FATTAHI, B., 1975, Proc. 9th Conf. on Severe Local Storms, A.M.S
JISCHKE, M. C. and PARANG, M., 1974, J. Atmos. Sci., 31, 506
LESLIE, F. W., 1977, J. Atmos. Sci., 34, 1022
LESSEN, M., SINGH, P. J. and PAILLET, F., 1974, J. Fluid Mechs., 63, 753
LETZMANN, J., 1931, Gerland's Beitr. Geophs., (Köppen-Band II), 33, 130
LETZMANN, J., 1933, Gerland's Beitr. Geophs., 39, 167
LEWELLEN, W. S., 1977, See Peterson (1977), p. 107

LILLY, D. K., 1965, Geofisica Internacional, 5, 43
LILLY, D. K., 1977, See Peterson (1977), p. 145
LONG, R. R., 1961, J. Fluid Mech., 11, 611
LUGT, H., 1959, Ph.D. Thesis, Tech. Univ. Stuttgart, F.D.R
LUGT, H., 1962, British Hydro. Res. Assoc., Rept. T716
MAXWORTHY, T., 1972, Astronautica Acta, 17, 363
MAXWORTHY, T., 1973, J. Atmos. Sci., 30, 1717
MORTON, B. R., 1963, Nature, 197, 840
MORTON, B. R., 1966, "Prog. in Aeronautical Sciences," 7, 145
MULLEN, J. B. and MAXWORTHY, T., 1977, Dyn. of Oceans and Atmos., 1, 181
PETERSON, R. E. Ed., 1977, "Symposium on Tornadoes", Texas Tech. Univ., Lubbock
SARPKAYA, T., 1971, J. Fluid Mechs., 45, 545
SNOW, J. T., 1978, J. Atmos. Sci., 35, 1660
SNOW, J. T., and CHURCH, C. R., 1981, 3rd Conf. on Atmos. and Oceanic Waves, San Diego
SNOW, J. T., CHURCH, C. R., BAKER, G. L. and AGEE, E. M., 1977, 10th Conf. on Severe Local Storms, A.M.S
SNOW, J. T., CHURCH, C. R. and BARNHART, B. J., 1979, 11th Conf. on Severe Local Storms A.M.S
SNOW, J. T. and PAULEY, R. L., 1981, Subm. to J. Atmos. Sci
TURNER, J. S., 1966, J. Fluid Mechs., 25, 377
TURNER, J. S. and LILLY, D. K., 1963, J. Atmos. Sci., 20, 468
WAN, C. A. and CHANG, C. C., 1972, J. Atmos. Sci., 29, 11L
WARD, N. B., 1972, J. Atmos. Sci., 29, 1194
WILCKE, J. C., 1785, K. Vet. Acad. nye Hand., 6, 290
YING, S. J. and CHANG, C. C., 1970, J. Atmos. Sci., 27, 3

Dedication After the completion of this paper, I was deeply saddened to learn of the death of Professor Jule Charney of the Massachusetts Institute of Technology. It was my great fortune to have known Professor Charney during the last few years of his life. He was a devoted, prolific and imaginative research scientist, who during his whole life, and even while fatally ill, made pioneering contributions to the atmospheric sciences. His counsel, wisdom and wit will be greatly missed by those who knew, admired and respected him. This contribution is an inadequate expression of my sense of loss and is dedicated to his memory.

Vortex Breakdown and the Criterion for Its Occurence

M. P. Escudier

Introduction

The addition of swirl to an axial flow leads to flow phenomena which are not yet well understood even for simple geometric configurations. An essential feature of many such flows is that they are characterised by an inner vortical core of small diameter surrounded by nearly irrotational flow. Since the core can support waves, it may be either supercritical or subcritical in nature, and a relatively sudden transition from the former to the latter can occur in the form of vortex breakdown. The present paper represents a summary of the outcome to date of a continuing long-term study of vortex flows confined within cylindrical tubes in which vortex breakdown plays a significant role. The paper is organised around a number of specific aspects of such flows: the governing parameters, visualization of the core and the outer zone, the criterion for breakdown, measured velocity and circulation profiles.

The work discussed here has been motivated by the observation[1] that vortex flows play a crucial role in determining the performance characteristics of certain designs of exhaust hoods for axial-flow turbomachinery, and also by their relevance to the design of swirl combustors and cyclone separators. Other workers have also been motivated by applications to a variety of practical devices, including the vortex diode, vortex whistle, Ranque-Hilsch tube, hydraulic turbine draft tube and gas-core nuclear rocket. In such engineering equipment it is generally the case that intense vortices occur as the result of the intentional introduction of swirl into the flow. In the atmosphere, such vortices occur by chance when there is an appropriate combination of ambient vorticity and an entraining updraft. Nevertheless, such natural vortices have much in common with the forced vortices of engineering interest, and no attempt has been made here to omit those aspects which might be thought to have relevance only to the latter.

Experimental Arrangement

The observations discussed here were made in the experimental apparatus shown schematically in Figure 1. This setup has been described more thoroughly elsewhere[2] and only key features are mentioned here. The apparatus consists of a perspex tube of diameter D (55 mm) with a tangential inlet of adjustable width t (2-10 mm) running along its entire length (425 mm). In the arrangement shown in Figure 1, the tube is closed at one end and has an axisymmetric contraction section of entrance diameter D_e (10-47 mm) and flare angle α (0-25°) installed at axial location L, the inlet slit being block-

Topics in Atmospheric and Oceanographic Sciences
Intense Atmospheric Vortices, ed. by Bengtsson/Lighthill

ed off beyond this point. The upstream section of length L is referred to as the generator section Similar experimental apparatuses have been used to simulate atmospheric vortices,[3,4,5,6] though generally with multiple inlet slits. An alternative arrangement of our system is to have both ends of the tube open thereby eliminating the influence of the endwall boundary layer.

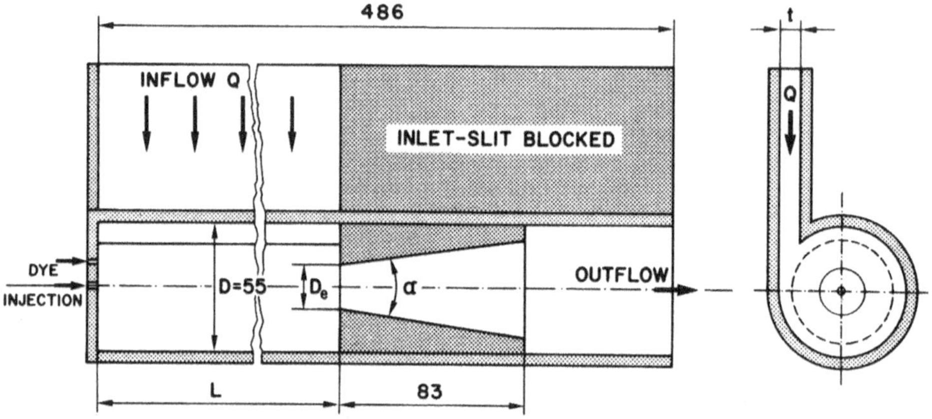

Figure 1 Schematic diagram of vortex tube

Water is used as the working fluid, and visual observations are made by injecting either conventional or fluorescent (fluorescein or rhodamine 6G) dyes into the flow. For fluorescent-dye photography, an Argon-ion laser is used to illuminate either the entire test section or a thin diametral plane thereby yielding a sectional view of the flow. Detailed measurements of the swirl and axial velocity components are made using a BBC-Goerz dual-beam back scatter LDA system employing a 5 mW Helium-Neon laser and incorporating a Bragg cell after the beam splitter to shift the frequency of one beam by 40 MHz.

Governing Parameters

In order to establish the governing parameters for vortex flows, it is convenient to consider the idealized arrangement depicted in Figure 2 which is representative not only of our setup but also of the more common guidevane systems, as used in earlier confined-vortex investigations.[7,8,9] Fluid of kinematic viscosity υ enters the system radially at a volume rate Q through a generator section of length L within which circulation Γ is imparted to the flow. The flow is then turned into the axial direction and passes through a tube of inlet diameter D_e and flare angle α. To describe the flow, three non-dimensional dynamical parameters are required, which are defined generally and related to our particular geometry as follows:

a circulation or swirl number $\Omega \equiv \pi\Gamma D_e/4Q = \pi^2 D D_e/4Lt$

a measure of the inflow angle $R \equiv Q/\Gamma L \cong t/\pi D$

and a Reynolds number $Re \equiv 4Q/\pi D_e \upsilon$

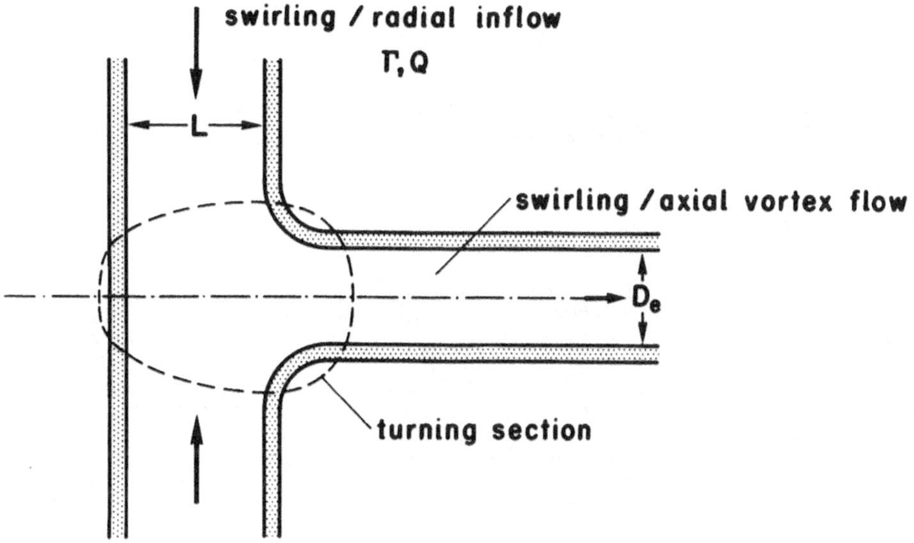

Figure 2 Idealized vortex-flow system

An advantage of our experimental arrangement in comparison with guidevane systems where D_e and L, and hence ΩR, are fixed, is seen to be that Ω and R are separately variable. A more fundamental advantage which will be demonstrated below is that over a wide range of conditions, only a single parameter (combining Ω, R and Re) is required.

Core-Flow Visualization

Dye injected through the endwall at a low flow rate is swept spirally inwards within the endwall boundary layer and then ejected axially close to the geometric centre of the generator tube. The core of dyed fluid is initially of small diameter (typically 0.1 - 1 mm) but undergoes a transition (possibly breakdown) to a much larger diameter a short distance (typically less than 10 mm) from the endwall. Clearly evident (Figure 3) as the fluid is convected downstream is the distinct division of the flow into two regions: an outer zone and a small-diameter inner core. Also appar-

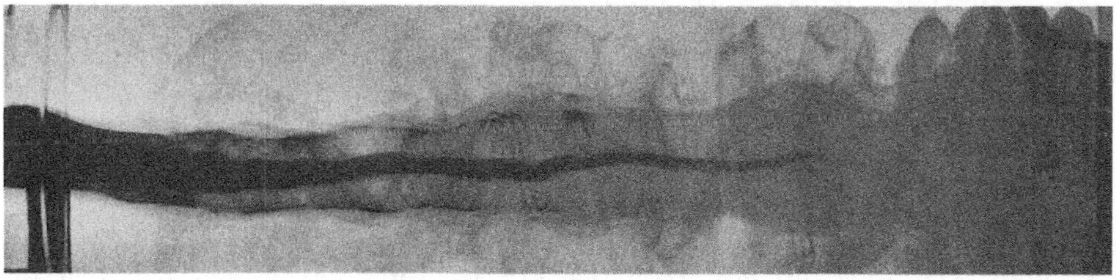

Figure 3 Generator-tube core-flow visualization

ent is the strongly anisotropic turbulent structure of the core at higher Reynolds numbers owing to strong suppression of radial mixing. Another observable feature is the decrease in core radius with longitudinal distance, and the existence of a velocity deficit at the core centre suggested by the slow-moving tail of dye which persists after the dye flow is cut off.

Outer-Zone Visualization

Dye injected into the fluid surrounding the core is rapidly dispersed at all but the lowest Reynolds numbers, and even then it is observed that the outer flow is easily disturbed. The fluorescent dye visualization technique reveals (Figure 4) that mixing of the outer flow occurs owing to the presence of slow-moving, large-scale structures which are probably Taylor-Görtler vortices generated by the tangential inflow over the inner concave surface of the tube[10], but which are also remarkably similar to the structures visible in the cross section of a turbulent spot[11]. The scale of these structures decreases with increasing Reynolds number, but is typically of the order of the tube radius. There seems to be no evidence of this large-scale mixing in previously reported work where the generator section is generally of a different design.

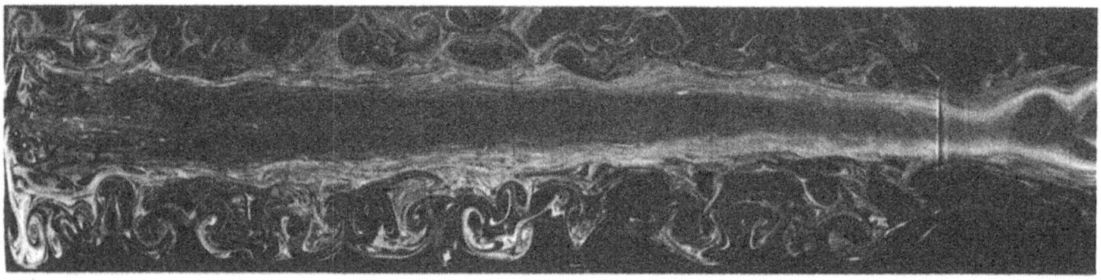

Figure 4 Generator-tube outer-flow visualization

Breakdown Criterion

For any given geometry (L, D_e, t, α) it is found that at a certain flow rate the core undergoes a distinct structural change at the contraction exit. With an increase in flow rate, this change moves upstream, becoming best defined about 10 mm downstream of the contraction inlet and then collapsing and disappearing at the inlet with a further increase in flow rate. Over a wide range of values of Ω and R, the Reynolds number Re_B for breakdown at the 10 mm location is well represented by the relation

$$\Omega^3 R Re_B = \text{constant},$$

as shown in Figure 5. For $L/D < 2$, departures from this relation occur, probably due to the influence of the endwall boundary layer. The apparatuses used in previous investigations of vortex breakdown in tubes are all subject to this limitation, which may explain why there is only limited agreement with our results[12], a point which is presently under investigation.

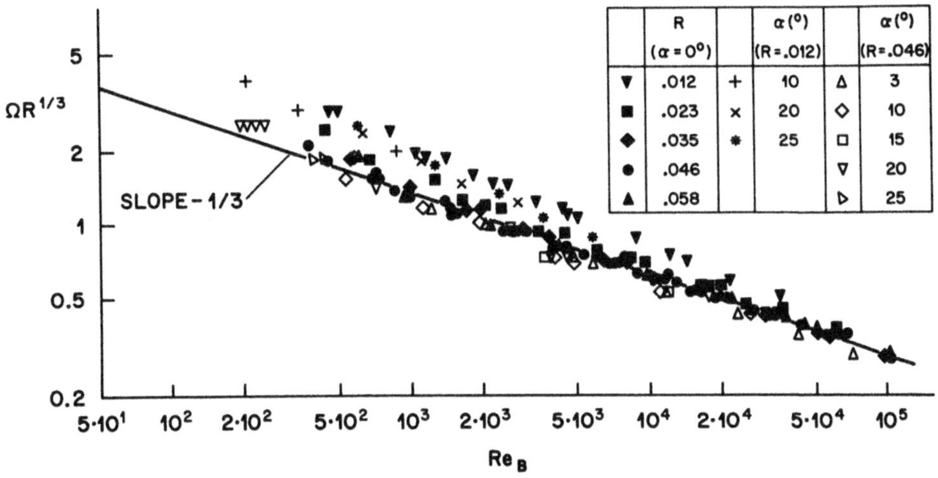

Figure 5   Correlation of data for vortex breakdown

The structure of the breakdown changes significantly with Re_B, versions of all forms (bubble, spiral, etc.) observed by previous investigators being found. Also, the scale of the breakdown increases strongly with increase in the flare angle α. Both points are illustrated in Figure 6 which consists of a series of photographs of vortex breakdown with values of the flow parameters as listed in the table below. All photographs were taken using the fluorescent dye technique and, with the exception of a) and c), show sectional views.

Table. Values of flow parameters for Figure 6

Photo	R	Ω	$Re_B \cdot 10^{-3}$	α
a,b	.046	7.25	0.22	20°
c		4.28	0.51	10°
d		2.03	3.80	20°
e		1.46	15.0	0°

Invariance, Criticality and Vortex Breakdown

We* demonstrate here that the breakdown criterion arrived at empirically can be established by combining a criticality criterion with some rather general arguments concerning the local invariance of the flow in and around the core. Attention is restricted to axisymmetric flow of an incompressible, low-viscosity Newtonian fluid and to geometrically similar flow systems (L/D_e = constant). Viscous effects are assumed to be confined within a region of small radius (<<L), outside of which the flow is irrotational.

---

* The author is indebted to his colleague Dr J.J. Keller for the arguments presented in this section.

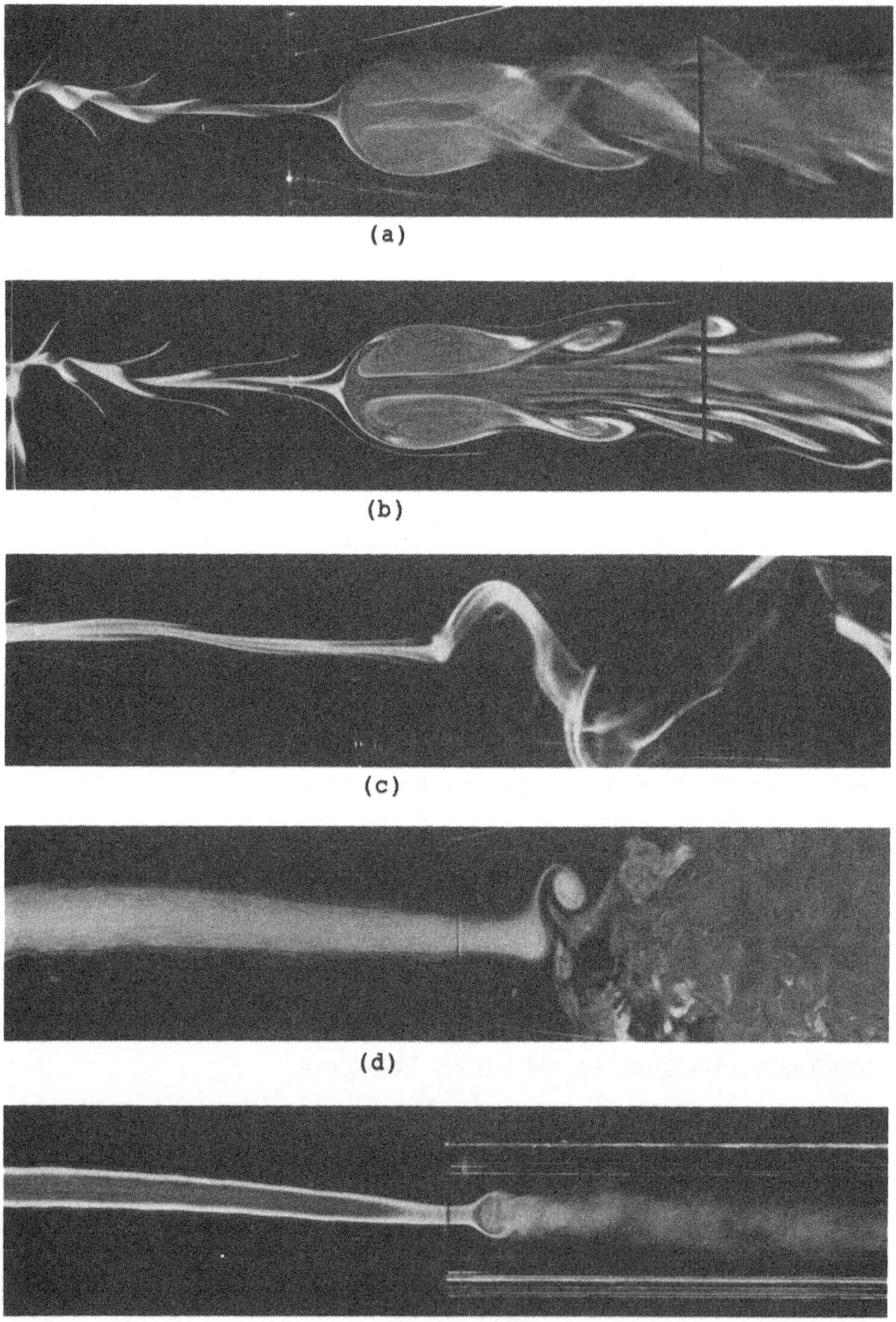

Figure 6 Examples of vortex breakdown

For the outer potential flow it can be shown that within the vicinity of the viscous core, the axial, radial and tangential velocity components are given by

$$w = 2\frac{Q}{L^2} g\left(\frac{z}{L}\right) \qquad u = \frac{-Qr}{L^3} g'\left(\frac{z}{L}\right) \qquad v = \frac{\Gamma}{2\pi r}$$

where the function g(z/L) is determined by the geometry of the flow system. In anticipation of their use for the analysis of the viscous core, we introduce normalized variables such that

$$w = \bar{w}.w_e,\ u = \bar{u}.w_e r_c/L,\ v = \bar{v}.\Gamma/2\pi r_c,\ r = \bar{r}.r_c \text{ and } z = \bar{z}.L,$$

where r_c is a measure (as yet undefined) of the core radius. Then the above results may be written in normalized form as

$$\bar{w} = 2f(\bar{z}) \qquad \bar{u} = -\bar{r}f'(\bar{z}) \qquad \bar{v} = 1/\bar{r}$$

Within the core region, the Navier-Stokes equations (subject to the restrictions mentioned above) may be written as

$$\left(\frac{\Gamma}{2\pi r_c w_e}\right)^2 \frac{\partial}{\partial \bar{z}}\left(\frac{\bar{v}^2}{\bar{r}}\right) + \frac{\partial}{\partial \bar{r}}\left[-\frac{1}{\bar{r}}\frac{\partial \bar{\psi}}{\partial \bar{z}} \cdot \frac{\partial}{\partial \bar{r}}\left(\frac{1}{\bar{r}}\frac{\partial \bar{\psi}}{\partial \bar{r}}\right) + \frac{1}{\bar{r}}\frac{\partial \bar{\psi}}{\partial \bar{r}} \cdot \frac{\partial}{\partial \bar{z}}\left(\frac{1}{\bar{r}}\frac{\partial \bar{\psi}}{\partial \bar{r}}\right)\right]$$

$$- \left(\frac{\upsilon L}{w_e r_c^2}\right) \frac{\partial}{\partial \bar{r}}\left[\frac{1}{\bar{r}}\frac{\partial}{\partial \bar{r}}\left\{\bar{r}\frac{\partial}{\partial \bar{r}}\left(\frac{1}{\bar{r}}\frac{\partial \bar{\psi}}{\partial \bar{r}}\right)\right\}\right] = 0$$

$$- \frac{1}{\bar{r}}\frac{\partial \bar{\psi}}{\partial \bar{z}}.\frac{\partial}{\partial \bar{r}}(\bar{r}\bar{v}) + \frac{1}{\bar{r}}\frac{\partial \bar{\psi}}{\partial \bar{r}}\frac{\partial}{\partial \bar{z}}(\bar{r}\bar{v}) - \frac{2\pi \upsilon L}{w_e r_c{}^2}\ \bar{r}\frac{\partial}{\partial \bar{r}}\left[\frac{1}{\bar{r}}\frac{\partial}{\partial \bar{r}}(\bar{r}.\bar{v})\right] = 0$$

wherein we have introduced a stream function $\bar{\psi}\cdot w_e r_c^2$ defined by

$$\bar{w} = \frac{1}{\bar{r}}\frac{\partial \bar{\psi}}{\partial \bar{r}} \qquad \bar{u} = -\frac{1}{\bar{r}}\frac{\partial \bar{\psi}}{\partial \bar{z}} \quad .$$

A convenient measure for the core radius is evidently

$$r_c = \sqrt{(\upsilon L/w_e)}$$

which may be recognized as the expression which is also given by similarity solutions.[13] The sole parameter which then emerges is $\Gamma^2/\upsilon L w_e$ and it is seen that both the core flow and the surrounding potential flow (for small radii) are invariant (in normalized form) to geometric scale for any given value of $\Gamma^2/\upsilon L w_e$. The restriction to geometrically similar flow systems turns out not to be of great importance, provided L/D_e>>1, however, the restriction to small radii is important: r_c/D_e would otherwise appear as a parameter. It is emphasized that the foregoing is in no way restricted to the phenomenon of vortex breakdown, to consideration of which we now return.

BENJAMIN[14] showed that steady axisymmetric disturbances superposed on an inviscid cylindrical flow first become unstable in the long-wavelength limit, and this leads to the following equation for the critical state

$$\frac{d^2\chi}{d\bar{y}^2} - [\frac{1}{\bar{w}}\frac{d^2\bar{w}}{d\bar{y}^2} - \frac{1}{2\bar{y}^2\bar{w}^2}\frac{d}{d\bar{y}}(\bar{v}^2\bar{y})\frac{\Gamma^2}{(2\pi)^2 \upsilon L w_e}]\chi = 0$$

subject to the boundary conditions

$$\chi(0) = 0 \quad \text{and} \quad d\chi/d\bar{y} \cong 0 \text{ at } \bar{y} = 1/2 \text{ (again provided } r_c/D_e \ll 1).$$

χ is the amplitude of the stream-function disturbance and $\bar{y} = \bar{r}^2/2$. No additional parameter appears in the criticality equation and for a given geometry the parameter $\Gamma^2/\upsilon L w_e$ is sufficient to determine both the core flow and also its criticality. The particular value of $\Gamma^2/\upsilon L w_e$ for which the flow in a given geometry is first critical is taken as the condition for "mild" vortex breakdown. It is seen at once that

$$\Gamma^2/\upsilon L w_e = 4\Omega^3 R Re/\pi.$$

We also note that $\upsilon L w_e/\Gamma^2$ can be regarded as a "Froude" number for a swirling flow.

Velocity and Circulation Profiles

Figure 7 shows a set of normalized velocity and circulation profiles, measured using the LDA technique, for L = 208.5 mm, D_e = 18 mm, t = 8 mm and $\alpha = 0$ ($\Omega = 1.46$, R = 0.046) and Re = 15,000, which corresponds to vortex breakdown just within the contraction ($\Gamma^2/\upsilon L w_e$ = 2,630). The axial locations are as follows: A, 100 mm from endwall; B, 10 mm upstream of contraction; C, 3 mm upstream of contraction; D, 8 mm within contraction; E, halfway along contraction; F, 8 mm before contraction exit; G, 3 mm downstream of contraction; and H, 25 mm downstream of contraction.

These profiles clearly confirm the division of a vortex flow into two distinct regions, as suggested by flow visualization. The change in character of the axial velocity profiles from a jetlike (supercritical) core upstream of breakdown, to a wakelike (subcritical) core downstream is typical. At lower Reynolds numbers, the core is jetlike throughout (even at the contraction exit), whilst at higher values it is wakelike throughout, all the way to the endwall.

The smooth curves fitted through the swirl velocity profiles correspond to a modification of Burgers profile to account for the fact that the outer flow is better represented[10] as a zone of constant than of zero vorticity $\bar{\omega}$:

$$\bar{v} = \alpha[1-\exp(-k\bar{r}^2)]/\bar{r} + \bar{\omega}\bar{r}/2$$

where α is the nondimensional core circulation and k is a constant. The vortical nature of the outer region is more readily apparent from the circulation distribution

$$\bar{\Gamma} = \bar{v}\bar{r} = \alpha[(1-\exp(-k\bar{r}^2)] + \bar{\omega}\bar{r}^2/2$$

which is plotted together with the experimentally determined profiles in Figure 7(c). It may be remarked that the level of the

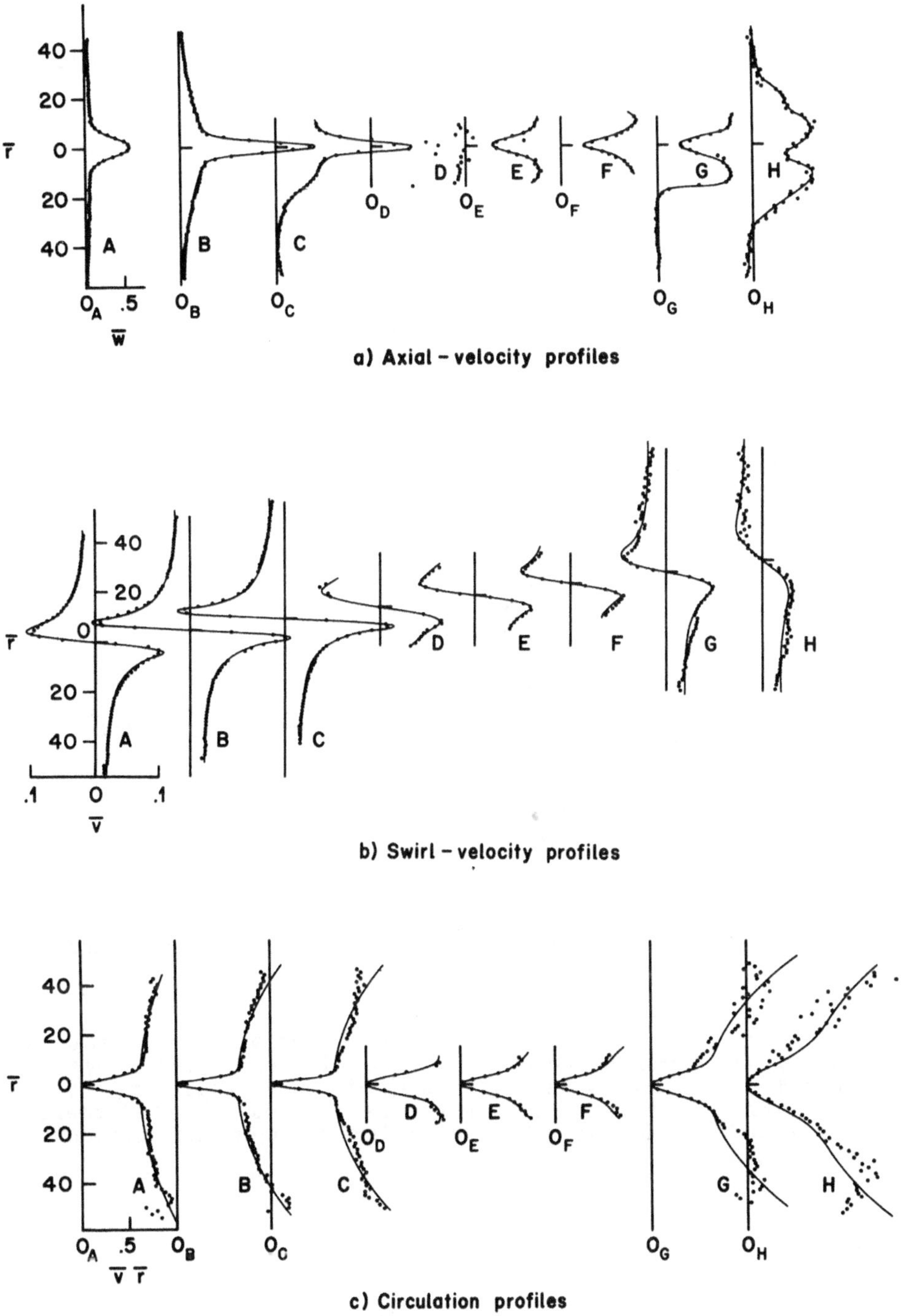

Figure 7 Normalized velocity and circulation profiles

vorticity in the outer zone is so low as to have a negligible influence on the core dynamics; only the fact that it acts over a very large area leads to a strong increase in circulation between the core and the outer wall.

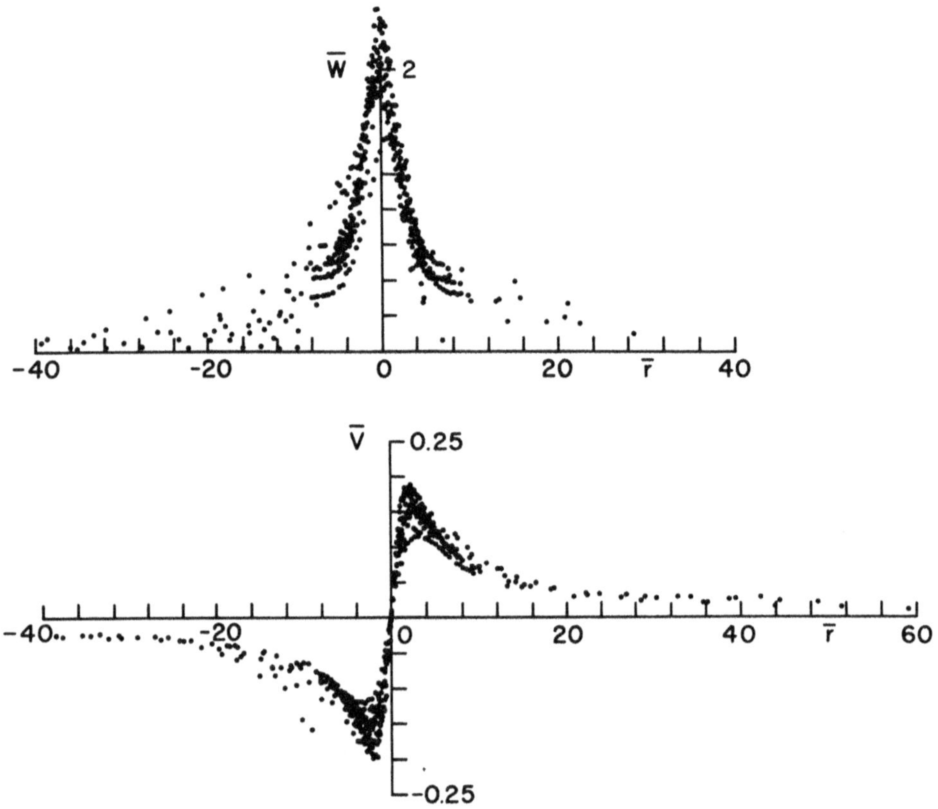

Figure 8 Normalized profiles just upstream of vortex breakdown

In Figure 8, eleven sets of normalized velocity profiles (for breakdown) at Reynolds numbers between 626 and 32,930 have been superposed. The corresponding range of core radii r_c = 0.03 - 0.24 mm. This figure, together with the collapse of the breakdown data in Figure 5, may be taken as confirmation of the arguments of the previous section.

Concluding Remarks

The observations and arguments discussed above reveal a fairly complete picture of vortex flow with special reference to the conditions under which vortex breakdown occurs. The essential parameter to describe vortex flows (at least in systems similar in design to ours) has been shown to be $\Gamma^2/\upsilon L w_e$ (= $4\Omega^3 R Re/\pi$) which corresponds to the inverse of a Froude number for swirling flow. It is evidently inadequate to characterize vortex flows through the circulation number Ω alone, as has sometimes been the case in the past, and it is clear that breakdown will occur even for flows with "low swirl" at a sufficiently high Reynolds number (which in practice may be unattainable owing to cavitation in a liquid, or the onset of compressibility effects in a gas).

Although many aspects of vortex-flow behaviour in general are similar to those discussed here, important differences lie in the manner in which the vortical core is formed. Implicit to our discussion has been the assumption that the core is formed as a result of the viscous torque required to decelerate fluid particles which approach the axis from the outer flow. This may be thought of as one limit, the other being the case when the core is dominated by (radially directed) vorticity generated within an endwall boundary layer by the outer swirling flow and subsequently turned into the axial direction. It seems to be the case that the guidevane flow systems, for which the aspect ratio L/D is generally of the order unity, lead to cores which are endwall dominated.

References

1. ESCUDIER, M.P. 1979. Estimation of pressure loss in ring-type exit chambers. J. Fluids Eng., 101, 511-516

2. ESCUDIER M.P., BORNSTEIN, J.B. and ZEHNDER, N. 1980. Observations and LDA measurements of confined vortex flow. J. Fluid Mech., 98, (1), 49-63

3. CHURCH C.R., SNOW, J.T. and AGEE, E.M. 1977. Tornado vortex simulation at Purdue University. Bull. Am. Meteor. Soc., 58, (9), 900-908

4. WAN, C.A. and CHANG, C.C. 1972. Measurement of the velocity field in a simulated tornado-like vortex using a three-dimensional velocity probe. J. Atm. Sci., 29, 116-127

5. WARD, N.B. 1972. The exploration of certain features of tornado dynamics using a laboratory model. J. Atm. Sci., 29, 1194-1204

6. MULLEN, J.B. and MAXWORTHY, T. 1977. A laboratory model of dust devil vortices. Dyn. Atmos. and Oceans., 1, 181-214

7. HARVEY, J.K. 1962. Some observations of the vortex breakdown phenomenon. J. Fluid Mech., 14, (4), 585-592

8. SARPKAYA, T. 1971. On stationary and travelling vortex breakdowns. J. Fluid Mech., 45, (3), 545-559

9. FALER, J.H. and LEIBOVICH, S. 1977. Disrupted states of vortex flow and vortex breakdown. Phys. Fluids, 20, (9), 1385-1400

10. ESCUDIER, M.P., BORNSTEIN, J., ZEHNDER, N. and MAXWORTHY, T. 1981. The dynamics of confined vortices. (In preparation)

11. PERRY, A.E., LIM, T.T. and TEH, E.W. 1981. A visual study of turbulent spots. J. Fluid Mech., 104, 387-405

12. ESCUDIER, M.P. and ZEHNDER, N. 1981. Vortex flow regimes. Accepted for publication, J. Fluid Mech.

13. DONALDSON, C. du P. and SULLIVAN, R.D. 1960. Behaviour of solutions of the Navier-Stokes equations for a complete class of three-dimensional viscous vortices. Proc. 1960 Ht. Trans. and Fluid Mech. Inst.

14. BENJAMIN, T.B. 1962. Theory of the vortex breakdown phenomenon. J. Fluid Mech., 14, 593-629

Acknowledgement

The author gratefully acknowledges the contributions to the work presented here of his colleagues Dr J. Bornstein, Dr J.J. Keller and Mr N. Zehnder, and also Prof. T. Maxworthy of the University of Southern California.

Pressure Fields Beneath Tornado-like Vortices

J. T. Snow

ABSTRACT

Wall static pressure fields beneath tornado-like vortices have been investigated using a vortex generator designed to model tornado-cyclone airflow. The data presented include (a) a series of radial profiles of the mean surface pressure beneath the vortex as a function of swirl, (b) measurements of the magnitude of the greatest pressure deficit to be found under the vortex as a function of swirl, and (c) a set of measurements of the pressure deficits associated with the individual subsidiary vortices in a multiple vortex system. These data are interpreted to show the development of the model tornado cyclone from the no-swirl state, and the evolution of its intense vortical core from a one-celled into a two-celled flow. It is shown that the greatest pressure deficits and largest pressure gradients are associated with single-celled vortices. The pressure deficits within the subsidiary vortices are found to be two to three times as great as that at the center of the parent flow. This work provides the background for better physical interpretation of barographic records obtained in actual tornado cyclones, and further supports a model of the evolution of these phenomena that is consistent with doppler radar observations of natural tornado cyclones.

1 INTRODUCTION

Large, long-lived tornadoes are relatively rare, accounting for only a few percent of all tornadoes reported. However, these intense events are of special interest because they produce almost all tornado-related injuries and deaths, and account for most of the property damage. In the last decade, it has come to be recognized that these especially strong tornadoes are usually produced adjacent to the intense updraft of a rotating thunderstorm. This type of storm appears to be internally well organized, and so is able to exist as an entity for periods of two to six hours, and occasionally even longer. During its lifetime, a rotating storm may produce a series of strong, long-lived tornadoes (a "tornado family"). The intense spiraling updraft that forms the core of such a storm is reflected at the surface as a region of reduced pressure and inward spiraling winds, or mesocyclone. Such a mesocyclone is also an area of enhanced vertical vorticity - when properly concentrated, this vorticity is sufficient to form the core of an intense tornado. Once a tornado is produced, the mesocyclone is identified as a tornado-cyclone.

Because of the relatively infrequent occurrence, and because of the violent winds associated with a large tornado, no direct measurements of winds or pressures are available from within the central core region of this phenomenon. Post event damage surveys have provided only crude estimates of the surface wind fields near and beneath such a tornado, while photogrammetric analyses have been restricted to investigations of the motions in the below-cloud, visualized portions of core. Further, radar is not yet able to resolve detail in the hidden, in-cloud portion of the vortex. Consequently present knowledge of the wind and pressure fields in and near a strong tornado vortex is very limited.

However, barographic traces are occasionally obtained in the near vicinity of tornadic events. While many of these records, particularly those associated with weak tornadoes (ones not generally associated with a tornado-cyclone), are of

Topics in Atmospheric and Oceanographic Sciences
Intense Atmospheric Vortices, ed. by Bengtsson/Lighthill

little more than passing interest, some clearly show the passage of a tornado-cyclone over the recording instrument. Particularly good examples are to be found in Fujita (1959, 1970) and Ward (1964). Such traces represent one of the few sources of directly measured quantitative data concerning these mesoscale events. While difficult to interpret, these records of the surface pressure continue to be subjects of interest for several reasons:

- they can provide clues about the overall dynamic structure of the tornado cyclone, and may allow inferences about those of the tornado vortex.
- they allow estimates of the magnitudes of velocities within the tornado cyclone.
- they may indicate the degree to which the destructive effects of the translating vortical system are due to rapid pressure falls.
- they provide a means by which laboratory and numerical models of tornado cyclones may be directly compared with natural events.

Also, doppler radar observations are available that show the evolution of the in-cloud portion of tornado-cyclones. These observations show that significant rotation appears first at mid-levels within the storm, and then builds downward. Tornadoes begin to be produced when the region of rotation extends to the lower surface. The dynamics behind this sequence of events are not yet clear.

This report discusses a series of quantitative measurements of the wall static pressure on the surface beneath a laboratory model of a tornado cyclone. When combined with a parallel set of qualitative observations of the visualized flow, these measurements provide a clear picture of the development of the swirling flow. Consequently, we are now able to better physically interpret the barographic traces obtained near actual tornado events, and to further support a proposed model for the evolution of the tornado cyclone.

2 TORNADO VORTEX SIMULATION

The observations reported herein were made in a vortex generator modeled after that of Ward (1972). This apparatus is designed to replicate the situation wherein tornado production is the result of low-level radial convergence of angular momentum into a rotating updraft. As shown in Fig. 1, its physical arrangement is such that dynamic and geometric similarity to a rotating thunderstorm updraft-tornado-cyclone is attained (Davies-Jones, 1973, 1976). The model tornado-cyclone forms in the central portion of a flat, disk-shaped volume (labeled the confluence and convergence regions in Fig. 1) at the base of the updraft. In this system, the tornado corresponds to the central, vortical core of the tornado-cyclone. It is stressed that this vertical core forms initially due to the convergence of the background vorticity within the tornado-cyclone, and is maintained by tilting of the horizontal vorticity present in the surface inflow layer. Further discussion of the physical basis for the Ward-type tornado simulator, and its relationship to the rotating thunderstorm, is contained in Church et al, (1979) and in the review by Davies-Jones, (1976).

A review of the analyses contained in Lewellen (1962), Davies-Jones (1973, 1976), and Church et al, (1977, 1979) indicates that the flow in both actual and model rotating thunderstorm - tornado-cyclone systems is characterized by two dynamic and (at least) four geometric non-dimensional parameters:

$$\text{Swirl parameter} = \Gamma/Q \tag{1}$$

$$Re_r = \text{radial Reynolds number} = Q/\upsilon \tag{2}$$

$$a = \text{internal aspect ratio of the convergence region} = h/r_o \tag{3}$$

$$\text{external aspect ratios} = \frac{r_s}{r_o}, \frac{r_w}{r_o}, \frac{l}{h} \tag{4}$$

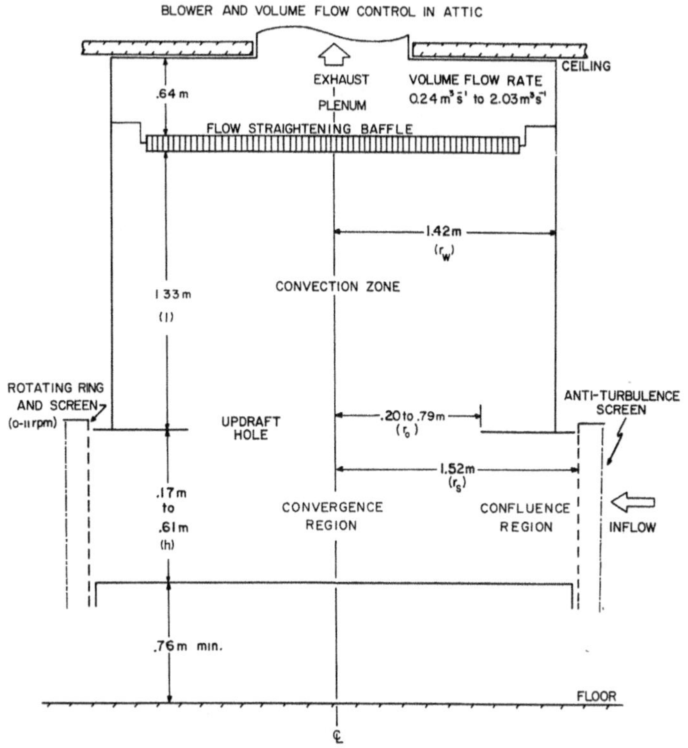

Figure 1. Cross-section through the Purdue vortex generator

(All symbols are defined in Fig. 1 and/or the Appendix.) In the governing equations of motion, the swirl parameter and the internal aspect ratio appear in combination as

$$S = \text{swirl ratio} = \frac{r_o' \Gamma}{2Qh} . \qquad (5)$$

The swirl ratio (describing the relative strength of the tangential velocity field compared to the through flow field) and the radial Reynolds number (a measure of the dissipation in the inflow layer) control the dynamics of the flow; the aspect ratios, its geometry. To properly scale the tornado cyclone, dynamic similarity requires that both the swirl ratio and the Reynolds number of the model flow be the same as those for the natural event. Geometric similarity requires that the geometry of the flow boundaries be scaled in proportion. Tables 1 and 2 in Church et al., (1979) list the values of these quantities likely to occur in nature and those obtainable in the Purdue vortex generator.

We note that in our present apparatus, the dynamic similarity requirements are not satisfied with respect to the radial Reynolds number, with the Reynolds number for the natural flow being several orders of magnitude larger than that for the laboratory model. However, both values are large enough that the flow is likely turbulent in each case. Davies-Jones (1973, 1976) and our own work has shown that for turbulent flows of this type, the core radius and key core transition points are strong functions of the swirl ratio, and are relatively insensitive to the Reynolds number provided it is sufficiently large.

Here we focus on phenomena in the convergence region. Consistent with previous work, we will show that the Reynolds number does affect the details of the flow, but that the swirl ratio is the parameter of primary importance in determining the central pressure within the vortex.

3 INSTRUMENTATION AND EXPERIMENTAL PROCEDURES

The volume flow rate Qh, the updraft hole radius r_o, the inflow depth h, and the screen rotation rate are the adjustable parameters in the experimental apparatus. These were set to obtain the various vortices investigated. The volume flow rate and the inflow angle θ (hence the swirl parameter Γ/Q) were measured as described in Church et al. (1979).

Each probe used to measure the wall static pressure consisted of a thin, flat plate with beveled edges. One or more sensing orifices were drilled through the flat face of each plate. These probes were mounted flat on the lower surface with pressure taps from the orfices extending downward through the surface. Each probe could be moved radially by means of a positioning assembly affixed to the bottom side of the lower surface.

In practice differential pressures were measured. The pressure differences (between a fixed reference point and the point(s) of interest) were sensed by a variable reluctance differential pressure transducer. The measurements reported in paragraphs 5, 6, and 7 were made using a single orifice probe and one pressure transducer. For the measurements of pressure deficits within subsidiary vortices described in paragraph 8, a three orifice probe and three transducers were utilized.

The radial profiles described in paragraph 5 were taken using as a reference pressure that at the outer edge of the confluence zone. The output of the transducer was read after being passed through a first order, variable time constant filter so that the data represents the time mean value at a radial position. This resulted in an underestimation of the central pressures in vortices with core diameter less than about 2 cm.

The reference pressure used for the measurements discussed in paragraphs 6, 7, and 8 was taken under the lower surface. No electronic filtering was used. The maximum pressure deficit values (paragraphs 6 and 7) were obtained by following the transducer output on an oscilloscope.

Three outputs were obtained during the investigations described in paragraph 8. These were recorded on a high-speed oscillographic recorder. The frequency response of each channel of the measuring system was known and appropriate adjustments made to each output signal. When the outputs from the two outside orifices were the same, it was considered that a subsidiary vortex had passed directly over the center orifice.

4 NON-DIMENSIONALIZATION

It is advantageous to non-dimensionalize the data to facilitate identification of those parameters controlling the distribution of surface pressure. This requires the selection of characteristic length and velocity scales. For flows in the convergence region, selection of the updraft radius r_o as the length scale is appropriate.

In the limit of a weak swirl, consideration of a simple model of inviscid convergent flow (see Rotunno, 1977, and Snow et al., 1980) suggests that the appropriate non-dimensionalization scheme is one based on a characteristic radial velocity V_{ro}. Such a scheme results in the distribution of dimensionless surface pressure being dependent only on the swirl parameter (Γ/Q). It is thus suitable for examining the development of the swirling flow from the no-swirl state and is used in paragraph 5.

Recognizing that in strongly swirling flow the functional dependence of the pressure deficit on radius will likely be significantly different than that for weak swirl, a non-dimensionalization based on a characteristic axial velocity V_{ro} is more appropriate. Such a scheme leaves the non-dimensionalized flow in the convergence region a function of the two parameters S and a. It appears to be the best scheme to use in studies of phenomena within large, intense vortices.

5 RADIAL PRESSURE PROFILES

Shown in Fig. 2 is a typical family of mean surface pressure profiles measured using the techniques previously described. The values of the parameters for each curve in this family are given in Table 1. The data were non-dimensionalized using the characteristic radial velocity. The updraft radius r_0 and inflow depth h were held constant at 30.48 and 60.96 cm, respectively, and the volume flow rate (hence the radial Reynolds number Re_r) held fixed. The screen speed was varied incrementally to change the swirl parameter between determinations of the profiles. This family of curves thus shows the response of the surface mean pressure field to a progressive increase in swirl parameter for fixed Re_r and a.

The curve labeled a presents the no-swirl situation. It shows a broad central region of slightly higher pressure, with the lowest pressure in a circular trough beneath the lip of the updraft hole. This profile was associated with boundary layer separation in the outer portion of the convergence zone. Separation gave rise to a stagnation region over the lower surface and around the central axis.

The weakly swirling flow of curve b did not have a concentrated core in the convergence zone. Rather, the shallow pressure well within the central region of higher pressure was a response to the appearance of a core zone well aloft. Air with significant angular momentum was prevented from approaching the axis at low levels since the magnitude of the adverse gradient presented by the high pressure ring to the inflow was still sufficient to trigger separation.

As the swirl parameter was increased, the dynamic pressure deficit induced by the intensifying swirling flow aloft offset the pressure rise due to the decelerating radial component. As the concentrated core built downward, the greatest induced pressures were around the axis so that the central pressure well gradually deepened and widened.

As the swirl strengthened, the favorable gradient presented to the inflow by the deepening central pressure well caused the center of the stagnation region to thin out. Eventually, this favorable gradient strengthened to where it forced reattachment of the separated inflow at some radius inside the line of separation. (Note that the ring of high pressure was eroded from the inside out, its outer circumference being fixed by the geometry of the convergence region.) A small but intense central core in contact with the lower surface formed as the reattached fluid approached the central axis. Further increases in input swirl reduced both the vertical thickness and radial width of this ring. At $S \simeq 0.20$, the component of the pressure field induced by the increased convergence of angular momentum strengthened to where it prevented boundary-layer separation altogether; fluid flowed directly from the confluence region into the core. A well-developed single-celled vortex extended to the lower surface (curve c). Further increases in swirl resulted in a rapid deepening of the pressure well (curve d).

For $0.1 < S < 0.5$, a series of vortex breakdowns developed in the core in the convection region and moved upstream with increasing swirl. By $S \simeq 0.5$, these had begun to reach the lower surface. In response, the pressure profile first flattened out (curve e) and then begun to reflect a small, central high (curve f). This represents the transition from a one- to a two-celled vortex. For very large swirl ($S > 1.0$), the inner core contained a well-developed central high (curve g), with the lowest mean pressures being found in an annular ring at the edge of the core. The central high is due to the deceleration of the downflow coming in from above. During the transition from a one- to a two-celled vortex, the central pressure rose, but for very high swirl, mean pressures in the central zone began to fall as the core broadened out significantly. We note that a two-celled vortex usually contained a pattern of two or more subsidiary vortices that develop in the zone of strong shear of the outer edge of the core. Since these measurements were made with a long time constant first-order filter, the high frequency fluctuations due to the subsidiary vortices were filtered out.

Additional profiles and further discussion can be found in Barnhart (1979) and Snow et al. (1980).

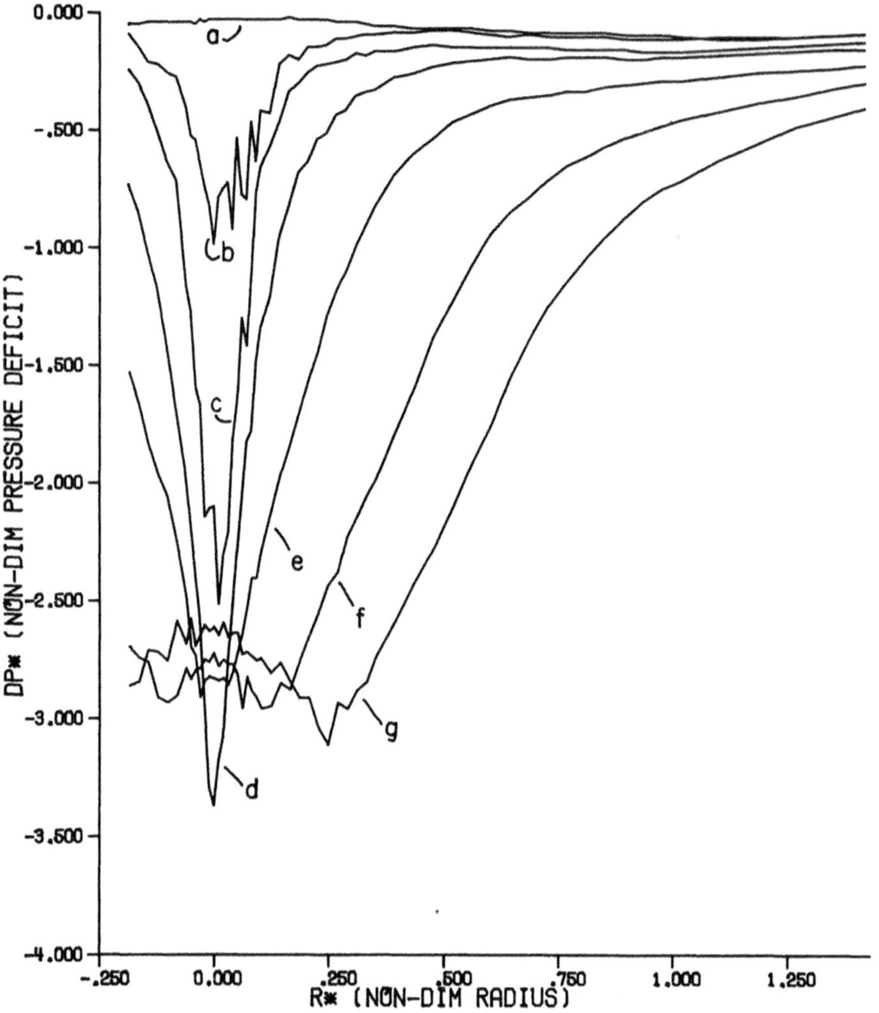

Figure 2. Profiles of mean surface pressure for the conditions listed in Table 1. Values have been non-dimensionalized using a characteristic radial velocity. Small drop at $r_* = 0.26$ in profile g is the time averaged effect of the secondary vortices

TABLE 1. Conditions used to produce the curves in Fig. 2

Curve	S	a	Re_r (x 10^4)	θ (deg)	Qh (m^3s^{-1})
a	---	0.5	2.384	(2.25)	0.662
b	0.14	0.5	2.384	8.00	0.662
c	0.29	0.5	2.385	16.25	0.662
d	0.42	0.5	2.411	22.74	0.669
e	0.60	0.5	2.400	31.02	0.666
f	1.17	0.5	2.384	49.48	0.662
g	1.79	0.5	2.398	60.87	0.666

6 VARIATION IN MAXIMUM PRESSURE DEFICIT

Fig. 3 is a plot of the maximum surface pressure deficit to be found in the vortex, regardless of radial location, as a function of swirl ratio. These data were taken for fixed aspect ratio a and constant flow rate Qh. The data have been nondimensionalized with ρV_{zo}^2. The general nature of the flow and the number of subsidiary vortices present are also noted. This plot is similar to Fig. 9 in Snow et al. (1980), except that here, for S >0.8, the sensing probe is moved off center to stay under the point of lowest pressure. Also, these data were taken without filtering, so that the deficit values for S > 0.8 are those occurring within the subsidiary vortices.

Consistent with previous discussion, in the interval $0 < S < 0.8$, four distinct regimes of flow are indicated

- for $0 < S < 0.2$, the flow is controlled by separation of the inflow at large radius. A core was present aloft, but did not extend persistently down to the surface.

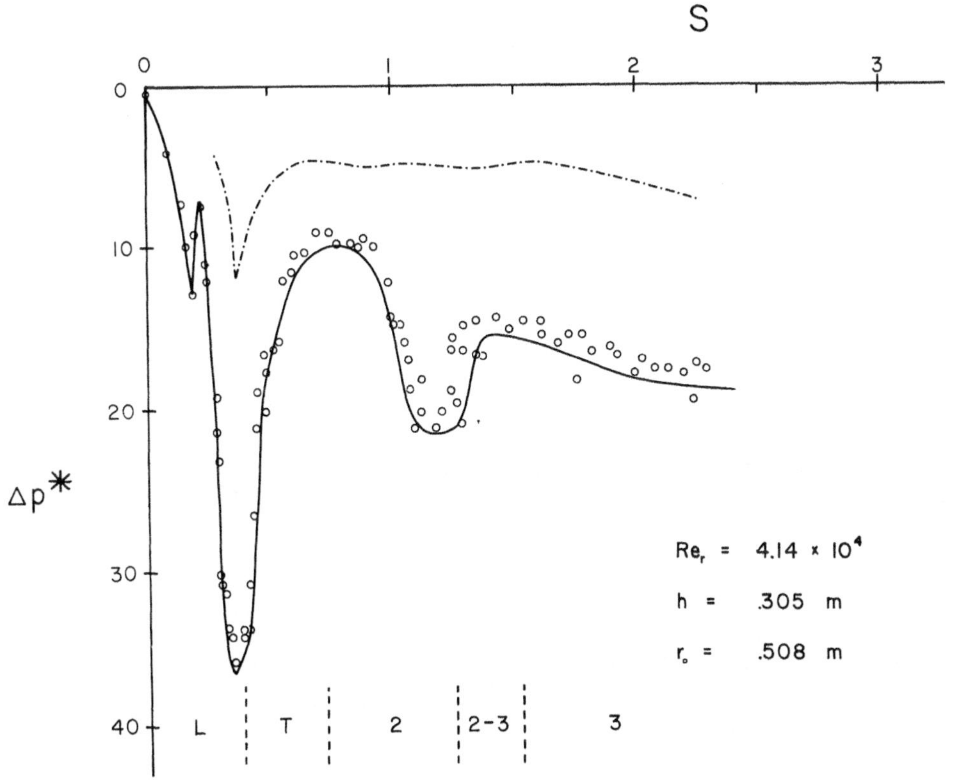

Figure 3. Plot showing dependence of maximum pressure deficit on swirl ratio for $Re_r = 4.14 \times 10^{-4}$, a=0.6. Values have been non-dimensionalized using a characteristic axial velocity. Lower row of characters denotes vortex structure: L=laminar, T=turbulent; 2,3=number of subsidiary vortices present, (2-3)=two-to-three subsidiary vortex transition zone. Circles are some of the actual data points, solid curve is hand-drawn envelope to the data. Dot-dash line is time-mean value of pressure at same location. For S < 0.8, values are all at center line; for S > 0.8, at progressively larger radii

- for $0.20 < S < 0.25$, the central pressure increased as S increased. Flow visualization showed that the favorable external pressure gradient imposed on the surface boundary layer had become strong enough to force reattachment. A small, laminar core was in contact with the surface. The annular region of separated flow flattened, eroded from the inside outward, and disappeared as $S \to 0.25$.
- for $0.25 < S < 0.45$, a concentrated, single-celled vortex had penetrated to the surface, and progressively intensified. The innermost core flow throughout this stage appeared to remain laminar. The transition to a turbulent vortex occurred aloft in the convection zone in the form of a "breakdown" of the core. This breakdown point moved upstream as S was increased. The maximum central pressure deficit occurred at $S \approx 0.45$, just before the breakdown reached the lower surface. (See also curve d in Fig. 2.)
- for $0.45 < S < 0.8$, the vortex was transitioning from a one- to a two-celled structure. The core was now turbulent over its full length. As $S \to 0.8$, the core expanded rapidly - the rise in central pressure in this interval reflects the loss of a concentrated core.

For $S > 0.8$, two secondary vortices were present. The lowest pressure in the vortex was found off axis, beneath these secondary flows. They provided two centers of localized convergence at the outer edge of the parent core, so as the input swirl was further increased ($S \to 1.1$), the maximum pressure deficit also increased. For $1.1 < S < 1.3$, there was a transition from two to three subsidiary vortices. The localized centers were lost, and the pressure rose. For $S > 1.3$, with three subsidiary vortices present, the pressure again decreased with increasing swirl, but at a much slower rate.

7 DEPENDENCE OF MAXIMUM SURFACE PRESSURE DEFICIT ON RADIAL REYNOLDS NUMBER AND ASPECT RATIO

To confirm that the radial Reynolds number was of secondary importance in determining the dynamics of the flow, a series of measurements were made to establish the dependence of the maximum central pressure deficit on Reynolds number and aspect ratio. The geometry was fixed, and the flow rate varied, changing Re_r. For each Re_r at which a data point was taken, the swirl ratio was adjusted until the pressure drop was maximized. The resulting family of curves, non-dimensionalized by ρV_{zo}^2, is shown in Fig. 4.

Flow visualization confirmed that the greatest central pressure deficits observed were coincident with the penetration of the vortex breakdown to the surface. It should be emphasized that the swirl ratio at which the vortex breakdown reaches surface is not constant for all Re_r. More swirl must be put in at low Re_r to achieve this transition. However, for the larger values of Re_r, the critical swirl ratio value for this event approaches a constant value of $S = 0.3$.

We note that while the dimensional data ranged over two orders of magnitude, the dimensionless data in Fig. 4 fall within the narrow range $\Delta p_* = 40 \pm 8$. This strongly suggests that ρV_{zo}^2 determines the magnitude of the maximum pressure deficit that can be achieved for different flow rates and geometries. Further, the clustering of all the curves about the $\Delta p_* = 40$ value shows that the dependence on Re_r is weak.

8 PRESSURE DEFICITS IN SUBSIDIARY VORTICES

For a swirl ratio $S = 2.45$, a Reynolds number $Re_r = 2.9 \times 10^4$, and an aspect ratio $a = 0.6$, surface pressure distributions within subsidiary vortices have been measured. This set of parameters was chosen as it gave a visually steady pattern of three subsidiary vortices. These orbited at a mean radius of 19 cm with an approximate period of 1.0 s. The three orifice probe was centered beneath the

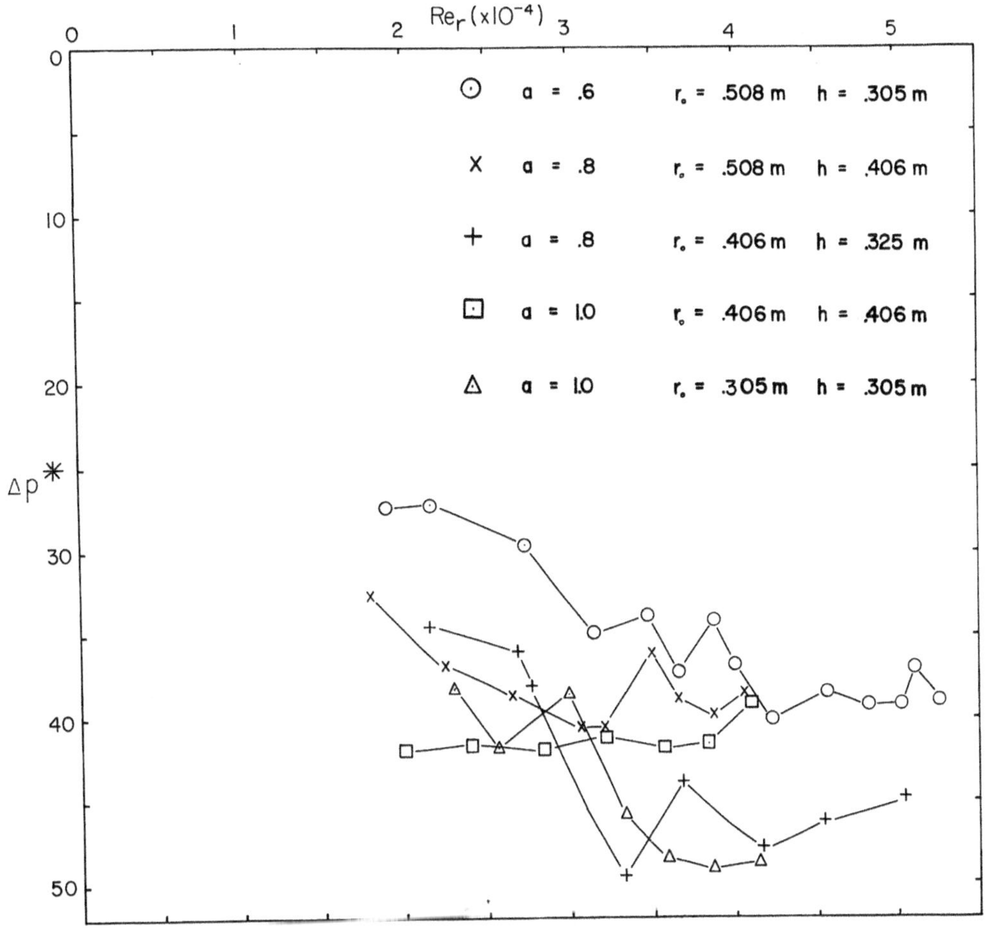

Figure 4. Dimensionless maximum surface pressure deficit *vs*. radial Reynolds number for five different flow geometries

track of the subsidiary vortices, and used with three transducers to make simultaneous observations of the pressure at r = 17, 19 and 21 cm as the subsidiary vortices passed overhead.

Employing Taylor's hypothesis, the temporal records from the transducers' outputs were transformed into a spatial representation of the pressure field. Assuming that each of the three traces represented the pressure deficit along a constant radius in the parent system, the limited-area plan form pressure fields for individual subsidiary vortices shown in Fig. 5 were synthesized.

Even though the flow conditions were selected to give an apparently steady pattern of subsidiary vortices, the measured fields showed considerable diversity. The measured pressure deficits ranged from 16 to 22 Pa. (Pressure deficits at the center of the parent circulation and in between the subsidiary vortices (at r = 19 cm) were in the range 7 to 9 Pa. ---the subsidiary vortices contained pressure drops two to three times those elsewhere in the parent pressure well.) Also frequently observed was a slightly stronger pressure gradient on the side of the subsidiary vortex closer to the center of the parent vortex.

Analysis of nearly direct transversals over the center tap on the three port probe indicate that the core radii of the stronger subsidiary vortices (defined as the radius (about the local center) of the inflection point in the pressure profile) were in the range 2 to 3 cm. This corresponds to 10 to 15% of the radius of the track of the subsidiary vortices around the system center.

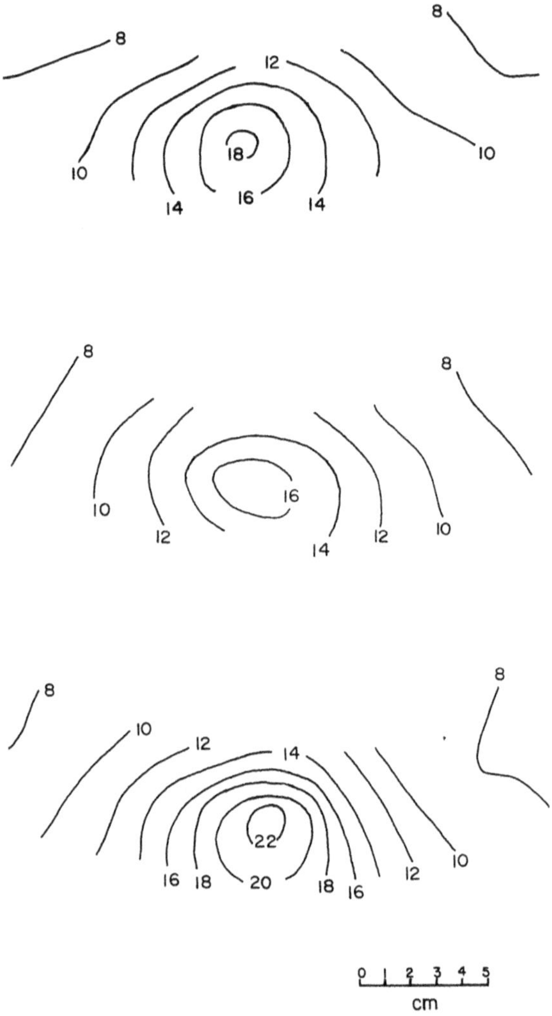

Figure 5. Limited-area surface pressure fields for three individual subsidiary vortices for $S = 2.45$, $Re_r = 2.90 \times 10^4$, $a = 0.6$. Units of deficit pressure are Pascals

9 DISCUSSION

The present study was undertaken to provide a better physical understanding of the mechanisms producing the surface pressure fields observed beneath actual tornado cyclones. As expected from the outset, the surface pressure field has been found to strongly reflect the evolution of the vortical flow as a function of increasing swirl. The laboratory observations reported herein show the potential complexity of the tornado-cyclone system. These interpretations indicate that in many cases the traditional simple model of the tornado as a single-celled vortex must be discarded and a much more complicated two-celled model used.

In examining the evolution of the surface pressure field in the outer portion of the convergence region as a function of swirl (Fig. 2), the radial profiles of mean surface pressure show a shifting balance between

- radial inertial forces acting to decelerate the inflow by establishing a region of higher pressure about the centerline (the presence of the resulting adverse gradient triggers separation of the surface boundary layer) ; and
- the radial pressure gradient force induced by conservation of angular momentum which acts to produce a region of lower pressure about the centerline (the resulting favorable pressure gradient acts to inwardly accelerate fluid in the surface boundary).

The modification of the pressure field by the conservation of angular momentum in the overlying flow thus acts to offset the inertial effects, and as swirl increases, to force reattachment of the separated boundary layer. Eventually the gradient within the dynamically induced pressure field becomes strong enough to prevent separation altogether. From this, both the cause of and the role played by the outer high pressure ring (the "humps") observed in microbarograph traces can now be recognized.

Figs. 2 and 3 also show that the dynamics of the central core are complicated. Whereas in the outer portion of the convergence zone it appears reasonable to assume that the pressure field is determined by the overlying flow and impressed through the surface boundary layer without great modification, beneath the core this is not the case. The data presented here reflect the development of a strong, downwardly directed pressure gradient working to turn and accelerate the inflowing fluid upward.

Another significant finding of this work is that the greatestdeficit pressures are associated with the single-celled vortex just upstream of the breakdown, and not with the subsidiary vortices in a multiple vortex system. The breakdown represents an abrupt increase in core radius. Hence, from angular momentum conservation arguments, it is expected that the pressure deficits should decrease once this feature is on the surface since the swirling air can not approach the central axis as close as was previously possible. Why the maximum pressure deficit associated with the vortex breakdown should be greater than the deficits in the subsidiary vortices of a multiple vortex system is less obvious. Certainly these subsidiary vortices provide localized centers of convergence. Work in progress at Purdue suggests that this result may be best explained in terms of the differences in the inertial (turning) processes between the two stiuations and the resulting vertical pressure gradient across the turning region. (We recall that it was also found that the pressure deficits within the multiple vortices were 2 to 3 times those elsewhere in the parent vortex)

The reader is directed to Pauley (1980) for additional mappings of pressure fields and further discussion.

The finding that the maximum pressure drop for given flow rate scales approximately with ρV_{z0}^2 is not too suprising. As shown by Morton (1969), there is close coupling between the flow force of the gross upflow (which is characterized by ρV_{z0}^2) and the strength of the axial flow in the vortex core, and so also to the strength of the azimuthal flow and the surface pressure deficit.

ACKNOWLEDGEMENTS

I wish to acknowledge Ms. B.J. Barnhart and Mr. R.L. Pauley, both of whom showed great patience and persistence in collecting the data described in this report. I also thank Dr. C.R. Church and Mr. G.L. Baker for many fruitful discussions concerning the physical interpretation of the data. Thanks are also due to Ms. D.A. Burks, who typed the manuscript.

REFERENCES

Barnhart, B.J., 1979: An experimental investigation of surface pressure distributions in tornado-like flows. Masters Thesis, Purdue University, W. Lafayette, IN. 144 p

Church, C.R., J.T. Snow, and E.M. Agee, 1977: Tornado vortex simulation at Purdue University. Bull. Amer. Meteor. Soc., 58, 900-908 + cover photo

Church, C.R., J.T. Snow, G.L. Baker and E.M. Agee, 1979: Characteristics of tornado-like vortices as a function of swirl ratio: A laboratory investigation. J. Atmos. Sci., 36, 1755-1776

Davies-Jones, R.P., 1973: The dependence of core radius on swirl ratio in a tornado simulator. J. Atmos. Sci., 30, 1427-1430

Davies-Jones, R.P., 1976: Laboratory simulations of tornadoes. Proc. Symp. on Tornadoes. Dept. of Civil Engr., Texas Tech University, Lubbock, TX. p. 151-173

Fujita, T.T., 1959: A detailed analysis of the Fargo tornadoes of June 20, 1957. Tech. Rep. No. 5, Severe Local Storms Project, Univ. of Chicago, 129 pp

Fujita, T.T., 1970: The Lubbock tornadoes: A study of suction spots. Weatherwise, 23, 161-173

Lewellen, W.S., 1962: A solution for three-dimensional vortex flows with strong circulation. J. Fluid Mech., 14, 420-432

Morton, B.R., 1969: The strength of vortex and swirling core flows. J. Fluid Mech., 38, 315-333

Pauley, R.L., 1980: Laboratory measurements of surface pressure minima in simulated tornado-like vortices. Masters Thesis, Purdue University, West Lafayette, IN. 91 p

Rotunno, R., 1977: Numerical simulation of a laboratory vortex. J. Atmos. Sci., 34, 1942-1956

Snow, J.T., C.R. Church, and B.J. Barnhart, 1980: An investigation of the surface pressure fields beneath simulated tornado cyclones. J. Atmos. Sci., 37, 1013-1026

Ward, N.B., 1964: The Newton, Kansas tornado cyclone of May 24, 1962. Preprints Eleventh Weather Radar Conf., Amer. Meteor. Soc., Boston, 410-415

Ward, N.B., 1972: The exploration of certain features of tornado dynamics using a laboratory model. J. Atmos. Sci., 29, 1194-1204

APPENDIX

The following symbols have been used in this report (see also Fig. 1):

a = internal aspect ratio of the convergence zone = h/r_o

h = depth of the convergence zone

l = depth of the convection zone

Q = average volume flow rate, per unit axial depth, across the confluence zone = $2\pi r_o V_{ro} = \pi r_o^2 V_{zo}/h$

Δp = pressure difference between the point of measurement and a fixed reference pressure

Re_r = radial Reynolds number = Q/υ

r_o = radius of the convergence zone

r_s = outer radius of the confluence zone

r_w = radius of the convection zone

r_* = non-dimensional radius = r/r_o

S = swirl ratio = $\dfrac{r_o \Gamma}{2Qh} = \dfrac{\Gamma}{2Qa} = \dfrac{\tan\theta}{2a}$

V_{ro} = height-averaged radial velocity at $r = r_o$

$V_{\theta o}$ = height-averaged tangential velocity at $r = r_o$

V_{zo} = area-averaged vertical velocity through the updraft hole

Γ = average circulation at the outer edge of the convergence zone = $2\pi r_o V_{\theta o}$

υ = kinematic viscosity of air = 0.153 $cm^2\ s^{-1}$

θ = inflow angle = $\tan^{-1}(\Gamma/Q)$

Convection and Mixing at High Rossby Numbers in Rotating Systems

A. D. McEwan

Summary: Laboratory experiments have revealed substantial reorganization of the mean motion in enclosed uniformly rotating fluid when subjected to intense stirring. The potential importance of these effects is reviewed in relation to the earliest stages of tropical cyclone genesis. A laboratory experiment in which localized mixing simulates the kinematics of a tropical disturbance cluster indicates that turbulently induced recirculations can bring about significant and rapid vortical organization.

1 INTRODUCTION

By the use of the 'compositing' technique, cumulative observations have provided a good synoptic account of the mean and eddy transport of momentum and heat in developing tropical disturbances (Gray, 1979, McBride, 1981). Not resolved by such techniques is the eddy contribution on scales of individual cloud clusters at which scale the intense vortex signature first appears.

The indications of some laboratory experiments to be described here are that when there exists a weak but finite environmental vorticity, turbulence possessing the kinematic character of intense convection is capable of efficiently concentrating this vorticity, importing angular momentum in so doing.

The present work is an evolution of earlier and more general attempts to identify experimentally the properties of rotationally modified turbulence. These are described in section 2. The present experiments described in section 3 are an attempt to imitate more realistically the conditions of a tropical cloud cluster, wherein the intense convection is confined in radius. The results are discussed in the context of cyclone genesis in section 4.

2 TWO SIGNIFICANT EXPERIMENTS

In recognition of the geophysical relevance of experiments on turbulence with rotation, a method of turbulence generation was sought which would possess the kinematical character of thermal convection while overcoming the scaling difficulties inherent in a laboratory scale thermal experiment.

Topics in Atmospheric and Oceanographic Sciences
Intense Atmospheric Vortices, ed. by Bengtsson/Lighthill

The method adopted in described in McEwan (1976). Water in a cylindrical container was recirculated through a regular array of sources and sinks in the base of the container by way of fine tubes, distributing manifolds and a pump. Without rotation no significant organized circulation was established on scales greater than the spacing of the sources and sinks. Since the source fluid emerged as a jet while returning more or less uniformly, the sinks being in most cases more numerous than the jets, the vertical motion had a structure resembling that of deep moist convection, with entrainment and recirculation in the rising column, the return flow representing low level convergence.

For the first experiment, described in McEwan (1976) the sources and sinks were distributed over the whole base of a closed cylindrical container, and observed through a transparent lid from above. When pumping was commenced in a previously uniform rotation, persistent cyclonic vortices soon appeared. Similar effects have been observed under different conditions (Malkus, 1968; McEwan, 1973; Dickenson and Long, 1980), but this experiment provided convincing evidence of a general relative velocity (and vorticity) distribution in these vortices. A completely self consistent description of this distribution could be derived assuming that the turbulence was scaled by the area averaged pumping rate $\overline{w}'$ and jet spacing ρ; then if angular momentum was diffused by this turbulence against turbulent pseudo viscosity and Ekman-layer spindown, a solution of the mean angular momentum equation gave excellent agreement (see figure 1) with the experimental results when the angular diffusivity and the pseudo viscosity were in the ratio of 1:3. In conformity with this concept the velocity scaling Q was shown to be proportional to $\overline{w}'\rho/\delta$, where δ is Ekman boundary layer thickness, corrected for source-sink Reynolds stress. An upper limit to this scale appeared to be imposed by a Rossby number defined by source exit velocity and container depth.

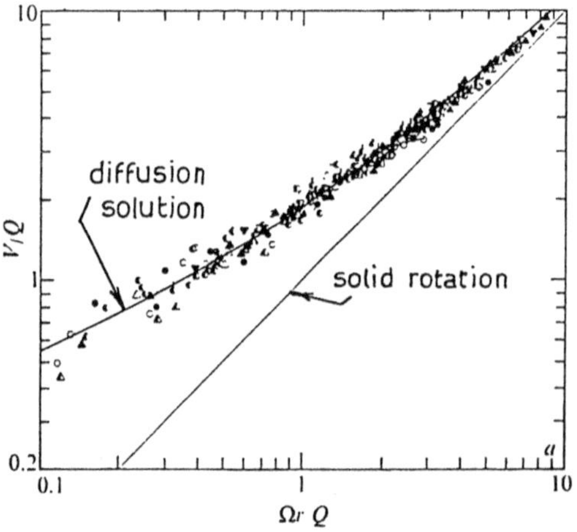

Figure 1

By inspection of figure 1, since the absolute azimuthal velocity $V = v+\Omega r$ tends toward the solid rotation value Ωr, v being relative azimuthal velocity relative to the vortex axis, any circuit embracing a whole system of vortices must have an average vorticity not departing significantly from 2Ω. Indeed since the vortices presumably exert a cyclonic torque on the ends there must exist a slight (although in these experiments imperceptible) anticylonic relative mean motion. On the other hand, most of the internal motion is cyclonic, thus the experiments indicate that *'uniformly distributed turbulence in rotation imports mean angular momentum and concentrates vorticity parallel to the rotation axis'*.

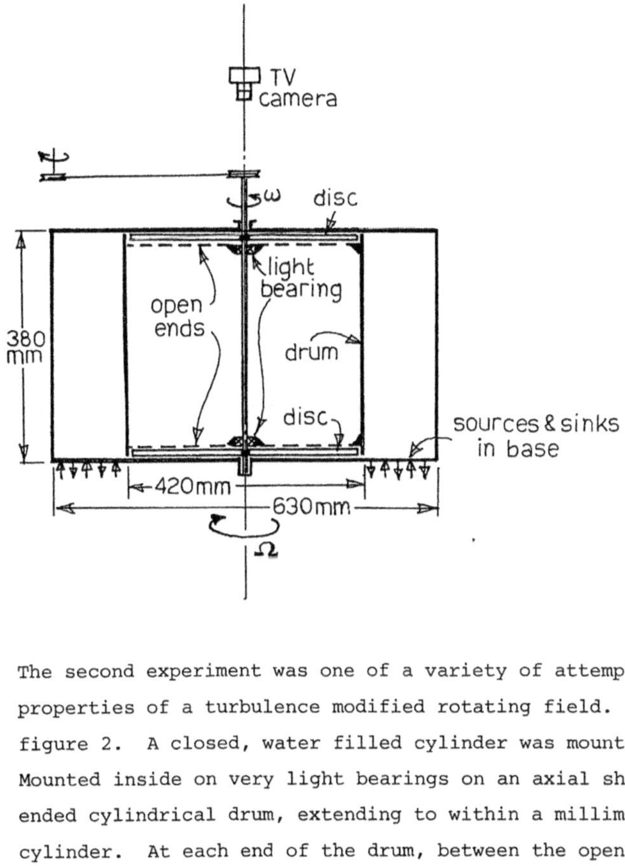

Figure 2

The second experiment was one of a variety of attempts to investigate the 'average' properties of a turbulence modified rotating field. Reference is made to figure 2. A closed, water filled cylinder was mounted coaxially on a turntable. Mounted inside on very light bearings on an axial shaft was a hollow, open ended cylindrical drum, extending to within a millimetre of the ends of the cylinder. At each end of the drum, between the open webs carrying the bearings were solid discs rigidly connected to the shaft. The shaft could be rotated at a remotely controllable speed. The whole was observed from above by a TV camera. When the discs were rotated at the same speed as the drum and with time allowed for equilibration, the drum therefore presented a *torque-free* inner boundary to the fluid occupying the annular space between it and the cylinder wall. An array of sources and sink under this annular space provided mixing in the same manner as in the former experiments.

In operation it was found that the drum rotated anticyclonically. There existed a consistent clockwise bias in the rate of rotation but as shown in figure 3 the rate of anticyclonic rotation of the drum, ω, over a wide range of pumping rates $\overline{w}'$, collapsed uniformly consistently against Ω when scaled by $\overline{w}'$.

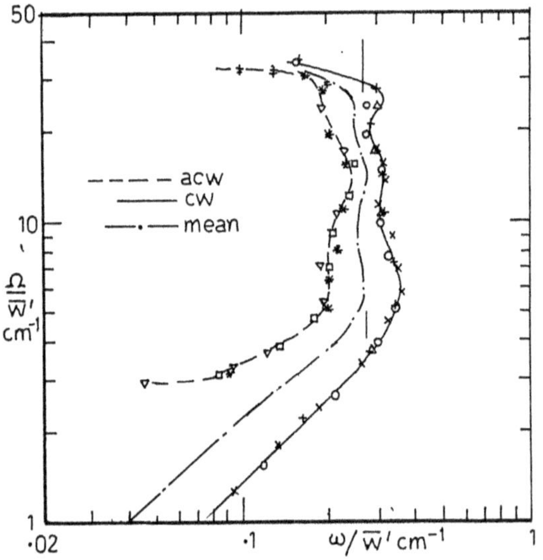

Figure 3

Taking the arithmetic mean to remove the bias three regimes were clearly defined:

(i) Low Rossby Number ($\overline{w}'/\Omega < 1/25$ cm) above which drum rotation fell to zero

(ii) Intermediate ($1/25 < \overline{w}'/\Omega < 1/5$ cm) when ω was independent of Ω

(iii) High Rossby number $\overline{w}'/\Omega > 1/5$) in which drum rotation became a constant proportion of background rotation, independent of turbulent intensity.

These results raise several interesting questions, but for the present discussion the most important observation is that where *there exists a radial gradient of azimuthal Reynolds stress, a flow rectification effect can persist as rotation reduces to very small values, apparently without limit.*

3 TROPICAL DISTURBANCE SIMULATION

The essential kinematic elements of a tropical disturbance cluster are a low but finite environmental vorticity, low vertical shear, an effective stratospheric lid, a virtually barotropic structure, and deep convection occasioned by some large-scale environmental convergence process, confined within a cluster of perhaps 50 km radius, or a group of clusters. To simulate these conditions in the simplest kinematically realistic way, the arrangement shown in figure 4 was adopted. A shallow cylindrical container on a turntable was provided with a regular triangular array of 18 sources and 42 sinks occupying the central 1/3 of the base. To simplify quantitative measurement, an array of circulation meters, of 130, 69 and 26 mm radius (the largest encompassing the source-sink region) was mounted on a central shaft at approximately the mid plane of the cylinder. To overcome friction the shaft was vibrated acoustically. Apart from the smallest meter, which had rectangular vanes, the meters were light wire spiders, bearing 8 mm spheres. Tests established that they measured mean azimuthal relative motion at their respective radii reliably and without bias. Motion was viewed from above by TV camera, and also recorded on cinè and still film. Pumping was commenced after solid rotation was established and the results given are steady state averages over at least several turntable rotation periods.

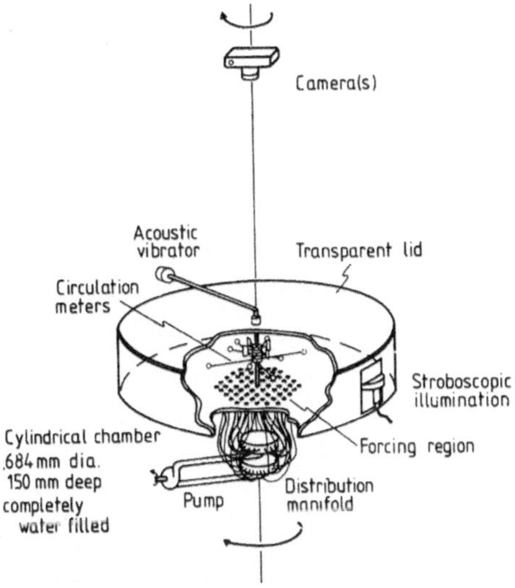

Figure 4

At high rates of turntable rotation Ω, the internal motion, revealed by suspended particles, resembled that of the original (McEwan, 1976) experiment, with coherent vortices forming, wandering and merging within the forcing zone. Only the smallest (26 mm radius) circulation meter registered any significant relative motion, an intermittent cyclonic rotation whenever one of the vortices lay within its' circuit. However, as the environmental rotation Ω was lowered there occurred a threshold below which all three circulation meters turned cyclonically, the rate increasing to several times Ω. The trend for one experimental sequence with pumping rate held constant is shown in figure 5.

The symbols (O,M,I) refer to the 130, 69 and 26 mm radius meters respectively and the height indicates the speed variability. As with the original experiment the

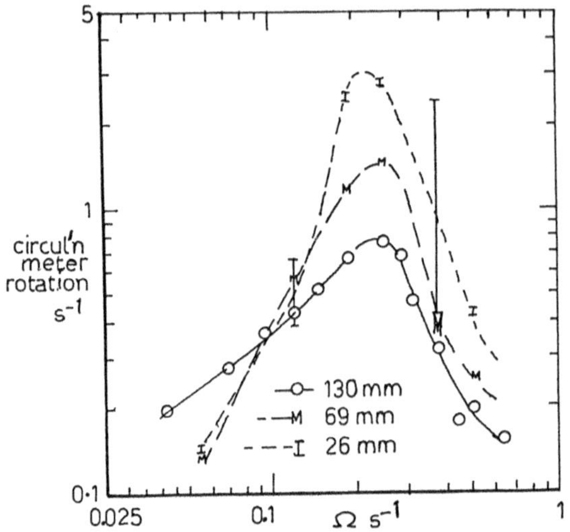

Figure 5

meter rotation increased inversely with radius to very low Ω indicating a corresponding degree of vorticity concentration towards the centre of the convection zone, and an importation of angular momentum.

Of particular interest is the circulation recorded by the outer meter. Figure 6 shows the angular velocity ω of this meter plotted against Ω. Both are normalized by pumping rate relative to a maximum pumping rate $\hat{w}'$ of 0.4 cm/sec. Also shown (broken line) is the value to be expected if absolute angular momentum were preserved from the outer boundary (342 mm radius). As shown, the angular momentum at the lowest angular velocities is almost completely conserved.

The similarity of profiles for different $\overline{w}'$ is indicative of a uniform Rossby scaling and therefore evidently insensitive to the only other time-scaling parameter, viscosity.

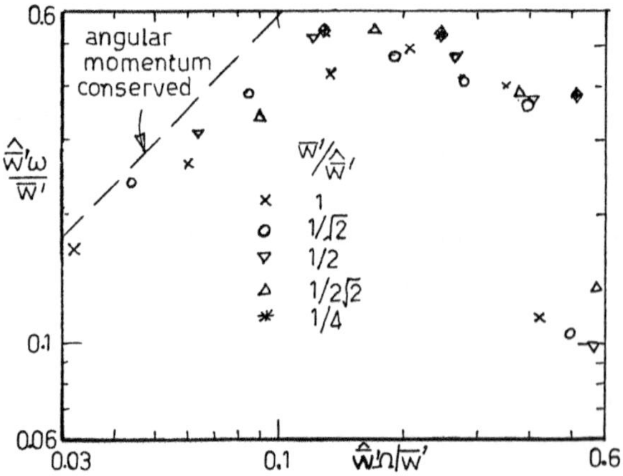

Figure 6

To accompany the explanation given in section 4 for the remarkable capacity of this type of spatially confined mixing to import angular momentum, some dye visualizations were performed.

Figure 7 shows sequences at three second intervals of dye injected into the source distribution manifold for three different rotation speeds. The arrows indicate the approximate extent of radial dye penetration, which is evidently suppressed by rotation, although within the penetration region, homogenization is rapid.

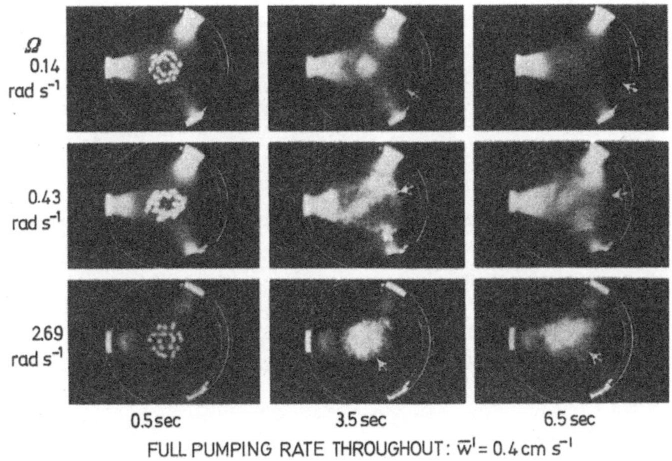

Figure 7

4 DISCUSSION: IMPLICATIONS FOR TROPICAL CYCLOGENESIS

To initiate the thermodynamic processes for warm-core formation and cyclone intensification a solenoidal circulation with upper level outflow needs to be established. Following Sawyer (1947) the equation for the radial acceleration of this circulation in cylindrical (r,θ,p,u,v,ω) coordinates is

$$\partial^2 u/\partial t^2 \;=\; \xi Z u - \partial/\partial t(\partial\phi/\partial r) + \omega\,\partial\alpha/\partial r + \xi F_\theta;$$

$\xi = f+2v/r = 2M/r^2$ is scaled absolute angular momentum, $Z = r^{-1}\,\partial M/\partial r$ is absolute vorticity and F_θ is the acceleration due to the azimuthal component of eddy body force. Other terms have their usual meaning.

If kinematical processes are to have effect in the early stages, F_θ must exceed Zu, since ξ and Z are both positive.

In axisymmetric models it has usually been assumed that F_θ derives mainly from vertical momentum flux gradients. More recently the importance of horizontal large-scale assymmetrical eddy terms has been recognized (McBride, 1981) but the indications of the present experiments are that even the fine-scale, azimuthally symmetrical eddy components require consideration. The horizontal flow contribution to F_θ is

$$F_{\theta_H} = -r^{-2}\, \partial/\partial r(r^2\overline{u'v'})$$

$$= -\overline{u'Z'}, \text{ the turbulent vorticity flux.}$$

Applying mixing length concepts, Taylors vorticity diffusion hypothesis (Rhines, 1977) yields

$$F_{\theta_H} = \kappa_v \partial\overline{Z}/\partial r$$

with κ_v a (positive) horizontal diffusivity; by this hypothesis unless $\overline{Z}$ increases with radius or ξ is negative, neither of which is likely, the contribution to u acceleration is negative.

If angular momentum is a diffused quantity, F_{θ_H} may contain a term of the form

$$r^{-2}\, \partial/\partial r(r\kappa_M(\partial M/\partial r)) = r^{-2}\, \partial/\partial r(r^2\kappa_M Z)$$

which differs in form from the previous expression by an additional term $r^{-2}Z\, \partial/\partial r(r^2\kappa_M)$. Providing κ_M increases with radius as occurs with decreasing rotational influence, or where horizontal transports are characterized by turbulent velocities whose magnitudes are independent of radial position, then this contribution can be positive.

This description may be appropriate for motion within a convective cluster. Motions beyond the cluster are more systematic by virtue of a prevailing stratification and the necessity for mass balance with cirrus outflow.

Kinematically the cluster can be regarded as a turbulent 'cloud' comprising rapid transport within vertical cells, surrounded by relatively quiescent fluid. The circumference represents an entraining boundary. Entrainment at this boundary and low level convergence in the boundary layer which gives the cells their existence is necessarily matched by outflow in the cirrus canopy. The situation is sketched in figure 8.

Figure 8

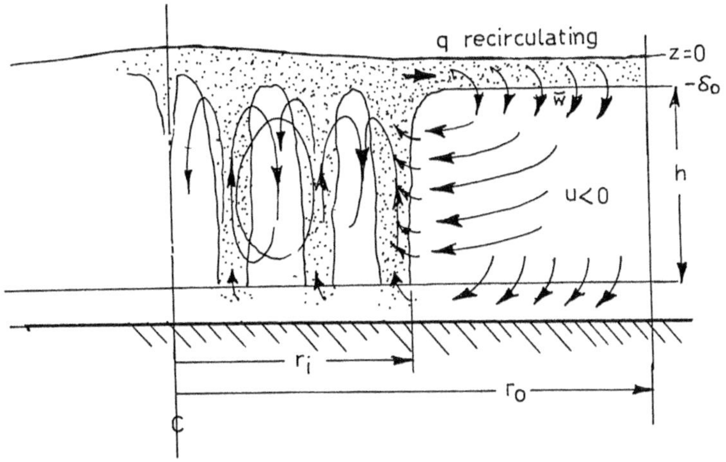

r_i is the cluster radius and r_o defines the radius to which the recirculation extends.

In an axisymmetric system the absence of a mean azimuthal pressure torque ($\partial\bar{p}/\partial\theta = 0$), requires that radial advection of angular momentum be matched by frictional torques alone. Either the returning flow is confined entirely within the boundary layer or angular momentum is virtually conserved through the whole depth of this region as it is drawn by entrainment into the cluster. The former possibility is not supported by laboratory experiments or by observation, for on compositing scales at least the radial inflow is found to be deep (McBride, 1981) in developing disturbances.

A consistent view, then, is that a confined convective cluster is surrounded by an annulus of fluid of uniform angular momentum. The vorticity enhancement within the cluster is then defined by the radius of the annulus r_o. Expressed as a velocity v relative to the environmental rotation Ω, where $v(r \geq r_o) = 0$,

$$v = \Omega(r_o^2 - r^2)/r, \tag{1}$$

The outflow region has the character of an Ekman wall jet. Theoretical description is not straightforward, but if characterized by boundary conditions in a manner similar to a conventional Ekman layer, the radial volume flux q(r) will be scaled by r and vertical thickness δ_o, i.e.:

$$q = 2\pi r \int_{-\delta_o}^{o} u dz = 2\pi r \delta_o v \Psi \tag{2}$$

where Ψ is a similarity constant of order unity.

So by continuity, with (1), the detrainment at the base of the outflow will be

$$\overline{w} = -2\Omega \int_{-\delta_o}^{o} \Psi dz, \tag{3}$$

independent of radius, providing Ψ is also independent.

Equating the radial outflow at the cluster perimeter to the detrainment over the annular area,

$$q(r=r_i) = \overline{w}\pi(r_o^2-r_i^2) \tag{4}$$

Entrainment dynamics imply that the interior radial velocity at the cluster perimeter $u(r_i)$ is scaled by the area-averaged volume flux in the convective columns, here represented by $\overline{w}'$. Thus

$$q \propto \overline{w}'hr_i$$

and by (4) above, with (3)

$$(r_o^2/r_i^2 - 1) = C\overline{w}'h/\Omega\delta_o r_i = CR_{o_R} \tag{5}$$

Where C is a constant of order unity. $\overline{w}'h/\Omega\delta_o r_i = R_{o_R}$ is a 'Recirculation Rossby Number'.

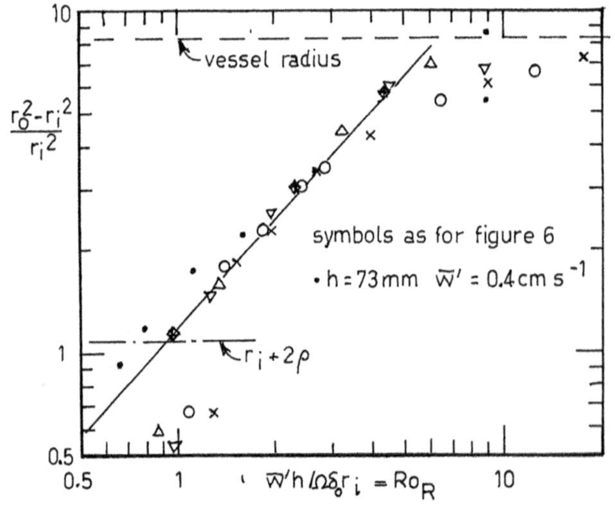

Figure 9

Equation (5) can be tested using the present experimental results. In figure 9 the quantity $(r_o^2/r_i^2 - 1)$ derived from the angular velocity of the 130 mm circulation meter using (1) is plotted against R_{o_R} for five different values of $\overline{w'}$ and one half-depth experiment. δ_o, being unknown, is taken as 10 mm.

The full-depth results (symbols) show a region where (5) is followed very closely, with C = 1.2. Departures occur where r_o approaches the container radius, or falls within distances from the cluster comparable with the source-sink spacing. The fit of half-depth results (solid symbols) is not so close, and is indicative of some geometrical dependence. However the general form of (5) is well confirmed, particularly since the scale factor C emerging experimentally is near unity.

To assess the foregoing in relation to the earliest stages of tropical cyclone genesis, a comparison of scales is given in table 1.

TABLE 1

	Tropical Disturbance	Experiment
Eddy vertical velocity $\overline{w'}$ at mid level, (intercloud, within cluster) cms^{-1}	20?	0.4
Convective cell spacing 2ρ cm.	$^1 10^6$	5.4
Outflow thickness δ_o cm.	$^1 10^5$	1.0?
Background vorticity $f \equiv 2\Omega$ s^{-1}	$^4 10^{-5}$	0.4
Depth h cm	$^1 10^6$	15
Cluster radius r_i cm.	$^5 10^6$	11.4
$R_{o_R} = \overline{w'}h/\Omega\delta_o r_i$	2.0	2.6

From the table R_{o_R} is comparable between examples, and if the proportionality of (5) is assumed, gives a 3.4 fold vorticity enhancement for the whole cluster, or with (1), an increase in the azimuthal velocity at the cluster perimeter amounting to 2.4 ms^{-1}. If a similar degree of internal vorticity enhancement to that indicated in figure 5 prevailed the average vorticity within one cell spacing distance 2ρ of the centre of the cluster would be twelve times the environmental value.

Encouraging through these results are, caution is needed in their extrapolation to the atmosphere, since they relate to an entirely homogeneous medium. Real systems, although virtually barotropic, possess a vertical stratification which would inhibit vertical exchange in the clear air surrounding of a cluster, and also in the inter-cloud regions within a cluster. Horizontal transports in these regions could therefore exhibit a significant degree of vertical structure, while in the present experiments vertical gradients were small and mean motions were almost depth independent away from the boundaries due to the high rate of vertical diffusion.

5 CONCLUSION

The kinematical character of deep thermal convection is simulated experimentally. It is found that intense mixing of this kind within a fluid initially in uniform solid rotation can bring about a localized concentration of vorticity and an importation of angular momentum from the surroundings. The net effect is to induce a relative mean circulation which persists even for very small rotation rates. When the mixing is confined in radius, strong cyclonic circulation can be induced by the creation of a peripheral region in which angular momentum is virtually conserved. The degree of intensification depends upon the radial extent of this periphery, which is directly dependent upon a 'Recirculation Rossby Number'. In the case of an active tropical disturbance cluster the process may explain the first formation of a closed vortex, and gives the opportunity for direct parameterization, providing some means for characterizing the vertical turbulent velocity $\overline{w'}$ in the cluster can be devised. Thermal stratification, absent in the experiment, may however exert considerable influence on the full scale.

This work was performed at the CSIRO Division of Atmospheric Physics in Aspendale, Victoria. The assistance of David Murray and George Scott is gratefully acknowledged.

6 REFERENCES

Dickenson, S.C. and Long, R.R. 1981 Oscillating grid turbulence including effects of rotation. J. Fluid Mech., (to appear)

Gray W.M. 1979 Hurricanes, their formation, structure and likely role in the tropical circulation. Meteorology over the tropical oceans (ed. D.B. Shaw) Royal Meteorological Society, p. 155

Holton, J.R. 1972 An introduction to dynamic meteorology. New York, Academic Press, pp. 319

Malkus, W.V.R. 1968 Precession of the Earth as a cause for geomagnetism. Science, 160, 259-264

McBride, J. 1981 Observational analysis of tropical cyclone formation. Part III: Budget analysis. J. Atmos. Sci. (to appear)

McEwan, A.D. 1973 A laboratory demonstration of angular momentum mixing. Geophys. Fluid Dynamics, 5, pp. 283-311

McEwan, A.D. 1976 Angular momentum diffusion and the initiation of cyclones. Nature, 260, 126-128

Rhines, P.B. 1977 The dynamics of unsteady currents. The Sea, 6, (ed. E.D. Goldberg, I.N. McCave, J.J. O'Brien and J.H. Stelle), pp. 189-318. Wiley

Sawyer, J.S. 1947 Notes on the theory of tropical cyclones. Quart. J. Roy. Met. Soc., 73, pp. 101-126

Intense Vortex Generation by Turbulence in a Rotating Fluid

E. J. Hopfinger and F. K. Browand[1]

Abstract Intense vortices are generated by turbulent forcing on a small scale of a fluid body initially in solid body rotation. The vortices extend over the entire depth and their core vorticity can be by a factor of about 50, and possibly more, larger than the background vorticity, 2Ω. The size of the vortices and their average spacing depend on the ratio of intensity of forcing to background vorticity. The concentrated vortex cores support waves consisting of helicoidal distortions and these waves induce strong axial flow which can cause local convergence and vortex intensification. Wave interaction frequently results in vortex breakdown. Complete distruction occurs on a time scale of about 10-20 rotation periods. The formation time of vortices, when the forcing is suddenly switched on, is of the order of 6-7 rotation periods.

1 Introduction

The idea of angular momentum mixing by turbulence in a fluid body initially in solid body rotation dates back to SCORER (1966). A consequence of angular momentum mixing is that vorticity is removed from the mixed region and expelled to the centre and the outer boundary. This results in vorticity transport up the gradient or in negative eddy viscosity. Experiments designed to demonstrate either the possibility of angular momentum mixing (GOUGH and LINDEN-BELL, 1968 ; BRETHERTON and TURNER, 1968 ; STRITTMATTER, ILLINGWORTH and FREEMAN, 1970) or to study the effect of rotation on homogeneous turbulence (IBBETSON and TRITTON, 1975) remained inconclusive. McEWAN (1976) was, however, able to show that vorticity can be concentrated on a smaller scale. He observed the formation of multiple vortices populating the cross section of the tank. Measurements of radial velocity distributions obtained from streak line photographs indicated vorticity concentration by a factor of 2-3 times the background vorticity, 2Ω. This feature is interesting but the vorticity concentration factor is too small for any relevance to tornados.

All of the rotating flow experiments have been performed in relatively shallow tanks and have the shortcomings that either the Reynolds number was too low or the forcing intensity and scale were unknown. In the present experiments a relatively deep tank is used, so that it was possible to have a zone of three dimensional turbulence adjacent to rotationally dominated turbulence. The turbulent flow is generated by a grid ossillating in a horizontal plane. The grid Rossby number is kept high, which assures a three dimensional turbulent field in a layer of about one to two grid meshes above the grid.

[1]On leave from the University of Southern California.

Topics in Atmospheric and Oceanographic Sciences
Intense Atmospheric Vortices, ed. by Bengtsson/Lighthill

The integral length scale and the rms turbulent velocity are known from HOPFINGER and TOLY (1976). The Reynolds number is of order 10^3 and away from the grid the Rossby number can reach very low values (order 10^{-2}).

Among the most striking phenomena observed are the extreme intensity of the vortices (section 4), comparable to what is observed in tornados, and the wave disturbances on the vortex core which can cause vortex breakdown and vortex intensification (section 5). The relevance to tornados is discussed in section 6.

2 Experimental Set-up

The experiments were conducted in a translucid cylindrical tank 40 cm in diameter and 80 cm in depth, shown schematically in figure 2.1.

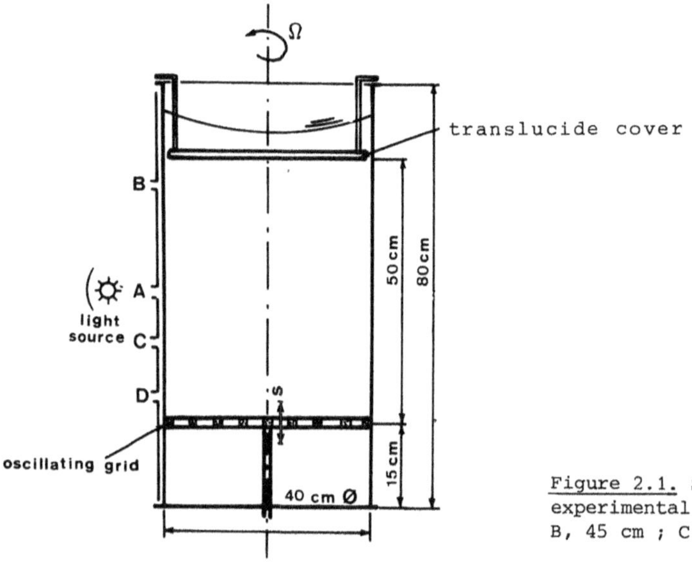

Figure 2.1. Schematic of the experimental set-up. A, z = 30 cm ; B, 45 cm ; C, 15.5 cm ; D, 5.5 cm

A turbulent velocity field is generated in this tank by means of a grid oscillating about a horizontal plane, located 15 cm above the lower end wall. This grid is made of 1 cm square bars and has a mesh size M = 5 cm. The stroke S and the frequency n are adjustable between $1 \leq S \leq 5$ cm and $2\pi \leq n \leq 20\pi$ rad S^{-1}. In all the experiments reported here we used S = 4 cm and n = 13.3 π rad S^{-1} and 6.6 π rad S^{-1}. Without rotation the rms horizontal turbulent velocity u and the integral scale l are given by (HOPFINGER and TOLY, 1976) :

$$u = 0.29\ nS\ (SM)^{1/2}/2\pi\ z \qquad (2.1)$$

$$l = 0.26\ z \qquad (2.2)$$

where z is measured from the midplane of the oscillating grid. This is more convenient than referring to the virtual origin which is about 0.5 cm off the midplane.

The whole system is placed on a rotating table, whose rotation rate can be adjusted continuously up to $\Omega = 3\pi$ rad S^{-1} and kept constant within 1 % or less. The rotation rates used were 2π, π and 0.2π rad S^{-1}, giving respective grid Rossby numbers n/f, ($f = 2\Omega$), of 3.32, 6.63 and 33.

The flow field in a cross section of the tank was obtained from streak line photographs. A 1 cm thick slice of fluid, containing trace amounts of wood particles (≈ 300 μm in diameter and of 1.15 gr/cm^3 density), was illuminated and time exposure photographs were taken. The structure parallel to the rotation axis was deduced from photographs and cinë films taken of a 2 cm thick fluid slice, illuminated parallel to the rotation axis. The vortex cores were made visual by introducing small air bubbles into the container which collected in the vortex cores.

3 Vortex Genesis

The most striking effect rotation has on the turbulent field is the generation of intense vortices. This is clearly seen from figure 3.1 which shows the instantaneous velocity field in a cross section of the tank for the case of $n = 13.3\pi$ rad S^{-1} and $\Omega = 2\pi$ rad S^{-1} (S = 4 cm in all experiments), corresponding to a grid Rossby number $Ro_g = n/f = 3.32$.

Figure 3.1. Streak line photographs of the flow field in a cross section 30 cm above the grid midplane.
$\Omega = 2\pi$ rad S^{-1}, $n = 13.3\pi$ rad S^{-1}; exposure time 1/4 s

The velocity field is similar to that observed by McEWAN (1976). Although the vortices occupy only a fraction of the total area, the flow field is dominated by them. Careful observation of films show that the sense of rotation is principally cyclonic - that is, in the same direction as the tank rotation. The velocity field in the vicinity of typical vortex cores is roughly 6-10 cm/sec at a radius of 1 centimeter. It was very difficult to obtain reliable velocity information from these observations of tracer particle trajectories. The major problem is the ambiguity in defining the beginning and end of individual particle paths (particles having a vertical velocity component can pass in or out of the light beam during the exposure time).

It is, however, noteworthy that the vortex cores generally contain few tracer paticles. We attribute this fact to the centrifuging of the heavier wood particles. Absense of particles in the cyclonic cores suggests that even larger velocities may exist here.

If the turbulence intensity of the grid is increased by increasing the oscillation frequency, or if the rotation rate is decreased for fixed oscillation frequency, then anticyclonic vortices are also observed. These are shown in figure 3.2a for a Rossby number $Ro_g = 6.63$, and in figure 3.2b corresponding to $Ro_g = 33.2$. At a Rossby number of 6.6 the anticyclonic vortices are not scattered randomly, but occupy a ring of fluid near the outer boundary. These vortices have much higher particle concentrations in the core regions which suggests they are much less intense than their cyclonic counterparts. For the largest Rossby number, $Ro_g = 33.2$ anticyclonic and cyclonic vortices are, perhaps, more interspersed.

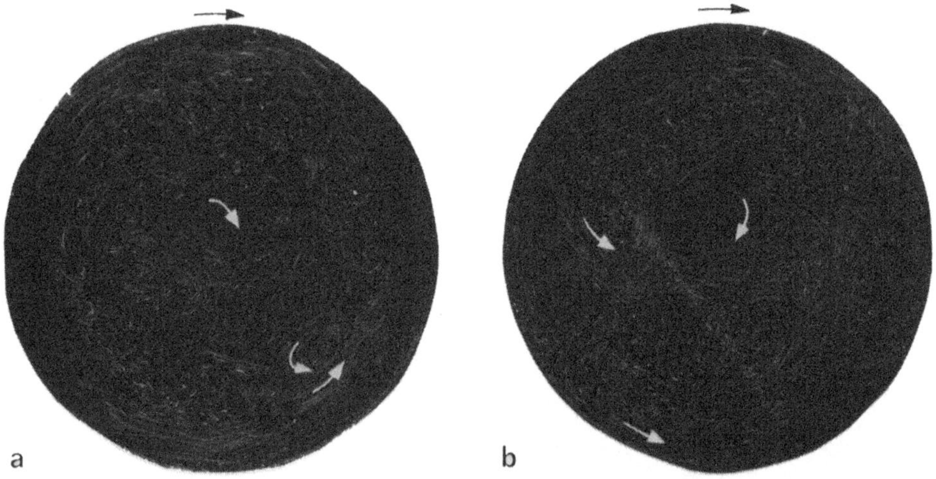

Figure 3.2. Streak line photographs of the flow field in a cross section 30 cm above the grid midplane. (a), $\Omega = \pi$ rad s^{-1}, $n = 13.3\,\pi$ rad s^{-1} ; (b), $\Omega = 0.2\,\pi$ rad s^{-1}, $n = 13.3\,\pi$ rad s^{-1}. The exposure time was 1/2 s

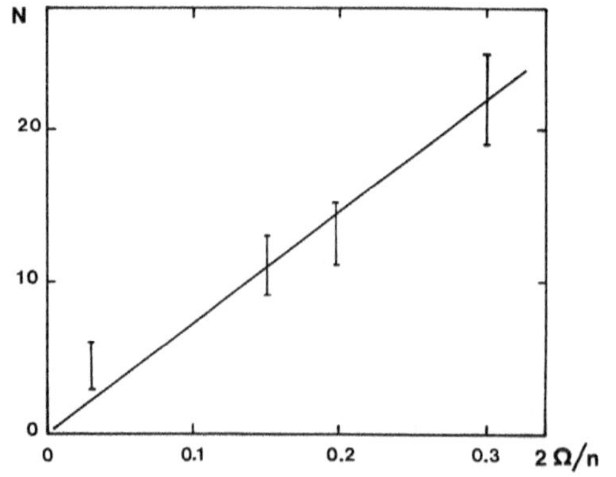

Figure 3.3. Number of vortices in a cross section plotted as a function of inverse grid Rossby number

The total population of vortices decreases with increasing Rossby number, and their size and average spacing increase. In figure 3.3. we have plotted the number of vortices, N, as a function of inverse Rossby number, $1/Ro_g$. The error bars on N are large because the selection criterion for vortices is somewhat subjective. Even so, figure 3.3 suggests a linear dependence between N and $1/Ro_g$, except perhaps for the largest value of Ro_g. A linear relation between the number of vortices and the inverse grid Rossby number would be supported by the assumption that a transition from three dimensional turbulence to a turbulent flow diminated by rotation (quasi-geostrophic turbulence), takes place at a certain height above the grid, represented by a fixed value of Rossby number based on local flow properties, and that the vortex spacing l_s is proportional to the turbulence length scale l_T at the edge of the turbulent Ekman layer. The relation obtained for the number of vortices is

$$N = K Ro_T / Ro_g, \tag{3.1}$$

where K is a constant and Ro_T is the transition Rossby number. The value of Ro_T could be obtained from measurements of the turbulent Ekman layer thickness. Flow visualisations in cross sections at z =15.5cm and 5.5cm (respectively C and D in figure 2.1) indicate that for Ro_g = 3.32 and 6.63 transition in flow structure takes place at $5 < z_T < 15$ cm which gives a value for $Ro_T \simeq 0.20$-0.5.

Television film observations show the vortices to be clearly present over times of a least 7 to 10 seconds (that is 7 to 10 rotation periods at Ro_g = 3.32) and occasionally even longer. During this time, the individual vortex strengths vary quasi-periodically until, at some stage, vortices become unrecognizable. The typical life-times are thus much longer than any eddy turnover time. It is not clear from these observations whether individual vortices are created and destroyed on this time scale or whether the vortices exist more or less permanently, but have highly modulated core strengths. Arguments in favor of distruction are, however, developed in section 5.

The time it takes to establish concentrated vortices when the forcing is suddenly switched on is also of interest. The non-dimensional time scale determined from visual observations is approximately

$$T_f \, \Omega / 2\pi \simeq 6\text{-}7, \tag{3.2}$$

where T_f is the formation time. Vortices appear throughout the tank on this time scale.

When the grid Rossby number is small, the mean flow is nearly solid body rotation, but for larger grid Rossby numbers, significant deparures from constant vorticity (solid body rotation) are observed. These conclusions can be obtained directly from the previous figures. For example, contrast figure 3.1. with figure 3.2a where a mean drift is clearly indicated near the wall. In figure 3.2b, the mean flow meanders considerably, but still, exists. In the words of previous researchers, an "expulsion" of vorticity takes place from the intermediate region to the region nearer the outer wall and to the region nearest the centre. The deficit in vorticity coincides with the centres of the anticyclones located near the boundary in figure 3.2a. The origin of these anticyclonic vortices is unclear, and will have to be studied further. Our present feeling is that they are generically different from the cyclonic vortices about which more will be said in later sections (although their initiation is not understood either).

4 Vortex Strength

The photographs discussed earlier have suggested that vorticity is highly concentrated in cores of relatively small extent. In order to visualize these vortex regions, we have illuminated a vertical slice of approximately 2 cm width through the center of the tank. Tiny gas bubbles were introduced into the bottom of the tank, and these collected in the low pressure wortex cores to provide a means of visualization. Such a vortex core visualization is shown in figure 4.1a for a Rossby number, $Ro_g = 3.32$ and 4.1b for $Ro_g = 6.64$.

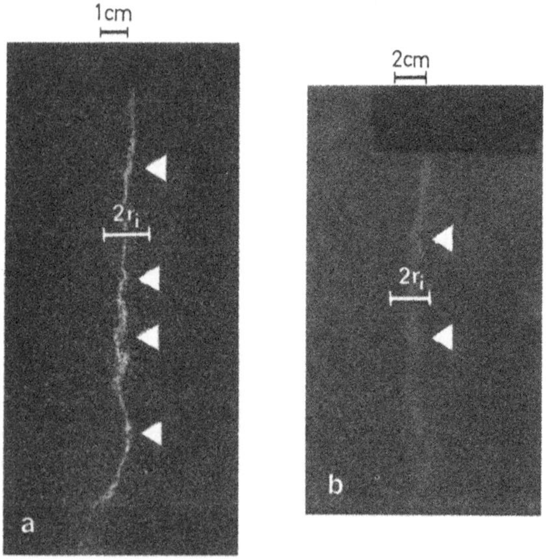

Figure 4.1. Photographs of vortex core made visual by small air bubbles introduced into the tank.
(a), $n = 13.3\,\pi$ rad s^{-1} and $\Omega = 2\,\pi$ rad s^{-1}, exposure time 1/60 s ;
(b), $n = 13.3\,\pi$ rad s^{-1}, $\Omega = \pi$ rad s^{-1}, exposure time 1/4 s

The cores are extremely thin and extend, coherently, over the entire depth of the rotationally dominated fluid column. They also tend to penetrate into the turbulent Ekman layer. Generally speaking, cores are aligned with the rotation axis, but of course they are not precisely vertical. They undergo large scale distortions, and appear to wander slowly about the tank. There are also many smaller scale distortions (several are marked in figure 4.1) which propagate along the core. These disturbances are discussed in the next section. One might describe such a flow field as being quasi two-dimensional, that is, having significant - but not perfect - correlation in the vertical direction.

Visual estimates of the illuminated core diameters give a number like 1-3 mm. The relation between the vortex core and the illuminated core size is uncertain, and another method was used to estimate the actual distribution of vorticity. The cover was removed from the tank and the water level lowered to 30 cm (for photographic reasons). The free surface was now populated with tiny craters, each one reflecting a concentrated vortex below. By filming the cross section of such craters, it was possible to determine the distribution of velocity which gave rise to the local changes in water surface elevation. The observed surface crater and that which would exist if surface tension were absent are shown in figure 4.2.

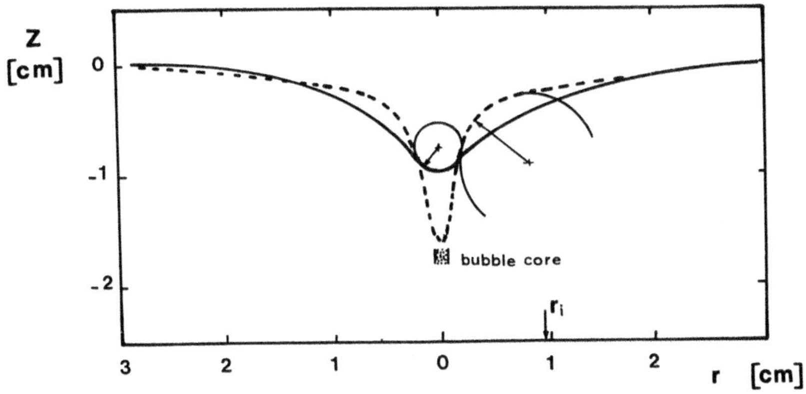

Figure 4.2. Free surface deformation produced by a vortex when $\Omega = 2\pi$ rad s^{-1}. ——— observed ; ------- corrected for surface tension

The velocity field required to produce the observed crater is shown as the dotted line in figure 4.3. The solid line is the velocity distribution obtained with a modified Rankine vortex model and approximates the surface deformation in figure 4.2 corrected for surface tension.

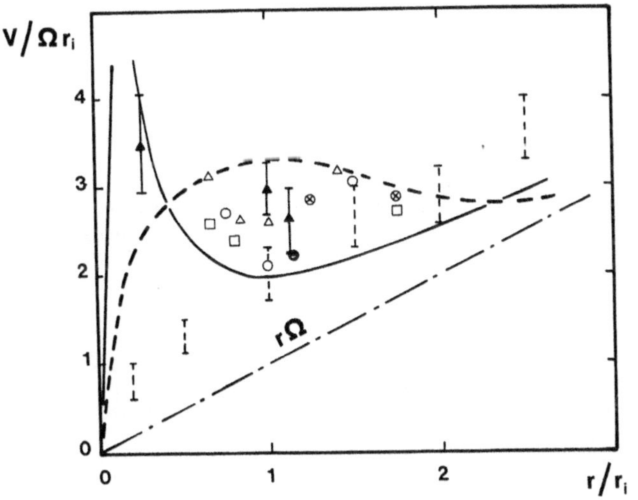

Figure 4.3. Radial velocity distribution in a vortex relative to the inertial frame for $\Omega = 2\pi$ rad s^{-1} and n = 13.3 π rad s^{-1}. ---- velocity field corresponding to the observed crater in figure 4.2 ; ——— velocity field obtained from the vortex model (which approximates the crater when surface tension is taken into account). r_i is the inertial radius $r_i = (\Gamma/2\pi\Omega)^{1/2}$. The open symbols correspond to velocities determined from selected vortices in figure 3.1 and ▲ refers to velocities determined from ciné films. Ι, indicates velocities measured by McEWAN (1976)

The following model velocity distribution was chosen for simplicity,

$$V = (\omega_c + \Omega)\, r \quad , \qquad r \le 2\epsilon/3 \quad ,$$
$$V = (\omega_c + \Omega)\, 2\epsilon/3 \quad , \qquad 2\epsilon/3 \le r \le 3\epsilon/2 \quad ,$$
$$V = \frac{\Gamma}{2\pi r} + \Omega r \quad , \qquad r \ge 3\epsilon/2 \quad ,$$

where V is the absolute azimuthal velocity, and ϵ, ω_c, and Γ are parameters to be determined. The "best fit" values are,

$\epsilon = 1.5$ mm,

$\omega_c = 257$ rad/s ,

$\Gamma = 36{,}4\ \mathrm{cm^2/s}$.

The concentration of vorticity is considerable. The core rotation rate, ω_c, is about 40 times larger than the ambient, rigid body rotation. The circulation contained in one core is only about 0.23 % of the circulation round the tank, however. Since there are approximately 20 such vortices (figure 3.3), the total circulation bound up in the vortex cores represents approximately 5 % of the tank circulation. It can be shown, by invoking similarity arguments, that the circulation in a vortex is independent of rotation rate and depends on intensity of forcing only. The reason for this interesting result is that as the size of the vortices increases with increasing Ro_g, vorticity is gathered from a larger region and the circulation remains relatively constant. This result is born out directly by observations of the velocity field at the three grid Rossby numbers (figure 3.1, 3.2a and 3.2b).

5 Travelling Wave Disturbances, Vortex Breakdown and Vortex Concentration

One of the most interesting dynamical aspects of this rotating turbulent flow are the wave disturbances which are observed to travel along the cores. Figure 4.1, for example, shows several waves travelling on a single core. The positions of the waves are indicated in the figure. The disturbances have the form of a helical distortion of the core. The distortions propagate as a result of self induction. The possibility of such vortex core motions was discussed in general terms by BATCHELOR (1967) and by HASIMOTO (1972). Simple dynamical arguments suggest four possible wave types. These possibilities are sketched in figure 5.1.

In all cases, the core vorticity vector is oriented downwards (clockwise rotation, as in the experiment). Example (a) is a left-handed, open helix which might for simplicity be called a "left twist". The induced motion causes the left twist to propagate upward in the opposite direction to the vorticity vector. The pattern rotates as it travels, with a sense of rotation opposite to the rotation of the core. A helix may also double back upon itself to form a loop or kink. Example (b) is a "left kink", which also travels in a direction opposite to the vorticity vector, but the pattern rotates in the same sense as the core rotation. From symmetry, the other two possibilities are a "right twist", and a "right kink", both of which propagate in the same direction as the vorticity vector. Again, the right twist rotates in a sense opposite to the core rotation, while the right kink rotates in the same sense as the core rotation. All of these wave types have been observed in our rotating, turbulent flow. Typical waves have

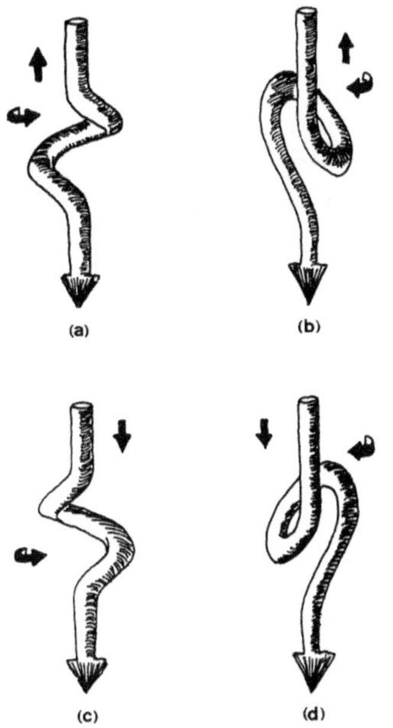

Figure 5.1. Basic helicoidal vortex core wave disturbances. The vorticity vector is directed downward. (a), left twist ; (b), left kink ; (c), right twist ; (d), right kink. The direction of propagation and the sense of rotation of the disturbances are indicated in the figure

wave envelopes which extend over several centimeters (wave lengths along core axis) and wave amplitudes of a fraction of a centimeter. Same waves satisfy the soliton solution of HASIMOTO (1972) as is shown in a separate paper (HOPFINGER and BROWAND, 1981).

Typical vortex cores contain many propagating waves. For the case $Ro_g = 3.32$ we observed on our 45 cm vortex core, on average, 3-4 upgoing waves and 1-2 downgoing waves. Wave interactions are frequent and often dramatic. Because there is so much activity on the vortex core, it is extremely difficult to follow the details of individual interactions. The following conclusions about wave interactions are therefore somewhat sketchy and preliminary. Additional, controlled experiments on wave interaction are planned.

Generally speaking, wave interactions involve energy dissipation, small scale turbulence production, and considerable local disruption of the vortex core region. There may be cases in which waves interact without modification of the core, but we were not able to demonstrate such interaction. We have attempted to classify the interactions leading to core breakdown into three categories, an example in each category is shown in figure 5.2 for $Ro_g = 3.32$. The first category (figure 5.2a) is believed to be the result of reflection interaction between an upgoing and a downgoing wave. The second category (figure 5.2b) is believed to represent an overtaking interaction of two waves travelling in the same direction, but having different wave speeds. The third category, which produces the most violent disruption (figure 5.2c), appears to be caused by a single wave of large torsion and may not be an interaction at all. The wave amplitude grows substantially as the wave progresses, and we cannot, however, rule out the possibility of additional waves accumulating from behind.

One is struck by the similarity between the features observed (figure 5.2) and the more conventional observations of vortex break-

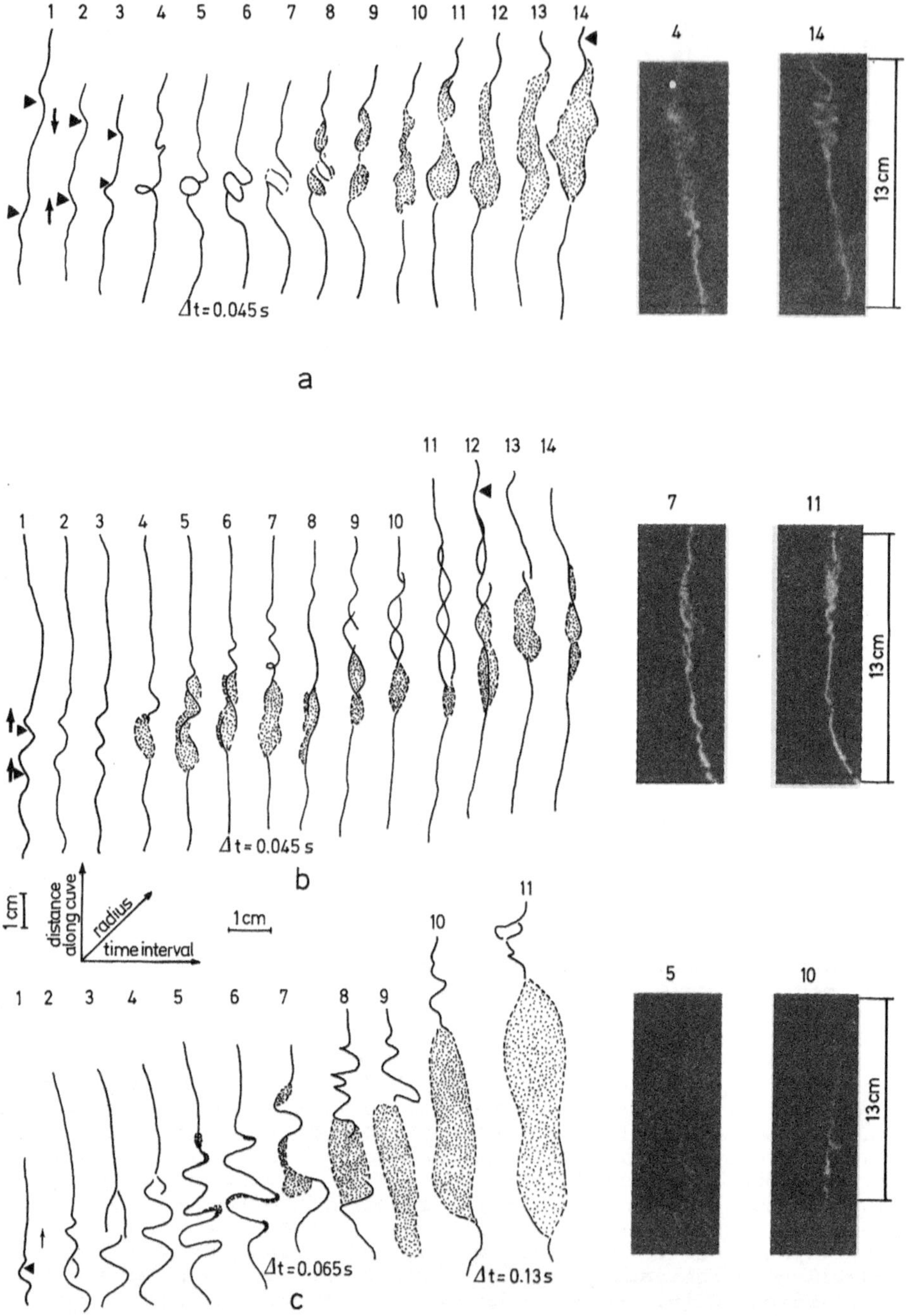

Figure 5.2. Most frequent local vortex breakdown modes. (a), loop mode ; (b), double helix ; (c), spiral mode. The vortex core is traced at regular time intervals from initiation of breakdown until occurance of small scale structure

down in various laminar, rotating flow apparatus. A double-stranded helical form analogous to 5.2b was observed by SARPKAYA (1971) and by FALER and LEIBOVICH (1977) to be a possible mode of stationary vortex breakdown ; and spiral breakdown (5.2c) has been identified to occur on delta wings by LAMBOURNE and BRYER (1961). It is plausible to conclude that waves of the form observed here are also responsible for these stationary breakdown phenomena. In these cases, a strong axial flow counters the forward propagation of the wave , effectively bringing the process to rest.

We have not found anything comparable to the axisymmetric bubble observed on delta wings and, for example, by FALER and LEIBOVICH (1977), and by ESCUDIER and ZEHNDER (1981) in their remarkable visualizations. Our flow lacks the symmetry which might be expected to be necessary to produce the axisymmetric bubble. On the other hand, the case involving head on collison of two waves doesn't seem to fit any descriptions of observed stationary vortex breakdown either. It may be that stationary breakdowns in most other devices start with a relatively "clean" upstream flow, and that wave disturbances in this case are produced only at the downstream end.

An interesting observation is that after a local breakdown the vortex generally reforms in about 1/2-1 rotation periods. Each time, a wave disturbance moves avay from the breakdown region (see for instance, configurations 10-12 in figure 5.2) and the core is reestablished. This would suggest that waves are also responsible for the maintenance of concentrated vortices. Wave disturbances induce a flow along the axis of the vortex which gives rise to convergence. Axial velocities in the vicinity of the vortex core have been determined by following particle motions on a ciné film frame by frame. Values of axial velocities measured in this way are of the same order as azimuthal velocities.

We have remarked on the possibility of complete core distruction by means of strong, high torsion waves. It is also possible for vortex cores to be destroyed by interaction with neighboring cores. Typically two neighboring cores approach one another over part of their length. They intertwine and produce considerable disturbance which redistributes the vorticity over a broader region, approximately the dimension shown in (11) of figure 5.2c. The intertwining may then continue up the core until both vortex cores are obliterated. The events which destroy complete cores happen much less frequently than "local" breakdowns. We have observed individual cores for as long as 30 rotation periods, but the average life time is probably 10-20 rotation periods. The important point to note is that the vortices disappear by way of catastrophic events and not by gradual, Ekman dissipation.

6 Conclusions and Relevance of the Results to Tornados

There are some important conclusions which emerge from the present study. First of all, it is shown that turbulent forcing, on a small scale, of a fluid body in solid body rotation can generate intense vortices, approximately parallel to the rotation axis. They wander about in a random fashion and individual vortices have a life time of about 10-20 rotation periods. At the lowest grid Rossby number studied, the core vorticity is approximately by a factor of 40 larger than the background vorticity, 2Ω . This factor can be exceded when the rotation rate is increased for the same forcing. The size of the vortices, characterized by the inertial radius r_i, is typically by a factor of 7-10 larger than the core size and about 5 times less than the average vortex spacing. For a given forcing all scales increase at lower rotation rates and vorticity is collected from a larger area. This implies that the circulation around a vortex depends on forcing only. Asymmetric wave disturbances are shown to travel along the vortex cores and it is conjectured that these play an important part in maintaining and possibly in intensifying vortices. The basis for this conjecture is that these waves induce on axial flow of velocity comparable with the azimuthal velocity and this can give rise to strong convergence. Waves are also a source of energy transport up the core from the turbulent Ekman layer (layer of forcing). Observations show that twice as many many waves travel upward than downward and the amplitude of upward travelling waves is generally larger.

Wave interaction generally leads to local vortex breakdown and production of smaller scales. Thus wave interactions are the mechanisms responsible for energy dissipation. Local breakdown and regeneration of the core occurs in about one rotation period. Complete disruption of a vortex takes place on average after 10-20 rotation periods.

Some interesting conclusions can be drawn when our experimental results are compared with tornados. A typical tornado has a core diameter of 150 m and a maximum velocity of 90 m S-1. They feed on thunderstorms of 3-5 km in extent which have a vorticity $f \sim 10^{-2}$ rad S^{-1}. Hence, vorticity concentration by a factor of 10^2 occurs, a value comparable with the experimental value of about 50. According to our experiments, the tornado scale 2 r_i (r_i is the radius where the induced velocity is equal to the background velocity) is about 7 times the core diameter and the tornado spacing would be about 5 times the tornado scale. This gives an average tornado spacing of 5 km. Therefore, only one tornado of strenght 2 rad S^{-1} can feed on a thunderstorm of 5 km extent. For the same thunderstorm, weaker forcing would result in more but weaker tornados (the number per unit area increases according to figure 3.3). On the other hand, too strong forcing would also give a weaker tornado, because low vorticity fluid would be drawn in from outside the storm region or, in any case, the scale would increase and vorticity concentration would be less intense. This suggests that for a given thunderstorm there exists an optimum forcing velocity. This velocity can be estimated from the conditions at the boundary of transition from three dimensional turbulence to rotationally dominated flow. If we take $Ro_T = 1/4$, the integral scale which is affected by rotation (integral scale at the transition boundary) is approximately equal to the tornado scale 2 r_i. Hence, $u = f\ 2\ r_i/4 \sim 5$ m S^{-1}.

The time of formation of a tornado can be estimated from equation (3.2) and is $T_f \simeq 6 \times 4\pi/f \sim 1$-2 hours (depending on f). A more precise knowledge of f would permit better estimates.

Acknowledgements

The authors gratefully acknowledge helpful comments by Yves Gagne, U. Frisch, P.-L. Sulen and J.-D. Fournier. Two of them, U.F. and P.-L. S. painted out the possible relevance of the paper by Hasimoto to the observed wave disturbances.

This research was partially financed by the Centre National pour l'Exploitation des Oceans under contract n° 80/2345.

References

BATCHELOR, G.K., 1967. An Introduction to Fluid Dynamics. Cambridge University Press

BRETHERTON, F.P. and TURNER, J.S., 1968. On the Mixing of Angular Momentum in a Stirred Rotating Fluid. J. Fluid Mech., 32, 449-464

ESCUDIER, M.P. and ZEHNDER, N., 1981. Vortex Flow Regimes. J. Fluid Mech. (to appear)

FALER, J.H. and LEIBOVICH, S., 1977. Disrupted States of Vortex Flow and Vortex Breakdow. Phys. Fluids, 20, 1385 -1400

GOUGH, D.O. and LYNDEN-BELL, D., 1968. Vorticity Expulsion by Turbulence : Astrophysical Implications of an Alka Seltzer Experiment. J. Fluid Mech., 32, 437-447

HASIMOTO, H., 1972. A Soliton on a Vortex Filament. J. Fluid Mech.,51,477-485

HOPFINGER, E.J. and TOLY, A.-J., 1976. Spatially Decaying Turbulence and its Relation to Mixing Across Density Interfaces. J. Fluid Mech., 78,155-175

HOPFINGER, E.J. and BROWAND, F.K. 1981. Solitary Waves in Rotating, Turbulent Flow. Nature (submitted for publication)

IBBETSON, A. and TRITTON, D.Y. 1975. Experiments on Turbulence in a Rotating Fluid. J. Fluid Mech., 68, 639-672

LAMBOURNE, N.C. and BRYER, D. W., 1961. The Bursting of Leading - EDGE Vortices : Some Observations of Discussion of the Phenomenon. Acro. Res. Counc. Reports and Memoranda, n° 3282

McEWAN, A.D. 1976. Angular Momentum Diffusion and the Initiation of Cyclones. Nature, 260, 126-128

SARPKAYA, T. 1971. On Stationary and Travelling Vortex Breakdowns. J. Fluid Mech., 45, 545-559

SCORER, R.S. 1966. ORIGIN OF CYCLONES. Sci. J. 2, 46-52

STRITTMATTER, P.A., ILLINGWORTH, E. and FREEMAN, K.G.,1970. A Note on the vorticity Expulsion Hypothesis. J. Fluid Mech., 43, 535-544

Laboratory Models of Tornado-like Vortices

B. A. Lugovtsov

Summary

A number of phenomena related to laboratory-produced tornado-like vortices are observed and discussed. These are the formation of a columnar vortex above evaporating fluid; appearance of multiple vortices in a single convergence system; a suppression of turbulence in fluid in rigid-body rotation; and the formation of a tornado-like vortex in rotating fluid heated from below. Qualitative explanations of these observed phenomena are given.

1 INTRODUCTION

The mechanism of the formation and structure of intense atmospheric vortices, which include hurricanes, tornadoes, water spouts and dust devils, seems not to have been fully elucidated, as yet. The difficulty of experimental investigations of these phenomena in natural conditions impels us to resort to model simulation of tornado-like vortices in the laboratory. It is clear, that small-scale laboratory studies cannot reproduce the whole complex of natural conditions accompanying the atmospheric vortices. Nevertheless, it appears likely that laboratory simulations, observations and investigations of tornado-like vortices may give the opportunity to gain an insight into the physics of these flows, develop intuition, stimulate reflections and suggest new and useful ideas for implementation in the atmosphere.

There are a number of papers devoted to the laboratory simulation and study of tornado-like vortices [1,8,11,19,25,26,28].

This paper presents miscellaneous, simple laboratory models of tornado-like vortices and describes the results of observations and investigations of these models carried out by the scientific workers of the Institute of Hydrodynamics, V.A. Vladimirov, V.V. Nickulin and V.F. Tarasov under the guidance of the present author.

Topics in Atmospheric and Oceanographic Sciences
Intense Atmospheric Vortices, ed. by Bengtsson/Lighthill

2 FORMATION OF COLUMNAR VORTICES ABOVE EVAPORATING FLUID

The following simple experiments demonstrate the formation of a columnar vortex in the presence of rising flow above evaporating fluid [20]. The cylinder 30 cm in diameter with warm water rotates with angular velocity about 1-2 rad/sec, (Fig. 1). The line OO' is the water level. The temperature of the water is about boiling-point so that intense evaporation takes place from the surface. Under appropriate conditions of rotation in rising flow there is formed a lofty columnar vortex well visible due to condensation of vapor (Fig.2).

A similar flow pattern is observed with a motionless cylinder if there is a jet of air tangential to the wall of the cylinder so that above the surface of water the volume of rotating air rises (Fig. 3).

In Fig. 4 is shown one more means for the formation of columnar vortex. Here a motionless inclined cylinder with warm water is placed in an external uniform flow of air. The velocity vector is directed normally to the plane of Fig. 4.

In spite of external differences in means of the formation of these vortices the observed phenomena are similar.

An axial cross-section of the considered pattern of flow is shown schematically in Fig. 1. The part of flow visible due to fog (region B) is shaded. Along the axis of rotation (region A) and near the surface of water (region C) the air is transparent due to high temperature in these regions. In region A the temperature is higher by 20-30°C than that of the ambient air.

In region D the flow is swirling strongly while in region E there is a weak swirling turbulent jet. The qualitative explanation of these phenomena is straightforward. The rotating air in region D displaces the light vapor in region C towards the axis of rotation (point a). Due to conservation of angular momentum this produces intensification of rotation of vapor. Near point a the vertical swirling jet is formed. The stability is provided by the position of the jet along the axis of rotation and by the condensation (in region B), which causes the stretching of the columnar vortex due to buoyancy.

These experiments demonstrate the possibility of the formation and stability of a tornado-like vortex in cases when there is only a thin swirling unstable layer of fluid. The experiment with an inclined cylinder demonstrates the possibility of the formation of columnar vortex without swirling in ambient flow.

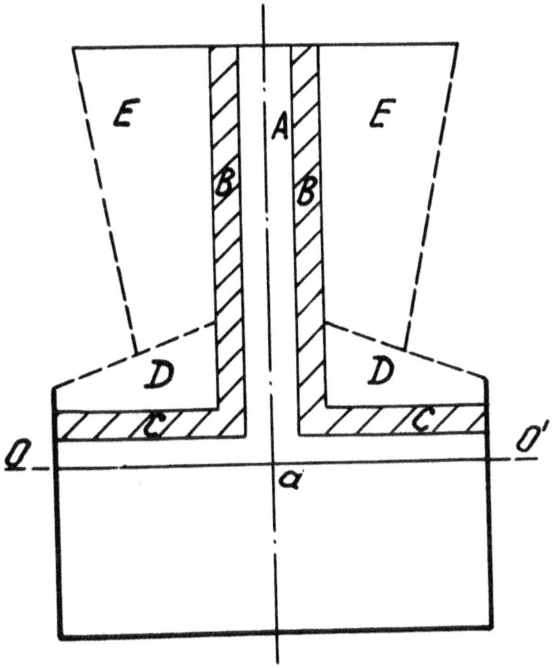

Figure 1

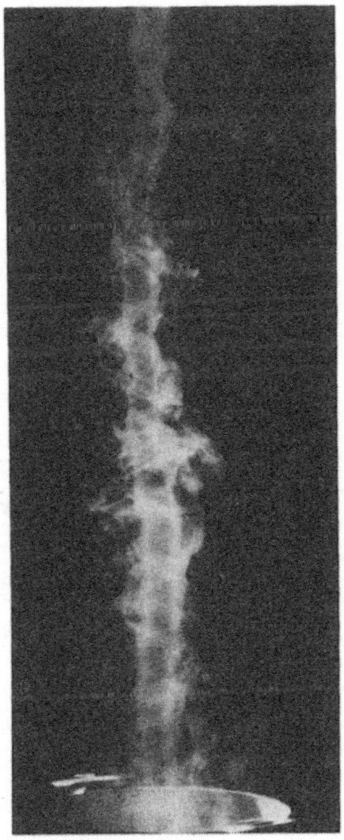

Figure 2

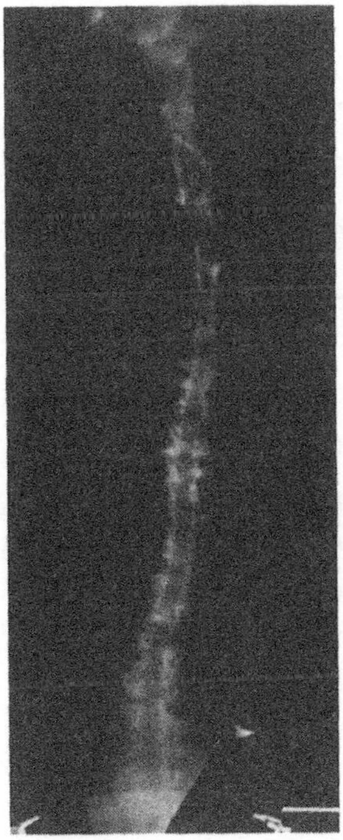

Figure 3

Figure 4

In this case the rotating volume of air is formed due to interaction of ambient flow with the wall of the cylinder. It may be suggested that a similar mechanism takes place in natural conditions under fortunate combinations of direction of wind and features of topography.

3 MULTIPLE VORTICES

In Fig. 5 is shown a vertical section of the laboratory device which was used for observation and qualitative investigation of the phenomenon of the formation of multiple vortices [23]. The rotating fluid flows out from cylinder 1 through the hole 0 in the bottom of the cylinder into the tank 2, from which the fluid flows out through the tube 4. The experiment is carried out as follows. The fluid in the cylinder is driven in nearly rigid-body rotation with angular velocity ω and then the hole 4 is opened. Under these conditions we observe the formation of some vertical vortices above the hole 0.

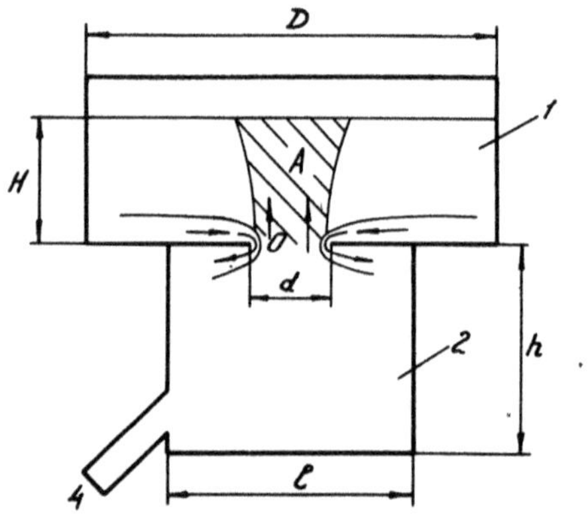

Figure 5

The device has the following dimensions (see Fig. 5): D=50 cm, d=8 cm, l=15 cm, h=30 cm. In these experiments the values ω, Q, H, (here Q is volume flow rate and H is the initial depth of fluid in cylinder 1) vary within the limits of $\omega=0{,}04 - 0{,}4\ \frac{\text{rad}}{\text{sec}}$, Q=80-1500 cm^3/sec, H=1-20 cm. Visualization of flow is achieved by dye (Fig.6) or aluminium powder on surface of fluid (Fig.7) and the form of free surface of fluid (Fig.9).

An analogous phenomenon was observed before [3,26]. In the paper [16] it was elucidated that the formation of multiple vortices resulted from the instability of a shear layer in rotating flow. This instability was studied before, both experimentally and theoretically [6,14,27].

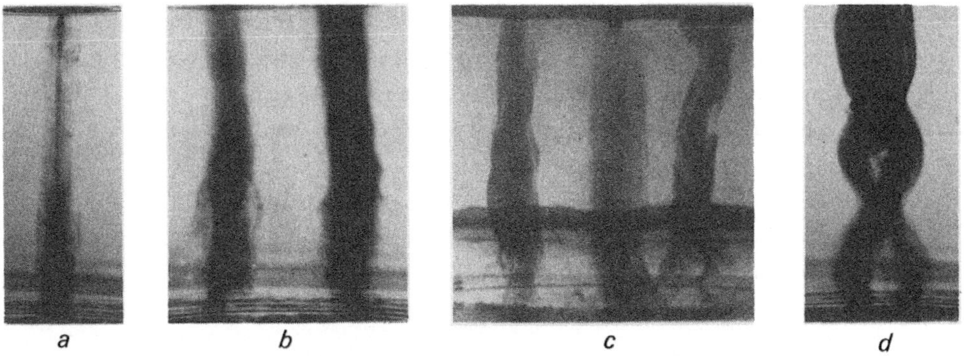

Figure 6

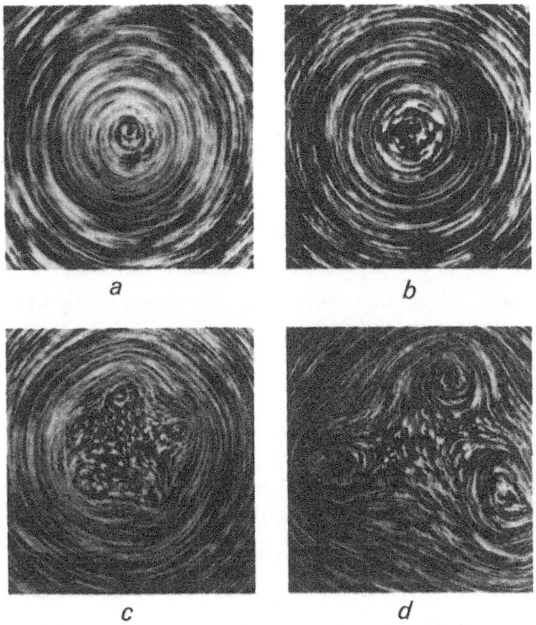

Figure 7

In this case the situation is similar. In Fig. 6 are shown some consecutive stages in the formation of multiple vortices. First one columnar vortex appears (Fig. 6a), which is separated into two vortices (Fig. 6b) and then into a triple system of vortices, rotating around the axis of symmetry; and so on. When H is small the number of vortices reaches to ten and more. The fluid in the space between such columnar vortices, provided that the number of vortices is more than two, is practically motionless (Fig. 7).

The observations reveal that reverse flow is the most critical feature in the mechanism for the formation of multiple vortices. In Fig. 7a we see a concentrated vortex, which appeared after the hole 0 was opened. When centrifugal gradient of pressure attains a certain critical value there is suction of motionless fluid from the tank 2 into the central part of the columnar vortex, tending to form a volume of motionless fluid of cylinder-like shape. This volume has a rather clear-cut boundary (Fig. 7b,7c).

In Fig. 7c the wave instability, observed near this boundary, is shown. After this wave has broken down the observed vortices are formed. This process is considerably nonlinear and the number of formed vortices is as a rule less than the number of waves characteristic of the instability.

In cases when H is large the development of the instability has a more complicated three-dimensional character. In this case the instability develops upwards from the lower end of the columnar vortex in the tank. With decrease of H the vortices are untwisted, from below upwards; and when H becomes almost equal to d two vortices are formed (Fig. 6d, 6b).

Qualitative explanations of the observed phenomena can be obtained by considering the stability of a model flow, in which the axial vorticity is as shown in Fig. 8. Here the continuous curve is to be the real distribution of vorticity and the stepped curve is to be the model distribution. In case $\omega_1 = \omega_3 = 0$, $\omega_2 \neq 0$ the necessary and sufficient criterion of instability with respect to disturbance in form of normal modes $e^{im\phi}$ is given [14] by

$$F(m) = [m(1-x) - 2]^2 - 4x^m < 0, \quad x \equiv (^a/_b)^2$$

From this criterion it follows that with increase of x the unstable range of m increases also. So, for $x < \frac{1}{4}$ all the modes are stable, for x slightly larger than $^1/_4$, only mode m=3 is unstable; while, provided $x > x_*$ where x_* is the critical value of x, mode m=4 becomes unstable and so on. For x=1 the vorticity distribution tends to a cylindrical vortex sheet and all the modes with m >2 become unstable.

For the model vorticity distribution the mode m=2 can be stable or unstable. However it can be proved that, if the vorticity distribution is such that $\omega_3 = \varepsilon\omega_2$ with $0 < \varepsilon < 1$, then mode m=2 is unstable for small x.

These results are in good qualitative agreement with the experiments.

It is possible to suggest that multiple vortices observed in this experiment and in experiments described in papers [3,26,27] may be similar to those observed in the atmosphere.

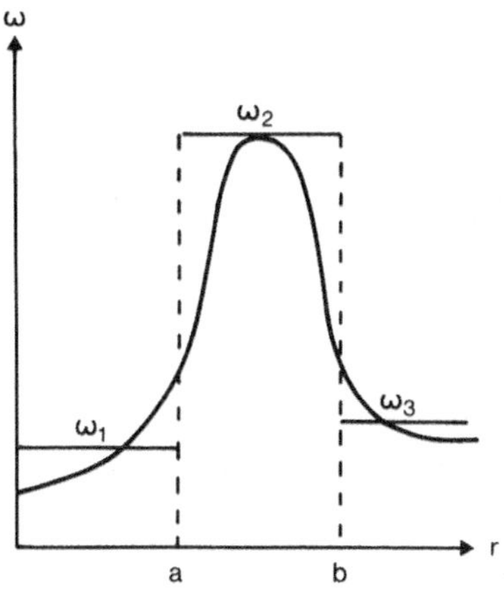

Figure 8

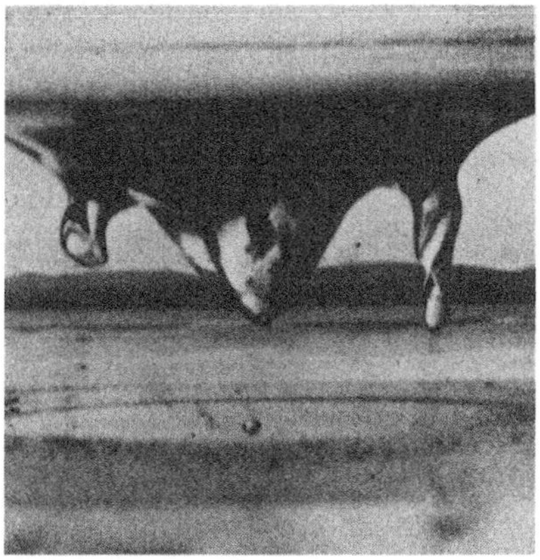

Figure 9

4 A SUPPRESSION OF TURBULENCE IN FLUID IN RIGID-BODY ROTATION

The following experiments demonstrate the phenomenon of suppression of turbulence in cores of concentrated vortices.

Influences of curvature of flow on turbulence are well known. Authors of papers on these problems usually speak about stabilizing and destabilizing effects of streamline curvature on stability of the mean flow, while noting the analogy between streamline curvature and stratification [2,13,18].

The experiments described below show that, in cores of concentrated vortices, turbulent mixing and turbulent transport are almost completely absent, even when intense turbulent mixing is present in the external region.

It is clear that this phenomenon may be important for the understanding and correct description of turbulent coefficients of transport in flows of tornado-like type.

This phenomenon shows itself clearly in transport of dye, smoke or other marker fluid by turbulent vortex rings. The ability of vortex rings to transport the marker fluid is well known and is explained by the fact that fluid is convected with the vortex ring through the ambient fluid. If the vortex ring is formed in a container filled with smoke, almost immediately after the formation of vortex we see a smoke ring, although much of the smoke may have been lost due to turbulent mixing across the boundary between atmosphere and vortex ring. But some amount of smoke within the core of the vortex ring is transferred over a great distance (Fig. 10).

On the other hand, if an initially "clean" vortex ring moves through a layer of dyed fluid, the vortex becomes dyed. Very quickly afterwards, the greater part of this dye is lost and it is seen that the core of the vortex ring remains "clean" while, on the boundary of the core, there appears a thin layer of dye (Fig.12). The boundary is fairly sharp [9,10,17,21]. Similar results were obtained in [12].

In the next experiment the dye was placed on the inner wall of the nozzle of the generator of vortex ring, near its end. A spiral line of dye appeared in the process of the formation (Fig. 13). A part of this spiral reaches into the core of the vortex ring. In Fig. 11 some frames from a film of the development in time of the dyed part of the core aa' are shown. The vortex ring moves along normal to the plane of Fig. 11. This film shows that turns of the spiral conserve their individuality and turbulent transport of dye along a radius is absent, though during exposure the vortex ring covers a distance of the order of fifty orifice-radii and initial Reynolds number is about 10^5, corresponding to developed turbulence in the atmosphere around the vortex ring.

For observation of this phenomenon in stationary conditions we used the following device. The Couette type of flow was created between cylinders. The inner cylinder was of lattice or 'grid' form. In this experiment the penetration of turbulence from the gap between the cylinders into the rotating fluid occupying the inner cylinder is observed. The outer solid cylinder had diameter 17 cm. The lattice was constructed of vertical rods, the number and cross-section of which varied within a wide range. Inner and outer cylinders can rotate independently with the rates ω_1 and $\dot{\omega}_2$. In the inner cylinder the fluid rotates almost in rigid-body manner with the rate ω. Visualization of the pattern of flow is achieved with aluminium powder suspended in the fluid and a slit source of light, which allows us to observe a horizontal cross-section of flow at a certain depth [4,5]. Such a view is shown in Figs. 14a,b,c,d.

These show a "laminar core" on the axis of rotation and "penetration" into it by turbulence from the gap. The values of ω_1, ω_2, ω were a) 1,3; - 6,3; 0; b) 2,4;- 5,3; 1,6; c) 5,4; - 2.4; 5; d) 8,0; 0; 7, 6 $\frac{\text{rad}}{\text{sec}}$. To provide equal intensity of turbulence in the gap, the difference of $\omega_1 - \omega_2$ was chosen approximately constant. The value of ω monotonically increases from Fig. 14a to Fig. 14d. In Fig. 14a the value of ω is zero and inner region is fully turbulent; in other words, "suppression" is absent. With increase of ω the radius of the "laminar core" increases monotonically. This shows that the degree of "suppression" becomes higher with the increase of ω [22,24].

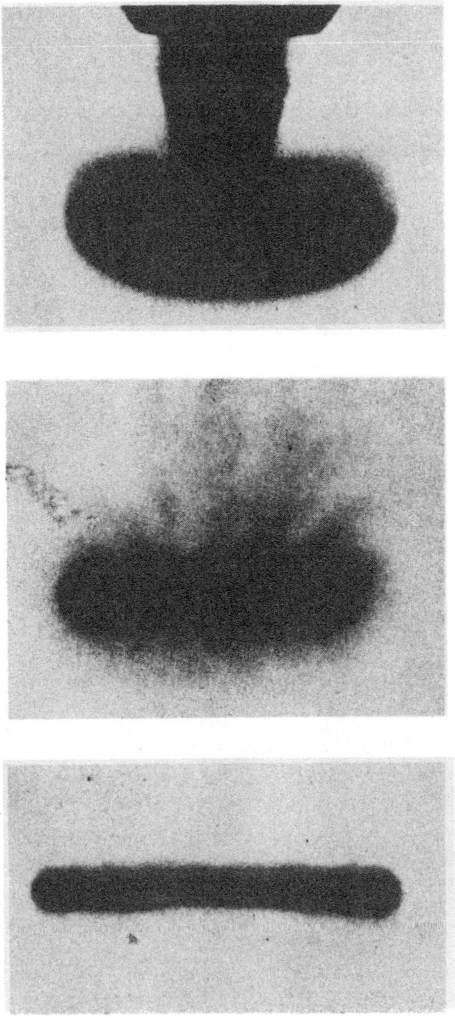

Figure 10

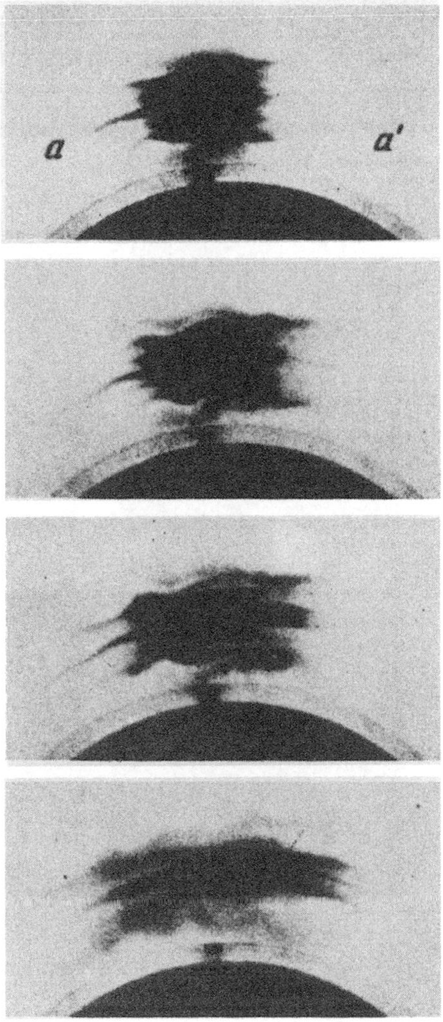

Figure 11

Figure 12

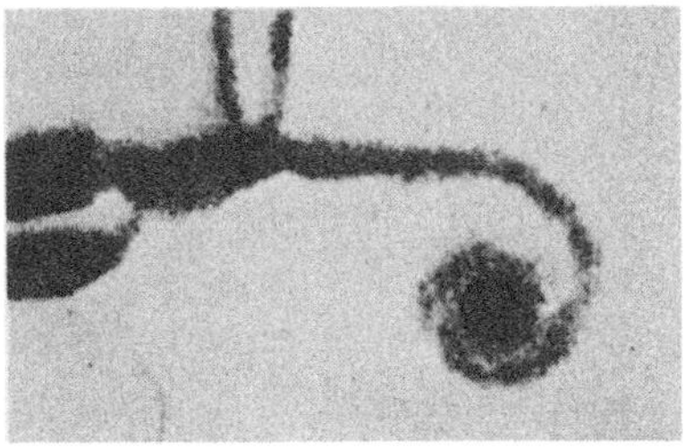

Figure 13

The flow in question was also visualized by dye. After introducing the dye into the gap one can observe quick colouring of the flow, except in the "laminar core". (Fig.15). In the case when $\omega=0$, however, the whole flow is coloured quickly. This experiment demonstrates that rigid-body rotation "suppresses" turbulent mixing [24].

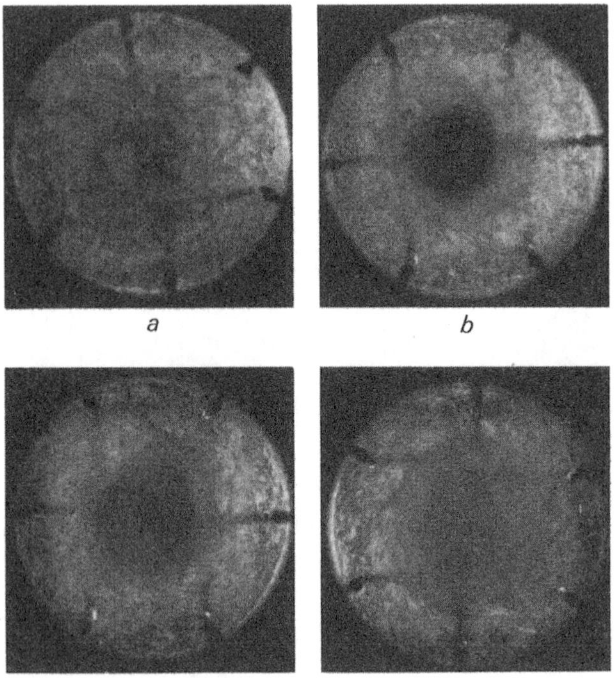

a b c d

Figure 14

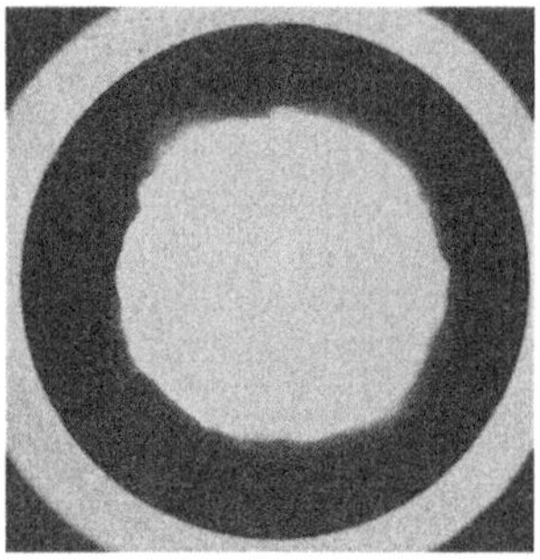

Figure 15

Qualitative explanations of the observed phenomena may be suggested, taking into account the well known property of elasticity of rotating fluid [5,7] and the analogy in the behaviour of small disturbances to fluids with rigid-body rotation and with stratification [22].

The degree of development of nonlinear processes in rotating fluid is determined by Rossby number $Rb = \frac{u}{\omega \ell}$ (here u and ℓ are characteristic scales of velocity and length of disturbances and ω is the vorticity in the mean flow). For $Rb \ll 1$ there is wave motion and, for $Rb \gg 1$, turbulence. The experiments under consideration show that in "laminar core" wave motion takes place, as here $Rb \ll 1$, and mixing is absent.

The analogy between stratification and rotation in complete form is developed in the paper [23] , including analogues to buoyancy frequency and Richardson number, and deviation of a sufficient criterion for stability of rotating fluid with respect to small disturbances.

This analogy allows us to interpret the results of the experiments under discussion in the following simple way. The core of the concentrated vortex, to first approximation, is in rigid-body rotation. As discussed above, this case corresponds to stable stratification and the suppression of turbulence in such conditions is well known [18] .

5 TORNADO-LIKE VORTEX IN ROTATING FLUID HEATED FROM BELOW

In the experiments described above no attempt has been made to measure flow parameters. For quantitative study of a tornado-like vortex the following simple device was used. The transparent beaker of fluid is heated from below. The diameter and height of the beaker are 10 cm and 13 cm. The fluid is driven into rotation by a transparent horizontal rotating top plate, 9 cm in diameter. The depth of fluid varies within 8 - 12 cm. The basic flow created without heating is almost rigid-body rotation at 0, 2ω (ω is

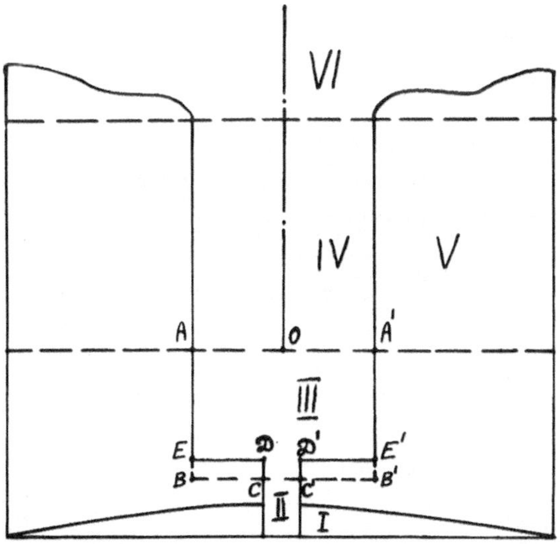

Figure 16

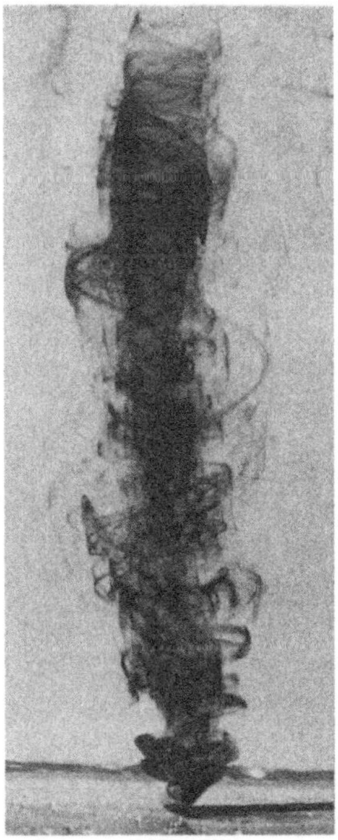

Figure 17

rotation rate of the upper plate). However, a columnar vortex is formed with heating in the centre of the beaker. This vortex (visualized by dye, which was introduced in the boundary layer) is shown in Fig. 17. The velocity field of the flow was measured by using the "hydrogen-bubble" technique [11] .

An axial cross-section of the considered pattern of flow is shown schematically in Fig. 16. The boundary layer is in the region I; in region II a rapidly swirling jet erupts from the boundary layer; in region III the jet jumps to a new diameter with lower axial and swirl velocities (vortex breakdown) [1,11]; region IV is the core of vortex, where the flow is almost rigid-body rotation while the vertical velocity is approximately constant within the core and very small in the outer flow; in region V the flow is almost irrotational. The difference in temperature between top plate and bottom was absolute $1^{\circ}C$. In all the experiments the flow is almost stationary.

The measurements showed that in the first approximation the radius of the core and the azimuthal component of velocity are independent of height. The approximate analytical solution, describing this flow within the framework of the theory of axial boundary layer, was given in the paper [15]. The influence of the bottom boundary layer and of vortex breakdown was taken into account by using the law of conservation of the axial component of momentum for the region EDD'E'B'C'CB. For core radius R (defined so that for r=R the aximuthal component velocity is maximal) the following formula was obtained:

$$R^2 = \frac{4\sqrt{5}\,\nu}{\sqrt{\alpha\beta g}} \left(C + \sqrt{1+C^2}\right), \quad C = K\frac{\Gamma^2}{\sqrt{\alpha\beta g}\; h^2 \nu}, \quad K \approx 4.10^{-4}$$

where Γ is circulation, ν is kinematic viscosity, h is depth of beaker, β is basic gradient of temperature, α is coefficient of thermal expansion and g is acceleration due to gravity. This formula explains the effect of R tending to increase with increase of Γ and decrease of h, as observed in the experiments, and yields reasonable quantitative agreement with experimental data.

REFERENCES

1. Barcilon A. 1967. A Theoretical and Experimental Model for a Dust Devil. J.Atmos.Sci., 24, 453-466

2. Bradshow P. 1969. The analogy between streamline curvature and buoyancy in turbulent shear flow. J.Fluid Mech., 36, 177-191

3. Church C.R., Snow J.T., Agee E.M. 1977. Tornado vortex simulation at Purdue University. Bull. Amer.Meteor.Soc., 58, 900-908

4. Coles D. 1965. Transition in circular Couette flow. J.Fluid Mech., 21, 385-425

5. Creenspan H.P. 1968. The theory of rotating fluids. Cambridge Univ.Press

6. Hide R., Titman C.W. 1967. Detached shear layers in rotating fluid. J.Fluid Mech., 29, 39-60

7. Howard L.N., Cupta A.S. 1962. On the hydrodynamic and hydromagnetic stability of swirling flows. J.Fluid Mech., 14, 463-476

8. Hsu C.T., Fattahi. 1976. Mechanism of tornado funnel formation. Phys.Fluids, 19, 1853-1857

9. Lugovtsov B.A. 1970. O dvihzenii turbulentnogo vihrevogo koltsa i perenose im passivnoi primesi. In the book: Nekotorye problemy matematiki i mehaniki: Izd. "Nauka", Leningrad, pp. 182-197

10. Lugovtsov B.A. 1979. Turbulentnye vihrevye koltsa. Dynamica sploshnoi sredy. Novosibirsk. v. 38, pp. 71-88

11. Maxworthy T. 1972 On the structure of concentrated, columnar vortices. Astron.Acta, 17, 363-374

12. Maxworthy T. 1977. Some experimental studies of vortex ring. J.Fluid Mech., 81, 465-495

13. Mellor G.L. 1975. A Comparative study of curved flow and density-stratified flow. J.Atmos.Sci., 32, 1278-1282

14. Michalke A., Timme A. 1967. On the inviscid instability of certain two-dimensional vortex-type flows. J.Fluid Mech., 29, 647-666

15. Nickulin V.V. 1980. Issledovanie vzaimodeistvija tornadopodobnogo vihrja s tverdymi granitsami. PMTF N 1, pp. 68-75

16. Rottunno R. 1978. A note on the stability of a cylindrical vortex sheet. J.Fluid Mech., 87, 761-771

17. Tarasov V.F., Yakushev V.I. 1974. Experimentalnye issledovanija perenosa primesi turbulentnym vihrevym koltsom. PMTF, N 1, pp. 130-136

18. Turner J.S. 1973. Buoyancy Effects in Fluids. Cambridge Univ. Press

19. Turner J.S. 1966. The constraints imposed on tornado-like vortices by the top and bottom boundary conditions. J.Fluid Mech., 25, 377-400

20. Vladimirov V.A. 1977. Formirovanie vihrevyh shnurov iz voshodjashyh potokov nad isparjahusheisja zidkostju. DAN SSSR, 236, 316-318

21. Vladimirov V.A., Tarasov V.F. 1979. Struktura turbulentnosti vblizi yadra koltsevogo vihrja. DAN SSSR, 245, 1325-1328

22. Vladmimirov V.A., Tarasov V.F. 1980. O svoistvah uprugosti zakruchennyh potokov. DAN SSSR, 253, 565-568

23. Vladimirov V.A., Tarasov V.F. 1980. Obrazovanie sistemy vihrevyh shnurov vo vrashajusheisja zidkosti. Izvest. AN SSSR, Mech. Fluid and Gas, N 1, pp. 44-51

24. Vladimirov V.A., Lugovtsov B.A., Tarasov V.F. 1980. Podavlenie turbulentnosti v yadrah concentrirovannyh vihrei. PMTF, 5, 69-76

25. Wan C.A., Chang C.C. 1972. Measurement of the velocity field in a simulated tornado-like vortex using a three-dimensional velocity probe. J.Atmos.Sci., 29, 116-127

26. Ward N.B. 1972. The exploration of certain features of tornado dynamics using laboratory model. J.Atmos.Sci., 29, 1194-1204

27. Weske J.R., Rankin T.M. 1963. Generation of secondary motions in the field of a vortex. Phys.Fluids, 6, 1397-1403

28. Ying S.J., Chang C.C. 1970. Exploratory model study of tornado-like vortex dynamics. J.Atmos.Sci., 27, 3-14

High Vorticity Regions in Rotating Thermally Driven Flows

R. Hide

Summary: The regular and irregular non-axisymmetric flow régimes of thermal convection in a rotating fluid annulus subject to differential heating in the horizontal are characterized by the presence of upper level jet-streams, where intense concentrations of vorticity and high concomitant horizontal temperature gradients are found. The main features of the upper-level flow pattern can be interpreted by straightforward arguments based on general thermodynamic considerations and the requirement that the flow should be quasi-geostrophic nearly everywhere. Thus, when the distribution of applied heating and cooling is such that the corresponding gradient of the impressed radial temperature field has the same sign at all radii, the most conspicuous feature of the upper-level flow pattern in the regular non-axisymmetric regime is a single jet-stream meandering in a wavy pattern between the bounding cylinders. When, however, the impressed radial temperature gradient changes sign near mid-radius (as in the case when heat is introduced throughout the body of the fluid and withdrawn at both side-walls), the corresponding upper-level flow consists of several closed eddies, each circulating "anticyclonically" with the horizontal flow largely confined to a narrow jet-stream at the periphery of each eddy. In some respects these stable closed eddies are dynamically similar to long-lived anticyclonic eddies (including the Great Red Spot) seen in Jupiter's atmosphere in the southern hemisphere. Previous work on stable baroclinic eddies is now being extended in various directions and supporting numerical work is also being carried out.

1 INTRODUCTION

The motion of a fluid that departs but little from solid body rotation with angular velocity $\underline{\Omega}$ is usually geostrophic nearly everywhere, with the relative Eulerian velocity $\underline{u}$ (as measured in a frame of reference that rotates with angular velocity $\underline{\Omega}$ relative to an inertial frame) satisfying

$$2\rho\underline{\Omega}\times\underline{u} = -\nabla p + \rho\nabla V. \qquad (1)$$

Here ρ denotes density, p pressure and ∇V is the acceleration due to gravity

Topics in Atmospheric and Oceanographic Sciences
Intense Atmospheric Vortices, ed. by Bengtsson/Lighthill

and centripetal effects. Equation (1) is the leading approximation to the full equation of motion

$$\rho\left\{\frac{D\underline{u}}{Dt} + 2\underline{\Omega}\times\underline{u} - \underline{r}\times\frac{d\underline{\Omega}}{dt}\right\} = -\nabla p + \rho\nabla V - \nabla\times(\nu\rho\nabla\times\underline{u}). \quad (2)$$

It is valid in regions where the Coriolis term $2\rho\underline{\Omega}\times\underline{u}$ greatly exceeds the relative acceleration term $\rho\, D\underline{u}/Dt \equiv \rho[\partial\underline{u}/\partial t + \underline{u}\cdot\nabla\underline{u}]$ (where t denotes time), the precessional term $\rho\,\underline{r}\times d\underline{\Omega}/dt$, the viscous term $\nabla\times(\nu\rho\nabla\times\underline{u})$ (where ν denotes kinematic viscosity) and any other term that must be added to equation (2) when further effects (e.g. magnetohydrodynamic forces) have to be taken into account.

Now equation (1) is lower in order than equation (2), so it cannot be solved under the complete set of boundary conditions. For this to be possible it is necessary to include every term in equation (2) in the analysis. It follows that the flow cannot be geostrophic everywhere; the system must exhibit boundary layers and detached shear layers where $\rho\, D\underline{u}/Dt + \nabla\times(\rho\nu\nabla\times\underline{u})$ is comparable in magnitude with $2\rho\,\underline{\Omega}\times\underline{u}$. Within these highly ageostrophic regions the magnitude of the relative vorticity $\nabla\times\underline{u}$ can be comparable with or even exceed $2\underline{\Omega}$. Many examples of such vorticity concentrations are found in laboratory systems and in nature. They are often associated with steep gradients of temperature (thermal fronts), as in the case of jet-streams and western boundary currents found in atmospheres and oceans.

Jet-streams are a pronounced feature of the non-axisymmetric flow régimes of thermal convection in a rotating fluid annulus, laboratory and theoretical studies of which have elucidated many aspects of the general circulation of the atmospheres of the Earth and other planets. This paper outlines the findings of work on annulus convection produced by internal heating and mentions one particularly interesting possible application (Hide 1980), to the interpretation of long-lived anticyclonic eddies (including the Great Red Spot) seen in Jupiter's atmosphere.

2 SLOPING CONVECTION IN THE LABORATORY

Laboratory experiments in thermal convection in a rotating fluid annulus which rotates about a vertical axis and is subject to axisymmetric applied differential heating were initiated by the writer over thirty years ago (for references see Hide and Mason 1975, Pfeffer, Buzyna and Kung 1980, Tritton and Davies 1981).

They show that when the rotation rate Ω exceeds a certain critical value Ω_R (which depends on the acceleration of gravity, the shape and dimensions of the apparatus, the coefficients of thermal expansion, thermal conductivity and viscosity of the fluid and its mean density, and the distribution and intensity of the applied differential heating, see section 4 below) Coriolis forces inhibit over-turning motions in meridian planes and promote a completely different kind of motion, which has been termed "sloping convection". The motion is then non-axisymmetric and largely confined to jet-streams, with typical trajectories of individual fluid elements inclined at only very small but essentially non-zero angles to the horizontal. The kinetic energy of the non-axisymmetric flow derives from the interaction of slight upward and downward motions in these sloping trajectories with the potential energy field produced by the action of gravity on the density variations maintained by the applied differential heating. The kinetic energy of the motion is dissipated by friction arising in boundary layers on the walls of the container and in the main body of the fluid.

Provided that Ω, though greater than Ω_R, does not exceed a second critical value Ω_I (see section 4 below), the main features of the non-axisymmetric motion are characterized by great regularity. This regular flow is either steady (apart from a steady drift of the horizontal flow pattern relative to the walls of the apparatus) or it exhibits periodic "vacillation" in the amplitude, shape and other characteristics. The number of "waves" m around the annulus is not uniquely determined by the impressed conditions; the flow is found to be "intransitive" owing to the occurrence of what have come to be called "multiple equilibrium states". But a quantity $\hat{m}$, defined as the most likely value of m at a point in the region of parameter space occupied by the regular flow régime (where $\Omega_R < \Omega < \Omega_I$, see Fig.7 below), tends to increase with increasing Ω, until at $\Omega = \Omega_I$ the quantity m has that value for which the azimuthal scale of the horizontal flow pattern (namely the mean circumference of the annulus divided by m) is about 1.5 times the radial scale. Then the motion undergoes a transition to the so-called régime of irregular flow, where $\Omega > \Omega_I$ in parameter space, which is an example of thermally-driven "geostrophic turbulence".

Many laboratory studies of various aspects of sloping convection have now been carried out. These include measurements of heat transfer, flow structure and régime transitions over a wide range of mechanical and thermal boundary conditions. Numerical studies are playing an increasingly important rôle in this work and significant if more limited analytical studies have been made based on the essentially non-linear governing mathematical equations.

These equations are the equations of motion, (2), together with the equations of continuity and state for a liquid, respectively

$$\nabla \cdot \underline{u} = 0 \tag{3}$$

and

$$\rho = \rho_0 [1 - \alpha (T - T_0)] \tag{4}$$

(where T denotes temperature, α thermal coefficient of cubical expansion and ρ_0 is the density at the reference temperature T_0), and the equation of heat transfer

$$\frac{\partial T}{\partial t} + (\underline{u} \cdot \nabla) T = \kappa \nabla^2 T + q \tag{5}$$

(where κ is the thermal diffusivity (equal to the thermal conductivity divided by the product of ρ and the specific heat c) and $q\rho c$ is the rate of diabatic heating per unit volume). Across any cylindrical vertical surface r = constant (where (r, ϕ, z) are cylindrical polar coordinates of a general point, $r = 0$ being the rotation axis (see Fig.4 below)), the rate of heat transfer is given by

$$H(r,t) = \int_0^d \int_0^{2\pi} \rho c \left[\kappa \frac{\partial T}{\partial r} + u_r T \right] r \, d\phi \, dz \tag{6}$$

where the fluid extends in the axial direction from $z = 0$ to $z = d$. It is important to notice that the geostrophic contribution to the advective heat flow term on the right-hand side of equation (6) would vanish if the flow were axisymmetric, since by equation (1) u_r, the r component of $\underline{u} = (u_r, u_\phi, u_z)$, satisfies $u_r \doteqdot (2\rho\Omega r)^{-1} \partial p/\partial\phi$. This result points to the <u>raison d'être</u> of the non-axisymmetric régime of flow found when $\Omega > \Omega_R$; geostrophic flow cannot convey heat perpendicularly to the axis of rotation unless it is non-axisymmetric!

The boundary conditions on $\underline{u}$ under which equations (2), (3), (4) and (5) must be satisfied are that $\underline{u} = 0$ at a rigid bounding surface and that the stress should vanish at a free surface. The thermal boundary conditions require continuity of heat flow which, at a bounding surface, is purely conductive and proportional to $\kappa \nabla T$. When the boundary conditions on the side-walls at $r = r^*$ where $r^* = a$ or b (say, with $b > a$, see section 4 below) are combined with the geostrophic relationship given by equation (1) and used in conjunction with the standard relationship for the radial flow in the Ekman boundary layers on $z = 0$ and $z = d$ to evaluate the radial heat flow at $r = r^*$ (see equation (6)), it is found (see

Hide and Mason 1970) that:

$$H(r^*,t) \doteq -\left(\frac{\nu}{\Omega}\right)^{1/2}\left[\hat{T}(r^*,d,t)-\hat{T}(r^*,0,t)\right]\left[\Gamma(r^*,d,t)-\Gamma(r^*,0,t)\right]. \quad (7)$$

Here

$$\hat{T}(r^*,z,t) \equiv \frac{1}{2\pi}\int_0^{2\pi} T(r^*,\phi,z,t)\,d\phi \quad (8)$$

and

$$\Gamma(r^*,z,t) \equiv \int_0^{2\pi} U_\phi(r^*,\phi,z,t)\,r\,d\phi, \quad (9)$$

where $U_\phi(r^*,\phi,z,t)$ is the value of u_ϕ evaluated just outside the viscous boundary layer on $r=a$ or $r=b$, as the case may be. Now it may be shown that $\hat{T}(r^*,d,t) > \hat{T}(r^*,0,t)$ (when $\alpha > 0$, see equation (4)) and that either $\Gamma(r^*,0,t) = -\Gamma(r^*,d,t)$ or $|\Gamma(r^*,0,t)| \ll |\Gamma(r^*,d,t)|$, according as the upper surface is in contact with a rigid lid or is free. Hence,

$$H(r^*,t) = (\text{negative definite quantity}) \times \Gamma(r^*,d,t). \quad (10)$$

This relationship between the heat flow at a side-wall and the line integral of the tangential velocity near the side-wall embodies the arguments used by Hide (1958) to provide a general interpretation of the upper-level flow pattern in the case when $q = 0$ everywhere (see equation (5)), heat being introduced into the system via one of the side-walls and removed via the other side-wall. The corresponding impressed radial temperature gradient has the same sign at all values of r and the upper level pattern of motion in the regular flow régime consists of a single jet-stream meandering in a wavy pattern between the bounding cylinders, with a positive (i.e. "westerly") azimuthal component when heat enters via the outer side-wall and leaves via the inner side-wall (so that the impressed radial temperature gradient is positive), and negative ("easterly") when the overall radial heat transfer is in the opposite direction, from the inner to the outer cylinder (see Figs. 1a, b).

As a further test of equation (10), Hide and Mason (1970) carried out experiments using internal heating, so that the term q in equation (5) is not equal to zero. This was done by passing an alternating electric current through the fluid. Heat could be removed via the inner side-wall, the outer side-wall, or both side-walls. The observed upper surface flow patterns (see Fig.2) were found to be in

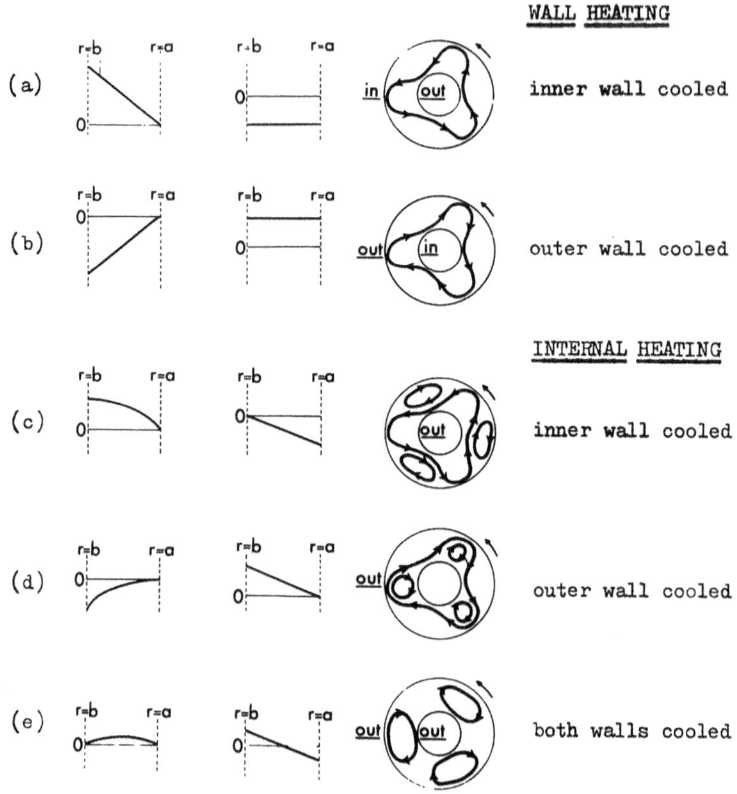

Fig.1. Schematic illustrations of the radial variation of the impressed temperature (left column), of the radial variation of the impressed radial temperature gradient (middle column); and of the corresponding upper surface relative flow pattern in the regular régime of thermal convection in a rotating fluid annulus (right column), as predicted on the basis of equation (10). (See Hide and Mason (1970) for further details)

good agreement with predictions for these three cases made on the basis cf equation (10) (see Figs. 1c, d, e). In the cases where heat is removed via one side-wall only, $H(r^*, t)$ vanishes at the other side-wall and, by equation (10), the quantity $\Gamma(r^*, d, t)$ must also vanish. For this to happen, $U_\phi(r^*, \phi, d, t)$ will be positive at some values of ϕ and negative at others (or zero at all values of ϕ, as in the axisymmetric regime that occurs when $\Omega < \Omega_R$). Figs 1c, d show how this requirement can be satisfied by adding closed eddies to the wavy pattern found in the cases when $q = 0$ (of Figs. 1a, b).

The most striking case of all studied in the experiments is the one illustrated by Fig.1e. Then, heat is removed via both side-walls, implying that the im-

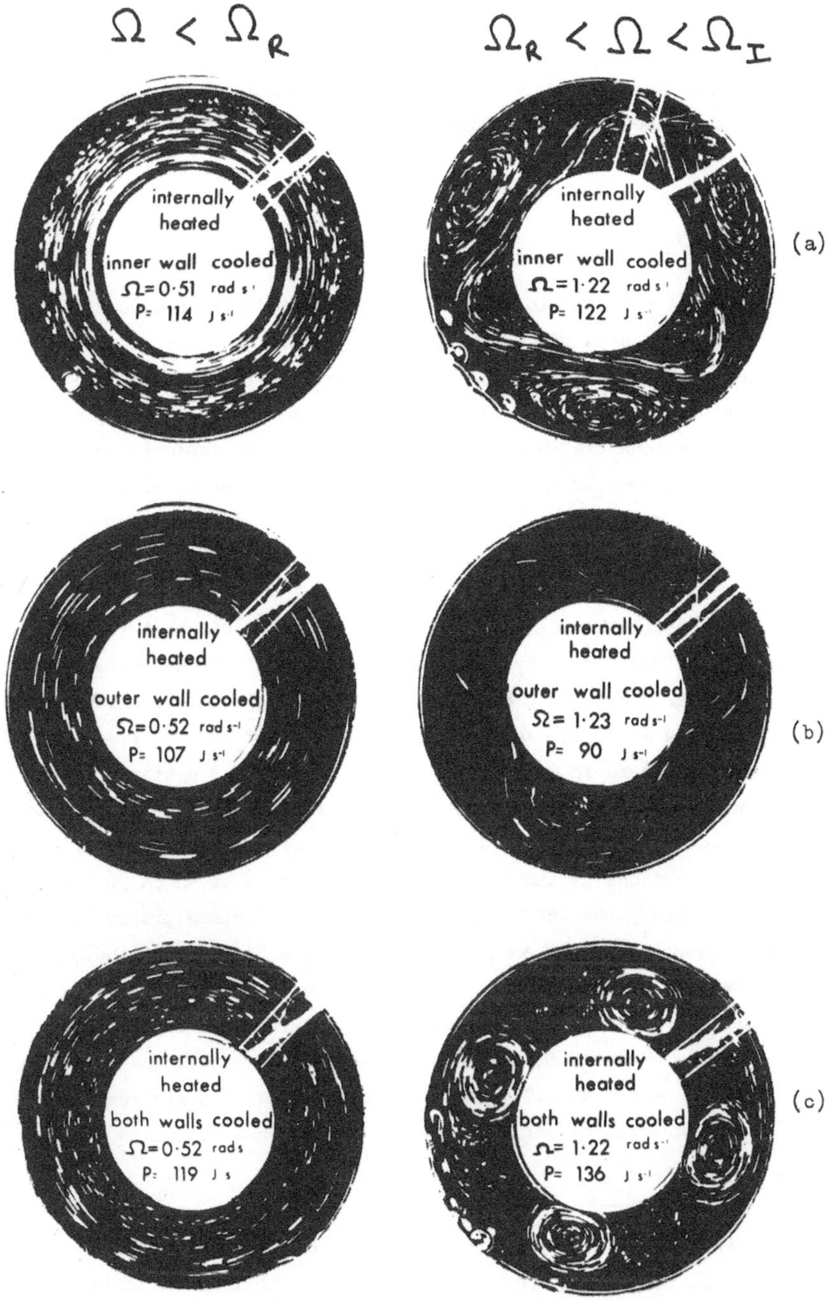

Fig.2. Streak photographs illustrating top-surface flow patterns of thermal convection in a rotating fluid annulus subject to internal heating in the axisymmetric regime (when $\Omega < \Omega_R$, see left column) and regular non-axisymmetric regime (when $\Omega_R < \Omega < \Omega_I$, see right column). They

(to be continued on opposite page)

pressed radial temperature gradient changes sign near mid-radius. The corresponding upper level flow consists of several separate closed eddies each circulating anticyclonically, in accordance with equation (10), with the horizontal flow confined to a narrow jet-stream at the periphery of each eddy (see Fig. 2c).

3 ATMOSPHERIC FLOWS

The meandering jet-streams within which the upper-level tropospheric air flow is mainly concentrated in the Earth's atmosphere are manifestations of sloping convection produced by differential solar heating, which maintains a systematic temperature contrast between tropical and polar regions in each hemisphere. These atmospheric motions are much less regular than those depicted in Figs. 1a, c, and 2a (cf. Fig.6), presumably because the Earth's angular speed of rotation exceeds the critical value Ω_I , although horizontal variations of surface conditions introduce complications which are not yet fully understood.

The atmosphere of the planet Jupiter is heated from below at about the same rate as its upper reaches are heated by solar radiation. Unlike the terrestrial case (where non-solar atmospheric heating is utterly negligible), north-south temperature gradients in Jupiter's atmosphere change sign several times between equator and pole and there is no evidence of any significant systematic temperature contrast between equatorial and polar regions (for references see Ingersoll, Dobrovolskis and Jakosky 1979). There are abundant observations of Jovian atmospheric motions at the upper cloud level, some of which go back many decades and even longer, and the "Pioneer" and "Voyager" space probes have added further details (see Peek 1958, Smith and Hunt 1976, NASA 1979, Mitchell et al. 1981). Our knowledge of what goes on below the cloud level is meagre and this produces difficulties with the interpretation of observations of upper-level atmospheric motions. Indeed, it has been argued elsewhere that the main task of the "Jovian meteorologist" should perhaps be to use these observations to improve our knowledge of the vertical structure of the planet (see e.g. Hide 1981).

---

show the dependence of the general characteristics of the flow pattern on the way heat is removed from the system and confirm predictions based on equation (10). The cases (a), (b) and (c) correspond to cases (c), (d) and (e) respectively of Fig.1. In the most striking case of all (see Figs.1e and 2c), sloping convection takes the form of closed anticyclonic eddies with the main motion concentrated in a jet-stream at the periphery of each eddy. (See Hide and Mason (1970) for further details.)

But here is not the place to discuss these observations and review the many interesting though largely controversial issues being debated by those of us who take an interest in the interpretation of the observations in terms of basic dynamical processes. There is however one striking phenomenon upon which laboratory experiments on sloping convection in a fluid annulus subject to internal heating might have some bearing. The highly stable closed anticyclonic eddies with the main motion concentrated in a jet-stream at the periphery of each eddy that are depicted in Figs.1e and 2c are remarkably similar dynamically to the long-lived eddies to be seen in Jupiter's atmosphere in the southern hemisphere. The largest, oldest and most conspicuous of these is the Great Red Spot in the South Tropical Zone, which may be at least three hundred years old. Next in size and age are the three White Ovals that formed in 1939 at the boundary between the South Temperate Belt and the South Temperate Zone, apparently as the residue of the highly irregular South Tropical Disturbance that was first seen in 1901. The smallest of the long-lived eddies are clearly seen in the magnificent "Voyager" pictures (NASA 1979) as about a dozen oval markings somewhat closer to the pole. The motion in each of these Jovian eddies is anticyclonic and largely confined to a narrow region at the edge of the eddy. It is tempting therefore to suppose (Hide 1980) that the eddies are manifestations of sloping convection in Jupiter's atmosphere, implying that they derive their kinetic energy directly from the potential energy due to the action of gravity acting on density variations produced by internal and solar heating, and that they transport heat from the interior to the edges of the latitudinal bands in which they arise. (See Fig.3.)

Preliminary calculations indicate that there is nothing unreasonable about this hypothesis so far as its implications for the vertical structure and other properties of Jupiter's atmosphere are concerned, but a detailed examination of these implications and a critical comparison of the hypothesis with other proposals as to the nature of the long-lived anticyclonic eddies will have to be considered elsewhere. The hypothesis raises a number of fluid-dynamical questions to be resolved by further laboratory and numerical work and some of this is now in hand. Amongst these questions is that of the instability of the strongly-sheared flow in the jet-stream itself. Experiments with a wall-heated annulus (Hide 1958) provide some evidence that when viscous effects are sufficiently small the jet-stream develops local instabilities on one side but not on the other. Pictures of Jupiter show highly irregular flow on a comparatively small scale just outside the Great Red Spot (and the other long-lived eddies), but not on the inside. It will be important to establish by experiment and theory (c.f. Narasimha 1980) whether this highly irregular flow arises as a result of a "one-sided" instability of the jet-stream at the edge of the main eddy.

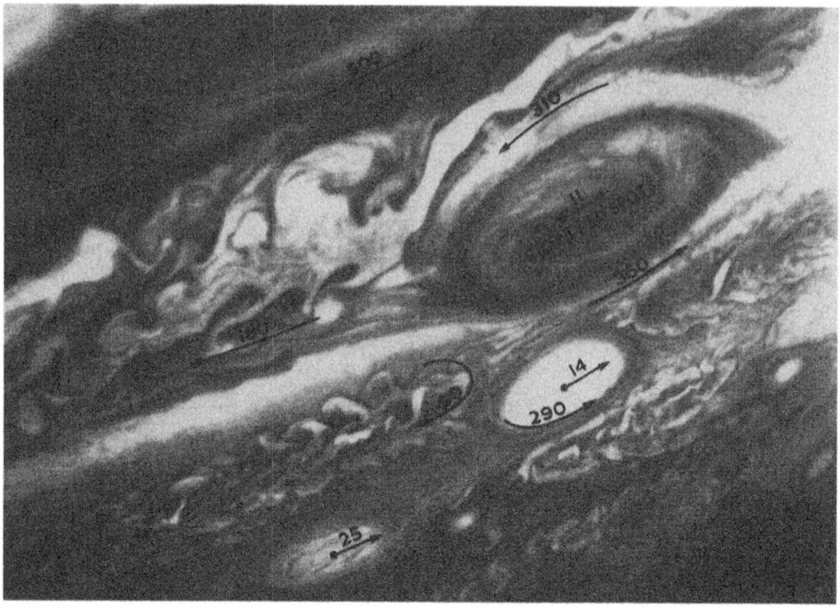

Fig.3. Long lived anticyclonic eddies in Jupiter's atmosphere, namely the Great Red Spot (which is about 20,000 km long) and one of the three White Ovals. The arrows indicate the sense of relative motion. The speeds (as determined by Dr R.F. Beebe of the New Mexico State University) are given in kilometres per hour; for comparison note that points on Jupiter's equator rotate at about 35,000 km per hour. (See NASA 1979 for further "Voyager" pictures of Jupiter)

4 APPENDIX: REGIMES OF THERMAL CONVECTION IN A ROTATING FLUID ANNULUS

The simplest system in which controlled and reproducible experiments on sloping convection have been carried out is the annular apparatus illustrated in Fig.4 when there are no internal heat sources of heat (i.e. $q = 0$, see equation (5)) but the bounding cylindrical side-walls in $r = a$ and $r = b$ are maintained at different temperatures T_a and T_b respectively, so that the impressed temperature field satisfies

$$T_i = \left[T_b \ln(r/a) - T_a \ln(r/b)\right] / \ln(b/a) \tag{11}$$

which simplifies to $T_i = \frac{1}{2}(T_b + T_a) + (T_b - T_a)\left[r - \frac{1}{2}(a+b)\right] / (b-a)$ when $(b-a) \ll \frac{1}{2}(b+a)$. Accurate determinations of the principal spatial and temporal characteristics of the fields of temperature and flow velocity over a wide range of precisely specified and carefully controlled experimental con-

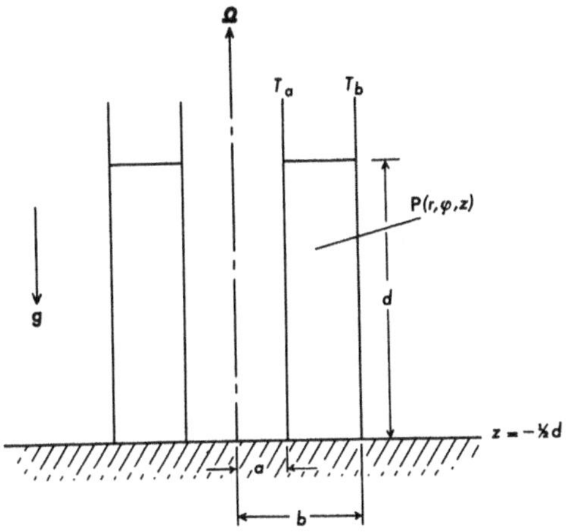

Fig.4. Schematic diagram of a rotating fluid annulus subject to a horizontal temperature gradient. (See Hide 1958 or Hide and Mason 1975 for further details)

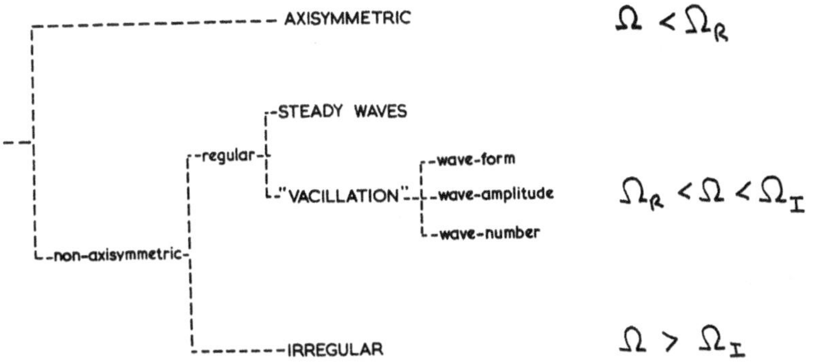

Fig.5. Broad classification of modes of free thermal convection in a rotating fluid annulus under axisymmetric boundary conditions. (See Hide (1958) or Hide and Mason (1975) for further details.)

ditions led to the discovery of several fundamentally different free types of flow, only one of which is symmetrical about the axis of rotation (see Figs. 5 and 6). The general character of the flow evidently depends largely on the values of certain external dimensionless parameters

$$\Theta \equiv g\, d\, \Delta\rho / \bar{\rho}\, \Omega^2 (b-a)^2 \tag{12}$$

and

$$\mathcal{T} \equiv 4\Omega^2 L^4 / \nu^2. \tag{13}$$

Here g denotes the acceleration of gravity (which is typically very much greater than $\Omega^2 b$), d is the depth of the liquid in the annular container, $\Delta\rho$ is

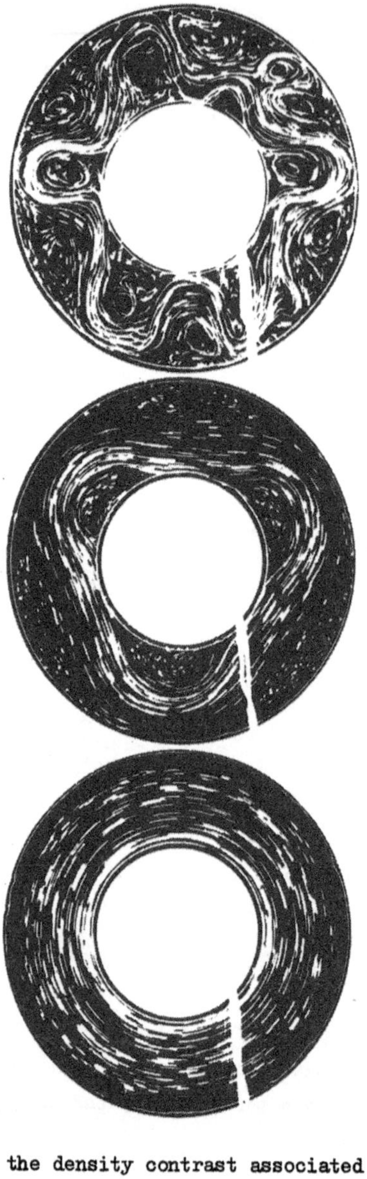

(I) symmetric ($\Omega = 0{\cdot}341$ rad s^{-1})

(II) steady waves ($\Omega = 1{\cdot}19$ rad s^{-1})

(III) irregular ($\Omega = 5{\cdot}02$ rad s^{-1})

Fig.6. Streak photographs giving one example of each of the main modes of thermal convection in a rotating fluid annulus subject to a radial temperature field of the form given by equation (11), namely (I) axisymmetric flow, (II) regular (steady) non-axisymmetric flow, and (III) irregular non-axisymmetric flow

the density contrast associated with the impressed temperature difference, i.e. $\Delta\rho \equiv |\rho(T_a) - \rho(T_b)|$, $\bar{\rho}$ is the mean density, Ω is the angular speed of rotation of the apparatus about a vertical axis, ν is the kinematic viscosity and L , which has the dimensions of a length, is equal to $(b-a)^{5/4}/d^{1/4}$ over a wide range of conditions.

When $\mathcal{J}$ is less than a certain critical value of about 2×10^5 (see Fig.7), viscosity ensures that the motion is essentially symmetrical about the axis of rotation for all values of Θ . However, when $\mathcal{J}$ exceeds this critical value

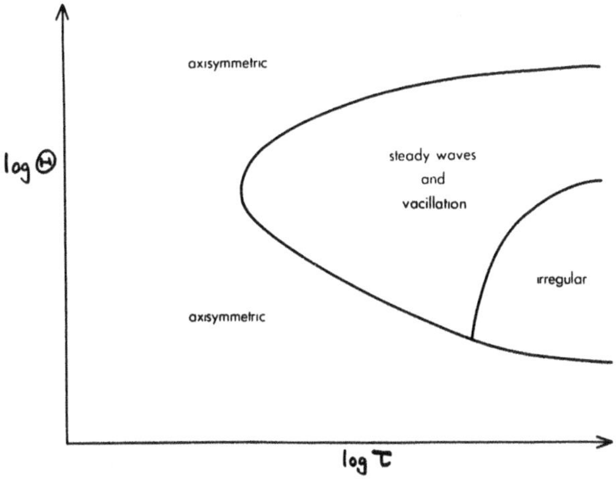

Fig.7. Schematic diagram illustrating the dependence of the mode of free thermal convection in a rotating fluid annulus under axisymmetric boundary conditions on the two principal dimensionless parameters found to specify the system, Θ and $\mathcal{J}$ (see equations (12) and (13))

there exists a range of Θ within which highly non-axisymmetric sloping convection occurs. These non-axisymmetric motions are either regular or irregular (see Fig.6), depending on the values of Θ and $\mathcal{J}$. The regular flows often exhibit periodic "vacillation" in amplitude, shape or wavenumber, but under certain conditions these periodic variations are so slight that, apart from a steady drift of the wavy pattern relative to the apparatus, the flow is virtually steady. In sharp contrast to this behaviour, the irregular flows exhibit complicated aperiodic fluctuations in both space and time.

Consider an experiment in which all the impressed conditions are kept fixed except Ω, which is increased in steps from low values to high values. This can be represented by a series of points on a straight line inclined at 45° to the $\mathcal{J}$ and Θ axes in the regime diagram of Fig.7, moving from the upper left part of the diagram to the lower right. The critical value Ω_R of Ω corresponds to the point where the transition from axisymmetric to regular non-axisymmetric flow occurs and the critical value Ω_I to the point where the transition from regular to irregular non-axisymmetric flow occurs.

REFERENCES

Hide, R. 1958 An experimental study of thermal convection in a rotating liquid. Phil.Trans.Roy.Soc.London, A250, 442-478; for further details see Ph.D. dissertation, Cambridge University, 1953

Hide, R. 1980 Jupiter and Saturn: Giant magnetic rotating fluid planets. Observatory, 100, 182-193

Hide, R. 1981 On the rotation of Jupiter. Geophys.J. Roy.Astron.Soc., 64 283-289

Hide, R. and Mason, P.J. 1970 Baroclinic waves in a rotating fluid subject to internal heating. Phil.Trans.Roy.Soc.London, A268, 201-232

Hide, R. and Mason, P.J. 1975 Sloping convection in a rotating fluid. Advances Phys., 24, 47-100

Ingersoll, A.P., Dobrovolskis, A.R. and Jakosky, B.M. 1979 Planetary atmospheres. Rev.Geophys.Space Phys., 17, 1722-1735

Mitchell, J.L., Beebe, R.F., Ingersoll, A.P. and Garneau, G.W. 1981 Flow fields within Jupiter's Great Red Spot and White Oval BC. J.Geophys.Res. (in the press)

Narasimha, R. 1980 The possible influence of curvature and rotation on ocean currents. Proc.Indian Acad.Sci.(Earth Planet.Sci.), 89, 267-275

NASA (National Aeronautics and Space Administration) 1979 Voyager encounters Jupiter. (NASA publication, Washington D.C.)

Peek, B.M. 1958 The planet Jupiter. (Faber, London)

Pfeffer, R.L., Buzyna, G. and Kung, R. 1980 Time-dependent modes of behaviour of thermally driven rotating fluids. J.Atmos.Sci., 37, 2129-2149

Smith, B.A. and Hunt, G.E. 1976 Motions and morphology of clouds in the atmosphere of Jupiter. pp 564-585 in Jupiter: Studies of the interior, atmosphere, magnetosphere and satellites (ed.T.Gehrels; Univ.of Arizona Press, Tucson)

Tritton, D.J. and Davies, P.A. 1981 Instabilities in geophysical fluid dynamics. pp.229-270 in Hydrodynamic instabilities and the transition to turbulence (ed. H.L.Swinney and J.P.Gollub, Springer-Verlag, Berlin, Heidelberg, New York)

Dynamic Meteorology: Data Assimilation Methods

Editors: **L. Bengtsson, M. Ghil, E. Källén**

1981. IX, 330 pages
(Applied Mathematical Sciences, Volume 36)
ISBN 3-540-90632-0

Contents: An Overview of Meteorological Data Assimilation. – A Review of Methods for Objective Analysis. – Normal Mode Initialization. – Assimilation of Asynoptic Data and the Initialization Problem. – Applications of Estimation Theory to Numerical Weather Prediction. – Convergence of Assimilation Procedures. – Some Climatological and Energy Budget Calculations Using the FGGE III-b Analyses During January 1979. – Appendix: Provisional Report on Calculation of Spatial Covariance and Autocorrelation of the Pressure Field.

Weather prediction with numerical models of the atmosphere has developed to such an extent that it is now used for operational forecasting all over the world. To specify the initial conditions for a numerical weather prediction (NWP) model, a vast amount of meteorological information has to be continuously monitored. This uniquely up-to-date study discusses all of the current and practical methods of analyzing these data, and how to make them optimally suitable as initial conditions for a numerical model, with suggestions for future developments. It functions both as a reference book for meteorologists and oceanographers, and as a solid introduction for anyone with a good background in applied mathematics or fluid dynamics.

Two aspects of the data assimilation problem are crucial: the dynamical and the statistical. While the equations governing NWP models have two time scales, only the slow one is meteorologically useful.

The book presents a common method for eliminating the undesirable fast waves present in the initial data: nonlinear normal-mode initialization as well as two recently developed alternative methods. The statistical aspects is currently treated by optimal interpolation techniques, which are also reviewed. A new approach is the application of Kalman-Bucy (KB) filtering methods to NWP models, described in detail here for the first time. The standard KB filter of estimation theory is modified to eliminate dynamical noise-fast waves, along with stochastic noise-observational errors.

Springer-Verlag
Berlin
Heidelberg
New York

J. Pedlosky

Geophysical Fluid Dynamics

1979. 180 figures, 2 tables. XII, 624 pages
ISBN 3-540-90368-2

Contents: Preliminaries. – Fundamentals. – Inviscid Shallow-Water Theory. – Friction and Viscous Flow. – Homogeneous Models of the Wind-Driven Oceanic Circulation. – Quasigeostrophic Motion of a Stratified Fluid on a Sphere. – Instability Theory. – Ageostrophic Motion. – Selected Bibliography. – Index.

This comprehensive and lucidly written introduction to the theory of ocean and atmospheric movement dynamics blends physical and intuitive reasoning with mathematical analysis. The book combines approximations of the basic fluid dynamical equations required for progress with the precise analysis necessary to make the progress meaningful. Emphasis is placed on the systematic derivation and application of quasi-geostrophic theory to central problems of oceanography and meteorology.

Designed for students and scientists, this book lays the basis for understanding and conducting modern research in oceanography and meteorology.

Springer-Verlag
Berlin
Heidelberg
New York

GPSR Compliance

The European Union's (EU) General Product Safety Regulation (GPSR) is a set of rules that requires consumer products to be safe and our obligations to ensure this.

If you have any concerns about our products, you can contact us on ProductSafety@springernature.com

In case Publisher is established outside the EU, the EU authorized representative is:

Springer Nature Customer Service Center GmbH
Europaplatz 3
69115 Heidelberg, Germany

Batch number: 10761697

Printed by Printforce, the Netherlands